Water
Pollution
Control

Water Pollution Control

A guide to the use of water quality management principles

Edited by
Richard Helmer and Ivanildo Hespanhol

Published on behalf of

UNEP
United Nations Environment Programme

Water Supply & Sanitation Collaborative Council

World Health Organization

Taylor & Francis
Taylor & Francis Group

LONDON AND NEW YORK

Published by Taylor & Francis
2 Park Square, Milton Park, Abingdon, Oxon, OX14 4RN
270 Madison Ave, New York NY 10016

First edition 1997

Transferred to Digital Printing 2007

© 1997 WHO/UNEP

ISBN 0 419 22910 8

A catalogue record for this book is available from the British Library

Publisher's Note
The publisher has gone to great lengths to ensure the quality of
this reprint but points out that some imperfections in the
original may be apparent

TABLE OF CONTENTS

FOREWORD

Publication of this book is a milestone for the Water Supply and Sanitation Collaborative Council. It demonstrates the Council's unique capacity to bring together water and sanitation professionals from industrialised and developing countries to formulate practical guidance on a key issue of the day.

Industrialised countries have extensive experience of the problems caused by water pollution and the strategies and technologies available to control it. In the developing world, although pollution is increasing rapidly with urbanisation and industrialisation, most countries have very limited experience of pollution control measures or of the institutional and legislative frameworks needed to make such measures effective. On the other hand, the Collaborative Council's developing country members have the specialist knowledge and skills with which to adapt the practices of the industrialised nations to their own circumstances.

This synergy among members is at the heart of the Council's approach to sector issues. By mandating specialist working groups to seek out good practices, to analyse them and to reach agreement on the best way forward, the Council is able to give its members authoritative guidance and tools to help them face their own particular challenges.

Water pollution control is clearly one of the most critical of those challenges. Without urgent and properly directed action, developing countries face mounting problems of disease, environmental degradation and economic stagnation, as precious water resources become more and more contaminated. At the Earth Summit in Rio de Janeiro in June 1992, world leaders recognised the crucial importance of protecting freshwater resources. Chapter 18 of Agenda 21 sees "effective water pollution prevention and control programmes" as key elements of national sustainable development plans.

At its second Global Forum, in Rabat, Morocco, in 1993, the Collaborative Council responded to the Rio accord by mandating a Working Group on Water Pollution Control, convened jointly with the World Health Organization and the United Nations Environment Programme. We were fortunate that Richard Helmer from the World Health Organization agreed to co-ordinate the Working Group. Richard had been a prime mover in the preparation of the freshwater initiatives endorsed in Rio de Janeiro and so was particularly well placed to ensure that the Group's deliberations were well directed. Experts from developing countries, UN agencies, bilaterals, professional associations, and academic institutions have all contributed over the last three and a half years. The Council is grateful to them, and I want to express my own personal appreciation for the voluntary time and effort they have devoted to the task.

The result is a comprehensive guidebook which I know will be a valuable tool for policy makers and environmental managers in developing and newly industrialised countries as they seek to combat the damaging health, environmental and economic impacts of water pollution. The council will play its part in advocacy and promotion. We all owe a duty to future generations to safeguard their water supplies and to protect their living environment.

Margaret Catley-Carlson,
Chair, Water Supply and Sanitation Collaborative Council

ACKNOWLEDGEMENTS

The co-sponsoring organisations would like to express their deep gratitude to all of those whose efforts made the preparation of this guidebook possible, through contributions to chapters, review of drafts, active participation in the working group process, or financial support to meetings, editorial work, etc.

The work was directed by a core group of staff from the World Health Organization (WHO), the United Nations Environment Programme (UNEP), the United Nations Centre for Human Settlements (UNCHS), the Food and Agriculture Organization of the United Nations (FAO) and experts from bilateral agencies who are members of the Water Supply and Sanitation Collaborative Council, WHO collaborating centres and experts from developing and newly industrialising countries. The activities have been implemented together with UNEP, the Danish Water Quality Institute (VKI), the Institute for Inland Water Management and Wastewater Treatment in the Netherlands (RIZA), the International Institute for Infrastructural, Hydraulic and Environmental Engineering of the Netherlands (IHE), the World Bank, the WHO Collaborating Centre for Water Quality Control, and the WHO European Centre for Environment and Health/Nancy Project Office. Other international organisations, in particular the International Association for Water Quality (IAWQ) and the International Water Resources Association (IWRA) have provided support to the Working Group. Additional support has also been received from bilateral and other external support agencies, particularly the Ministry of Foreign Affairs/DGIS of the Netherlands. Financial support for the activities undertaken by the Working Group has been provided by UNEP and by the Government of the Netherlands.

The Working Group brought together a group of experts who contributed individually or collectively to the different parts of the book. It is difficult to identify adequately the contribution of each individual author and therefore the principal contributors are listed together below:

Martin Adriaanse, Institute for Inland Water Management and Waste Water Treatment (RIZA), Ministry of Transport, Public Works and Water Management, Lelystad, The Netherlands (Chapter 9)

Guy J.F.R. Alaerts, The World Bank, Washington, D.C., USA formerly at International Institute for Infrastructural, Hydraulic and Environmental Engineering (IHE), Delft, The Netherlands (Chapters 3 and 8)

Mohamed Al-Hamdi, Sana'a University Support Project, Sana'a, Yemen currently Ph.D. fellow at the International Institute for Infrastructural, Hydraulic and Environmental Engineering, Delft, The Netherlands (Case Study XIII)

Humberto Romero Alvarez, Consultivo Técnico, National Water Commission, Mexico, D.F., Mexico (Case Studies VII and VIII)

Lawrence Chidi Anukam, Federal Environmental Protection Agency (FEPA), Abuja, Nigeria (Case Study IV)

Carl R. Bartone, Urban Development Division, World Bank, Washington, D.C., USA (Chapter 7)

Janis Bernstein, The World Bank, Washington, D.C., USA (Chapter 6)

M. Bijlsma, International Institute for Infrastructural, Hydraulic and Environmental Engineering (IHE), Delft, The Netherlands (Chapter 3)

Benedito Braga, Department of Civil and Environmental Engineering, Escola Politécnica da Universidade de São Paulo, São Paulo, Brazil (Case Study VI)

S. Andrew P. Brown, Wates, Meiring & Barnard, Halfway House, South Africa (Case Study V)

Peter A. Chave, Pollution Control, Bristol, UK formerly of National Rivers Authority, Bristol, UK (Chapter 5)

Renato Tantoco Cruz, River Rehabilitation Secretariat, Pasig River Rehabilitation Program, Carl Bro International a/s, Quezon City, Philippines (Case Study III)

Rainer Enderlein, Environment and Human Settlement Division, United Nations Economic Commission for Europe, Geneva, Switzerland (Chapter 2)

Ute Enderlein, formerly Urban Environmental Health, Division of Operational Support in Environmental Health, World Health Organization, Geneva, Switzerland (Chapter 2)

Roberto Max Hermann, Department of Hydraulic and Sanitary Engineering, Escola Politécnica da Universidade de São Paulo, São Paulo, Brazil (Case Study VI)

Ivanhildo Hespanhol, Department of Hydraulic and Sanitary Engineering, Escola Politécnica da Universidade de São Paulo, São Paulo, Brazil, formerly of Urban Environmental Health, World Health Organization, Geneva, Switzerland (Chapter 4)

Niels H. Ipsen, Water Quality Institute (VKI), Danish Academy of Technical Sciences, Hørsholm, Denmark (Chapters 1 and 10)

Henrik Larsen, Water Quality Institute (VKI), Danish Academy of Technical Sciences, Hørsholm, Denmark (Chapters 1 and 10)

Palle Lindgaard-Jørgensen, Water Quality Institute (VKI), Danish Academy of Technical Sciences, Hørsholm, Denmark (Chapter 9)

José Eduardo Mestre Rodríguez, Bureau for River Basin Councils, National Water Commission, Mexico, D.F., Mexico (Case Study VIII)

Ilya Natchkov, Ministry of Environment, Sofia, Bulgaria (Case Study IX)

Ioannis Papadopoulos, Agricultural Research Institute, Ministry of Agriculture, Natural Resources and Environment, Nicosia, Cyprus (Case Study XI)

Herbert C. Preul, Department of Civil and Environmental Engineering, University of Cincinnati, Cincinnati, USA (Case Study XII)

Yogesh Sharma, formerly National River Conservation Directorate, Ministry of Environment and Forests, New Delhi, India (Case Study I)

Lars Ulmgren, Stockholm Vatten, Stockholm, Sweden (Chapter 1)

Siemen Veenstra, International Institute for Infrastructural, Hydraulic and Environmental Engineering (IHE), Delft, The Netherlands (Chapter 3)

Vladimir Vladimirov, CPPI Water Component, c/o Centre for International Projects, Moscow, Russian Federation (Case Study X)

W. Peter Williams, Monitoring and Assessment Research Centre (MARC), King's College London, London, UK (Chapter 2)

Chongua Zhang, The World Bank, Washington, D.C., USA (Case Study II)

Chapter 7 draws heavily on the work and accumulated experiences of the Water and Sanitation Division of the World Bank, and of the environment team of the Urban Development Division and the UNDP/UNCHS/World Bank Urban Management Programme. The author is particularly indebted to John Briscoe, K.C. Sivaramakrishnan and Vijay Jagannathan for their comments and contributions.

Case Study I was an outcome of the initiative of Professor Dr Ir G.J.F.R. Alaerts of IHE, Delft who provided encouragement and invaluable guidance for which the author is grateful. The leadership and kind support of Mr Vinay Shankar, formerly Project Director of the Ganga Project, in allowing the case study to be produced is also gratefully acknowledged.

The advice and assurance of the Programme Coordination Unit for the Danube Programme based in Vienna and it's Team Leader Mr. David Rodda, is acknowledged in the preparation of Case Study IX. The views expressed in the case study are those of the author and do not necessarily represent those of the Task Force or any of its members.

The basic information and data for Case Study XII were gathered for the development of a Water Management and Conservation Plan for the country of Jordan by the author, in the year 1992, during a consulting assignment with the Chemonics International Consulting Division, Inc. of Washington, D.C. under a contract with the US Agency for International Development

(USAID). The assistance of others connected with the project is gratefully acknowledged. The views and opinions cited in this case study are those of the author and the named references and do not necessarily reflect the views and opinion or policies of USAID.

The draft text for this book was reviewed by the Working Group members through meetings and written comments and amendments. The broad range of issues and the wide geographical scope covered by the Working Group can best be demonstrated through complete listings of all members as given in the Appendix. In this way the co-sponsoring agencies and the editors would like to express their great appreciation for the dedication given by all participants to this project. The book would, however, not have been possible without the editorial assistance of Dr Deborah Chapman who undertook technical and language editing as well as layout and production management, in collaboration with the publisher. As the editor of the UNEP/WHO co-sponsored series of guidebooks dealing with various aspects of water quality management, she was responsible for ensuring compatibility with *Water Quality Assessments* and *Water Quality Monitoring*, two of the other books in the series.

Chapter 1[*]

POLICY AND PRINCIPLES

1.1   Introduction

During recent years there has been increasing awareness of, and concern about, water pollution all over the world, and new approaches towards achieving sustainable exploitation of water resources have been developed internationally. It is widely agreed that a properly developed policy framework is a key element in the sound management of water resources. A number of possible elements for such policies have been identified, especially during the preparation of Agenda 21 as well as during various follow up activities.

This chapter proposes some general principles for the policy making process and for policy document structure. Some examples of policy elements which support the overall sustainable management of water resources are also given.

1.2   Policy framework

Policy statements regarding water pollution control can be found within the legislative framework of most countries. However, the statements are often "hidden" in official documents, such as acts of government, regulations, action and master plans. Moreover, government statutes and constitutional documents often include paragraphs about environmental policies. Such statements are rarely coherent, and inconsistencies with other policies often exist because they have been developed separately with different purposes.

Water pollution control is usually specifically addressed in connection with the establishment of environmental legislation and action plans, but also within the framework of water resources management planning. Moreover, documents related to public health aspects may also consider water pollution. These three interacting areas are often administered in different line ministries — typically a Ministry of Environment, a Ministry of Water and a Ministry of Health. In addition, the policy making process, if it exists, may often take place independently.

---

[*]   *This chapter was prepared by H. Larsen, N.H. Ipsen and L. Ulmgren*

To reach a situation where the adopted political intentions can result in a real impact on the practical management of water resources, it is important to define policy statements clearly and in proper policy documents. It is recommended that the water pollution control policy statements either be placed within a water resources policy document or within an environment policy document, or the statements can form a document in themselves, referring to overall health–water and resources–environment policies. The approach selected will depend on the administrative organisation of water resources and environmental management in a particular country.

Some general principles that should be considered within the policy making process are as follows:

- A water pollution control policy, ideally, should be seen as part of a coherent policy framework ranging from overall statements such as can be found in government statutes, constitutions, etc., to specific policy statements defined for environment and water resources management as well as for particular sector developments.
- The policy making process should therefore incorporate consultations and seek consensus with all line ministries relevant for water resources management, including organisations responsible for overall economic development policies. In addition, when formulating new development policies for other sectors, water resources policy statements should be taken into account where appropriate.
- Policy statements must be realistic. Good intentions reflected in statements such as "No pollution of surface waters shall occur..." cannot be applied in practice and therefore become meaningless in the context of an operational policy.
- The statements in a policy document need to be relatively long-lived because they must pass a laborious political adaptation process. Thus, detailed guidelines, which may need regular adaptation to the country's actual development level, should be avoided and placed into the more dynamic parts of the legislation system, such as the regulation framework, that can be amended at short notice.

1.2.1 The policy document

A policy document should be formulated clearly and concisely, but at the same time it must be operational. This means that the statements should be easily understood and the document should form a guide for administrators formulating laws and regulations as well as those enforcing, and thereby interpreting, such texts. To fulfil these requirements the policy document should include, in addition to very general statements, well explained guiding principles for water pollution management as well as outlines for strategies for the implementation of the policy.

1.2.2 Overall policy statements

The overall policy statements, relevant for water pollution control, define a government's concept of the water resources as well as its long-term priorities for exploitation of the resource. These statements should, preferably, be derived from the country's general environment and water resources management policies. They should also document the government's willingness to let management instruments ensure the long-term protection and sustainable exploitation of water resources along with social and economic development.

Agenda 21 adopted some conceptual statements concerning water resources, but which apply to water pollution control as well as to other elements of water resources management. Two central statements were *"Fresh water should be seen as a finite and vulnerable resource, essential to sustain life, development and the environment"* and *"Water should be considered as a social and economic good with a value reflecting its most valuable potential use"*. The latter statement suggests an overall concept for prioritising water-related development activities.

1.3 Guiding principles for water pollution control

The guiding principles of the policy document put the political intentions into more practical terms by setting a more detailed conceptual framework that supports the overall policy objectives. It is recommended that these principles should be clarified by a short narrative interpretation. The following guiding principles provide a suitable basis for sound management of water pollution.

Prevent pollution rather than treating symptoms of pollution. Past experience has shown that remedial actions to clean up polluted sites and water bodies are generally much more expensive than applying measures to prevent pollution from occurring. Although wastewater treatment facilities have been installed and improved over the years in many countries, water pollution remains a problem, including in industrialised countries. In some situations, the introduction of improved wastewater treatment has only led to increased pollution from other media, such as wastewater sludge. The most logical approach is to prevent the production of wastes that require treatment. Thus, approaches to water pollution control that focus on wastewater minimisation, in-plant refinement of raw materials and production processes, recycling of waste products, etc., should be given priority over traditional end-of-pipe treatments.

In many countries, however, an increasing proportion of water pollution originates from diffuse sources, such as agricultural use of fertilisers, which cannot be controlled by the approach mentioned above. Instead, the principle of "best environmental practice" should be applied to minimise non-point source pollution. As an example, codes of good agricultural practice that address the causes of water pollution from agriculture, such as type, amount and time of application of fertilisers, manure and pesticides, can give guidance to farmers on how to prevent or reduce pollution of water bodies. Good agricultural practice is recognised by the United Nations Economic Commission for Europe (UNECE) as a means of minimising the risk of water pollution and of promoting the continuation of economic agricultural activity (UNECE, 1993).

Use the precautionary principle. There are many examples of the application and discharge of hazardous substances into the aquatic environment, even when such substances are suspected of having detrimental effects on the environment. Until now the use of any substance and its release to the environment has been widely accepted, unless scientific research has proved unambiguously a causal link between the substance and a well-defined environmental impact. However, in most cases it takes a very long time to establish such causal links, even where early investigations suggest clear indications of such links. When, eventually, the necessary documentation is provided and action can be taken to abandon the use of the substance, substantial environmental damage may already have occurred. Examples of such situations include a number of pesticides which are now being abandoned because contamination of groundwater resources has been demonstrated.

The examples clearly show that action to avoid potential environmental damage by hazardous substances should not be postponed on the grounds that scientific research has not proved fully a causal link between the substance and the potential damage (UNECE, 1994).

Apply the polluter-pays-principle. The polluter-pays-principle, where the costs of pollution prevention, control and reduction measures are borne by the polluter, is not a new concept but has not yet been fully implemented, despite the fact that it is widely recognised that the perception of water as a free commodity can no longer be maintained. The principle is an economic instrument that is aimed at affecting behaviour, i.e. by encouraging and inducing behaviour that puts less strain on the environment. Examples of attempts to apply this principle include financial charges for industrial waste-water discharges and special taxes on pesticides (Warford, 1994).

The difficulty or reluctance encountered in implementing the polluter-pays-principle is probably due to its social and economic implications (Enderlein, 1995). Full application of the principle would upset existing subsidised programmes (implemented for social reasons) for supply of water and removal of wastewater in many developing countries. Nevertheless, even if the full implementation of the polluter-pays-principle is not feasible in all countries at present, it should be maintained as the ultimate goal.

Apply realistic standards and regulations. An important element in a water pollution control strategy is the formulation of realistic standards and regulations. However, the standards must be achievable and the regulations enforceable. Unrealistic standards and non-enforceable regulations may do more harm than having no standards and regulations, because they create an attitude of indifference towards rules and regulations in general, both among polluters and administrators. Standards and regulations should be tailored to match the level of economic and administrative capacity and capability. Standards should be gradually tightened as progress is achieved in general development and in the economic capability of the private sector. Thus, the setting of standards and regulations should be an iterative and on-going process.

Balance economic and regulatory instruments. Until now, regulatory management instruments have been heavily relied upon by governments in most countries for controlling water pollution. Economic instruments, typically in the form of wastewater discharge fees and fines, have been introduced to a lesser extent and mainly by industrialised countries.

Compared with economic instruments, the advantages of the regulatory approach to water pollution control is that it offers a reasonable degree of predictability about the reduction of pollution, i.e. it offers control to authorities over what environmental goals can be achieved and when they can be achieved (Bartone *et al.*, 1994). A major disadvantage of the regulatory approach is its economic inefficiency (see also Chapter 5). Economic instruments have the advantages of providing incentives to polluters to modify their behaviour in support of pollution control and of providing revenue to finance pollution control activities. In addition, they are much better suited to combating non-point sources of pollution. The setting of prices and charges are crucial to the success of economic instruments. If charges are too low, polluters may opt to pollute and to pay, whereas if charges are too high they may inhibit economic development.

Against this background it seems appropriate, therefore, for most countries to apply a mixture of regulatory and economic instruments for controlling water pollution. In developing countries, where financial resources and institutional capacity are very limited, the most important criteria for balancing economic and regulatory instruments should be cost-effectiveness (those that achieve the objectives at the least cost) and administrative feasibility.

Apply water pollution control at the lowest appropriate level. The appropriate level may be defined as the level at which significant impacts are experienced. If, for example, a specific water quality issue only has a possible impact within a local community, then the community level is the proper management level. If environmental impacts affect a neighbouring community, then the appropriate management level is one level higher than the community level, for example the river basin level.

On a wider scale, the appropriate management level may be the national level for major water bodies where no significant water pollution impacts are anticipated for neighbouring states. Where significant impacts occur in several nations, the appropriate management level is international (e.g. an international river basin commission). The important point is that decisions or actions concerning water pollution control should be taken as close as possible to those affected, and that higher administrative levels should enable lower levels to carry out decentralised management. However, in considering whether a given administrative level is appropriate for certain water pollution control functions, the actual capacity to achieve these functions (or the possibility of building it) at that level should also be taken into account. Thus, this guiding principle intends to initiate a process of decentralisation of water pollution control functions that is adapted to administrative and technical feasibility.

Establish mechanisms for cross-sectoral integration. In order to ensure the co-ordination of water pollution control efforts within water-related sectors, such as health and agriculture, formal mechanisms and means of co-operation and information exchange need to be established. Such mechanisms should:
- Allow decision makers from different sectors to influence water pollution policy.
- Urge them to put forward ideas and plans from their own sector with impacts on water quality.
- Allow them to comment on ideas and plans put forward by other sectors.

For example, a permanent committee with representatives from the involved sectors could be established. The functions and responsibilities of the cross-sectoral body would typically include at least the following:

- Co-ordination of policy formulation on water pollution control.
- Setting of national water quality criteria and standards, and their supporting regulations.
- Review and co-ordination of development plans that affect water quality.
- Resolution of conflicts between government bodies regarding water pollution issues that cannot be resolved at a lower level.

Encourage participatory approach with involvement of all relevant stakeholders. The participatory approach involves raising awareness of the importance of water pollution control among policy-makers and the general public. Decisions should be taken with full public consultation and with the involvement of groups affected by the planning and implementation of water pollution control activities. This means, for example, that the public should be kept continuously informed, be given opportunities to express their views, knowledge and priorities, and it should be apparent that their views have been taken into account.

Various methods exist to implement public participation, such as interviews, public information sessions and hearings, expert panel hearings and site visits. The most appropriate method for each situation should take account of local social, political, historical, cultural and other factors. In many countries in transition, for example, only professional and scientific experts usually participate and other groups have mostly been excluded from the process. Public participation may take time but it increases public support for the final decision or result and, ideally, contributes to the convergence of the views of the public, governmental authorities and industry on environmental priorities and on water pollution control measures.

Give open access to information on water pollution. This principle is directly related to the principle of involvement of the general public in the decision-making process, because a precondition for participation is free access to information held by public authorities. Open access to information helps to stimulate understanding, discussions and suggestions for solutions of water quality problems. In many countries, notably the countries in economic transition and the developing countries, there is no tradition of open access to environmental information. Unfortunately, this attitude may seriously jeopardise the outcome of any international co-operation that is required.

Promote international co-operation on water pollution control. Transboundary water pollution, typically encountered in large rivers, requires international co-operation and co-ordination of efforts in order to be effective. Lack of recognition of this fact may lead to wasteful investments in pollution load reductions in one country if, due to lack of co-operation, measures are introduced upstream that have counteractive effects. In a number of cases (e.g. the Danube, Zambezi and Mekong rivers), permanent international bodies with representatives from riparian states have been successfully established, with the objective of strengthening international co-operation on the pollution control of the shared water resources.

A framework for international co-operation on water pollution control that has been widely agreed is the Convention on the Protection and Use of Transboundary Watercourses and International Lakes (UNECE, 1994). Although some countries have already started international co-operation on water pollution control, there is still a huge need for concerted planning and action at the international level.

1.4   Strategy formulation

Strategy formulation for water pollution control should be undertaken with due consideration to the above mentioned guiding principles, as well as to other principles for water resources management laid down in various documents, e.g. Agenda 21, that have been widely agreed. When formulating a water pollution control strategy, it should be ensured that various complementary elements of an effective water pollution control system are developed and strengthened concurrently. For example, financial resources would not be used very effectively by spending them all on the formulation of policies and the drafting of legislation, standards and regulations, if there is no institutional capacity to fill the established framework and enforce the regulations.

The main components of a rational water pollution control system can be defined as:

- An enabling environment, which is a framework of national policies, legislation and regulations setting the scene for polluters and management authorities.
- An institutional framework that allows for close interaction between various administrative levels.
- Planning and prioritisation capabilities that will enable decision-makers to make choices between alternative actions based on agreed policies, available resources, environmental impacts and the social and economic consequences.

All three components are needed in order to achieve effective water pollution control and it is, therefore, advisable to develop all three components hand-in-hand.

At the policy level the strategy must provide general directions for water quality managers on how to realise the objectives of the water pollution control policies and on how to translate the guiding principles into practical management. The strategy should provide adequate detail to help identify and formulate concrete actions and projects that will contribute to achieving the defined policies.

1.5   References

Bartone, C., Bernstein, J., Leitmann, J. and Eigen, J. 1994 *Toward Environmental Strategies for Cities: Policy Considerations for Urban Development Management in Developing Countries.* UNDP/UNCHS/World Bank, Urban Management Programme, Washington, D.C.

Enderlein, R.E. 1995 Protecting Europe's water resources: Policy issues. *Wat. Sci. Tech.,* **31**(8), 1–8.

UNECE 1993 *Protection of Water Resources and Aquatic Ecosystems.* Water Series No. 1, ECE/ENVWA/31, United Nations Economic Commission for Europe, New York.

UNECE 1994 *Convention on the Protection and Use of Transboundary Watercourses and International Lakes.* ECE/ENHS/NONE/1, Geneva, United Nations Economic Commission for Europe, New York.

Warford, J.J. 1994 Environment, health, and sustainable development: The role of economic instruments and policies. Discussion paper for the Director General's Council on the Earth Summit Action Programme for Health and Environment, June 1994, World Health Organization, Geneva.

Chapter 2[*]

WATER QUALITY REQUIREMENTS

2.1 Introduction

Control of water pollution has reached primary importance in developed and a number of developing countries. The prevention of pollution at source, the precautionary principle and the prior licensing of wastewater discharges by competent authorities have become key elements of successful policies for preventing, controlling and reducing inputs of hazardous substances, nutrients and other water pollutants from point sources into aquatic ecosystems (see Chapter 1).

In a number of industrialised countries, as well as some countries in transition, it has become common practice to base limits for discharges of hazardous substances on the best available technology (see Chapters 3 and 5). Such hazardous water pollutants include substances that are toxic at low concentrations, carcinogenic, mutagenic, teratogenic and/or can be bioaccumulated, especially when they are persistent. In order to reduce inputs of phosphorus, nitrogen and pesticides from non-point sources (particularly agricultural sources) to water bodies, environmental and agricultural authorities in an increasing number of countries are stipulating the need to use best environmental practices (Enderlein, 1996).

In some situations, even stricter requirements are necessary. A partial ban on the use of some compounds or even the total prohibition of the import, production and use of certain substances, such as DDT and lead- or mercury-based pesticides, may constitute the only way to protect human health, the quality of waters and their aquatic flora and fauna (including fish for human consumption) and other specific water uses (ECLAC, 1989; UNECE, 1992; United Nations, 1994).

Some water pollutants which become extremely toxic in high concentrations are, however, needed in trace amounts. Copper, zinc, manganese, boron and phosphorus, for example, can be toxic or may otherwise adversely affect

[*]   *This chapter was prepared by Ute S. Enderlein, Rainer E. Enderlein and W. Peter Williams*

aquatic life when present above certain concentrations, although their presence in low amounts is essential to support and maintain functions in aquatic ecosystems. The same is true for certain elements with respect to drinking water. Selenium, for example, is essential for humans but becomes harmful or even toxic when its concentration exceeds a certain level.

The concentrations above which water pollutants adversely affect a particular water use may differ widely. Water quality requirements, expressed as water quality criteria and objectives, are use-specific or are targeted to the protection of the most sensitive water use among a number of existing or planned uses within a catchment.

Approaches to water pollution control initially focused on the fixed emissions approach (see Chapter 3) and the water quality criteria and objectives approach. Emphasis is now shifting to integrated approaches. The introduction of holistic concepts of water management, including the ecosystem approach, has led to the recognition that the use of water quality objectives, the setting of emission limits on the basis of best available technology and the use of best available practices, are integral instruments of prevention, control and reduction of water pollution (ICWE, 1992; UNCED, 1992; UNECE, 1993). These approaches should be applied in an action-orientated way (Enderlein, 1995). A further development in environmental management is the integrated approach to air, soil, food and water pollution control using multimedia assessments of human exposure pathways.

2.2.   Why water quality criteria and objectives?

Water quality criteria are developed by scientists and provide basic scientific information about the effects of water pollutants on a specific water use (see Box 2.1). They also describe water quality requirements for protecting and maintaining an individual use. Water quality criteria are based on variables that characterise the quality of water and/or the quality of the suspended particulate matter, the bottom sediment and the biota. Many water quality criteria set a maximum level for the concentration of a substance in a particular medium (i.e. water, sediment or biota) which will not be harmful when the specific medium is used continuously for a single, specific purpose. For some other water quality variables, such as dissolved oxygen, water quality criteria are set at the minimum acceptable concentration to ensure the maintenance of biological functions.

Most industrial processes pose less demanding requirements on the quality of freshwater and therefore criteria are usually developed for raw water in relation to its use as a source of water for drinking-water supply, agriculture and recreation, or as a habitat for biological communities. Criteria may also

Box 2.1   Examples of the development of national water quality criteria and guidelines

Nigeria
In Nigeria, the Federal Environmental Protection Agency (FEPA) issued, in 1988, a specific decree to protect, to restore and to preserve the ecosystem of the Nigerian environment. The decree also empowered the agency to set water quality standards to protect public health and to enhance the quality of waters. In the absence of national comprehensive scientific data, FEPA approached this task by reviewing water quality guidelines and standards from developed and developing countries as well as from international organisations and, subsequently, by comparing them with data available on Nigeria's own water quality. The standards considered included those of Australia, Brazil, Canada, India, Tanzania, the United States and the World Health Organization (WHO). These sets of data were harmonised and used to generate the Interim National Water Quality Guidelines and Standards for Nigeria. These address drinking water, recreational use of water, freshwater aquatic life, agricultural (irrigation and livestock watering) and industrial water uses. The guidelines are expected to become the maximum allowable limits for inland surface waters and groundwaters, as well as for non-tidal coastal waters. They also apply to Nigeria's transboundary watercourses, the rivers Niger, Benue and Cross River, which are major sources of water supply in the country. The first set of guidelines was subject to revision by interested parties and the general public. A Technical Committee comprising experts from Federal ministries, State Governments, private sector organisations, higher educational institutions, non-governmental organisations and individuals is now expected to review the guidelines from time to time.

Papua New Guinea
In Papua New Guinea, the Water Resources Act outlines a set of water quality requirements for fisheries and recreational use of water, both fresh and marine. The Public Health Drinking Water Quality Regulation specifies water quality requirements and standards relating to raw water and drinking water. The standards were established in accordance with WHO guidelines and data from other tropical countries.

Viet Nam
In Viet Nam, the water management policy of the Government highlights the need for availability of water, adequate in quantity and quality for all beneficial uses, as well as for the control of point and non-point pollution sources. The Government is expected to draw up and to update a comprehensive long-term plan for the development and management of water resources. Moreover, an expected reduction in adverse impacts from pollution sources in upstream riparian countries on the water quality within the Mekong River delta will be based on joint studies and definitions of criteria for water use among riparian countries of the river.

A set of national water quality criteria for drinking-water use as well as criteria for fish and aquatic life, and irrigation have been established (ESCAP, 1990). Criteria for aquatic life include: pH (range 6.5–8), dissolved oxygen (> 2 mg l^{-1}), NH_4-N (< 1 mg l^{-1}), copper (< 0.02 mg l^{-1}), cadmium (< 0.02 mg l^{-1}), lead (< 0.01 mg l^{-1}) and dissolved solids (1,000 mg l^{-1}). More recently, allowable concentrations of pesticides in the freshwater of the Mekong delta have been established by the Hygiene Institute of Ho Chi Minh City as follows: DDT 0.042 mg l^{-1}, heptachlor 0.018 mg l^{-1}, lindane 0.056 mg l^{-1} and organophosphate 0.100 mg l^{-1}. According to Pham Thi Dung (1994), the actual concentrations of these pesticides during the period June 1992 to June 1993 were considerably below these criteria.

Sources: ESCAP, 1990; FEPA, 1991; Pham Thi Dung, 1994

Table 2.1   Definitions related to water quality and pollution control

Term	Definition
Water quality criterion (synonym: water quality guideline)	Numerical concentration or narrative statement recommended to support and maintain a designated water use
Water quality objective (synonyms: water quality goal or target)	A numerical concentration or narrative statement which has been established to support and to protect the designated uses of water at a specific site, river basin or part(s) thereof
Water quality standard	An objective that is recognised in enforceable environmental control laws or regulations of a level of Government[1]
Precautionary principle	The principle, by virtue of which action to avoid the potential adverse impact of the release of hazardous substances shall not be postponed on the ground that scientific research has not fully proved a causal link between those substances, on the one hand, and the potential adverse impact, on the other

[1]   Water quality standards are discussed in Chapter 3

Sources: Adapted from Dick, 1975; CCREM, 1987; Chiaudani and Premazzi, 1988; UNECE, 1992, 1993

be developed in relation to the functioning of aquatic ecosystems in general. The protection and maintenance of these water uses usually impose different requirements on water quality and, therefore, the associated water quality criteria are often different for each use.

Water quality criteria often serve as a baseline for establishing water quality objectives in conjunction with information on water uses and site-specific factors (see Table 2.1). Water quality objectives aim at supporting and protecting designated uses of freshwater, i.e. its use for drinking-water supply, livestock watering, irrigation, fisheries, recreation or other purposes, while supporting and maintaining aquatic life and/or the functioning of aquatic ecosystems. The establishment of water quality objectives is not a scientific task but rather a political process that requires a critical assessment of national priorities. Such an assessment is based on economic considerations, present and future water uses, forecasts for industrial progress and for the development of agriculture, and many other socio-economic factors (UNESCO/WHO, 1978; UNECE, 1993, 1995). Such analyses have been carried out in the catchment areas of national waters (such as the Ganga river basin) and in the catchment areas of transboundary waters (such as the Rhine, Mekong and Niger rivers). General guidance for developing water quality objectives is given in the *Convention on the Protection and Use of Transboundary Watercourses and International Lakes* (UNECE, 1992) and other relevant documents.

Water quality objectives are being developed in many countries by water authorities in co-operation with other relevant institutions in order to set threshold values for water quality that should be maintained or achieved within a certain time period. Water quality objectives provide the basis for pollution control regulations and for carrying out specific measures for the prevention, control or reduction of water pollution and other adverse impacts on aquatic ecosystems.

In some countries, water quality objectives play the role of a regulatory instrument or even become legally binding. Their application may require, for example, the appropriate strengthening of emission standards and other measures for tightening control over point and diffuse pollution sources. In some cases, water quality objectives serve as planning instruments and/or as the basis for the establishment of priorities in reducing pollution levels by substances and/or by sources.

2.3   Water quality criteria for individual use categories

Water quality criteria have been widely established for a number of traditional water quality variables such as pH, dissolved oxygen, biochemical oxygen demand for periods of five or seven days (BOD_5 and BOD_7), chemical oxygen demand (COD) and nutrients. Such criteria guide decision makers, especially in countries with rivers affected by severe organic pollution, in the establishment of control strategies to decrease the potential for oxygen depletion and the resultant low BOD and COD levels.

Examples of the use of these criteria are given in the case studies on the Ganga, India (Case Study 1), the Huangpu, China (Case Study 2) and Pasig River, Philippines (Case Study 3). Criteria for traditional water quality variables also guide decision makers in the resolution of specific pollution problems, such as water pollution from coal mining as demonstrated in the case study on the Witbank Dam catchment, South Africa (Case Study 5).

2.3.1 Development of criteria

Numerous studies have confirmed that a pH range of 6.5 to 9 is most appropriate for the maintenance of fish communities. Low concentrations of dissolved oxygen, when combined with the presence of toxic substances may lead to stress responses in aquatic ecosystems because the toxicity of certain elements, such as zinc, lead and copper, is increased by low concentrations of dissolved oxygen. High water temperature also increases the adverse effects on biota associated with low concentrations of dissolved oxygen. The water quality criterion for dissolved oxygen, therefore, takes these factors into account. Depending on the water temperature requirements for particular aquatic species at various life stages, the criteria values range from 5 to 9.5 mg l^{-1}, i.e. a minimum dissolved oxygen concentration of 5–6 mg l^{-1} for warm-water biota and 6.5–9.5 mg l^{-1} for cold-water biota. Higher oxygen concentrations are also relevant for early life stages. More details are given in Alabaster and Lloyd (1982) and the EPA (1976, 1986).

The European Union (EU) in its *Council Directive of 18 July 1978 on the Quality of Fresh Waters Needing Protection or Improvement in Order to Support Fish Life* (78/659/EEC) recommends that the BOD of salmonid waters should be $\leq$ 3 mg O_2 l^{-1}, and $\leq$ 6 mg O_2 l^{-1} for cyprinid waters. In Nigeria, the interim water quality criterion for BOD for the protection of aquatic life is 4 mg O_2 l^{-1} (water temperature 20–33 °C), for irrigation water it is 2 mg O_2 l^{-1} (water temperature 20–25 °C), and for recreational waters it is 2 mg O_2 l^{-1} (water temperature 20–33 °C) (FEPA, 1991). In India, for the River Ganga, BOD values are used to define water quality classes for designated uses and to establish water quality objectives that will be achieved over a period of time. For Class A waters, BOD should not exceed 2 mg O_2 l^{-1} and for Class B and C waters it should not exceed 3 mg O_2 l^{-1} (see section 2.4.1 and Box 2.3).

Water quality criteria for phosphorus compounds, such as phosphates, are set at a concentration that prevents excessive growth of algae. Criteria for total ammonia (NH_3) have been established, for example by the EPA, to reflect the varying toxicity of NH_3 with pH (EPA, 1985). Criteria have been set for a pH range from 6.5 to 9.0 and a water temperature range from 0 to 30 °C (Table 2.2). Ammonium (NH_4^+) is less toxic than NH_3. Similar values form the basis for the control strategy in the Witbank Dam catchment, South Africa (Case Study 5).

In a number of industrialised countries, as well as some countries in transition and other countries of the United Nations Economic and Social Commission for Asia and the Pacific (ESCAP) region, increasing attention is being paid to the development of water quality criteria for hazardous substances. These are substances that pose a threat to water use and the functioning of aquatic ecosystems as a result of their toxicity, persistence, potential for bioaccumulation and/or their carcinogenic, teratogenic or mutagenic effects. Genetic material, recombined *in vitro* by genetic engineering techniques, is also very often included in this category of substances. In

Table 2.2   Criteria for total ammonia (NH_3) for the protection of aquatic life at different water temperatures

pH	Ammonia concentration (mg l^{-1})						
	0 °C	5 °C	10 °C	15 °C	20 °C	25 °C	30 °C
6.50	2.50	2.40	2.20	2.20	1.49	1.04	0.73
6.75	2.50	2.40	2.20	2.20	1.49	1.04	0.73
7.00	2.50	2.40	2.20	2.20	1.49	1.04	0.74
7.25	2.50	2.40	2.20	2.20	1.50	1.04	0.74
7.50	2.50	2.40	2.20	2.20	1.50	1.05	0.74
7.75	2.30	2.20	2.10	2.00	1.40	0.99	0.71
8.00	1.53	1.44	1.37	1.33	0.93	0.66	0.47
8.25	0.87	0.82	0.78	0.76	0.54	0.39	0.28
8.50	0.49	0.47	0.45	0.44	0.32	0.23	0.17
8.75	0.28	0.27	0.26	0.27	0.19	0.16	0.11
9.00	0.16	0.16	0.16	0.16	0.13	0.10	0.08

Source: EPA, 1985

accordance with the precautionary principle, when developing water quality criteria, many countries are also taking into account substances (including genetically modified organisms) for which there is insufficient data and which are presently only suspected of belonging to the category of hazardous substances.

The elaboration of water quality criteria for hazardous substances is a lengthy and resource-expensive process. Comprehensive laboratory studies assessing the impact of hazardous substances on aquatic organisms often need to be carried out, in addition to a general search and analysis of published literature. In Canada, for example, the average cost of developing a criterion for a single substance by means of a literature search and analysis is in the order of Canadian $ 50,000. In Germany, the average cost of laboratory studies for developing a criterion for a single hazardous substance amounts to about DM 200,000 (McGirr *et al.*, 1991).

Some countries have shared the costs and the workload for developing water quality criteria amongst their regional and national agencies. For example, the Canadian Council of Resource and Environment Ministers (CCREM) has established a task force, consisting of specialists from the federal, provincial and territorial governments, to develop a joint set of Canadian water quality criteria. This has enabled them to produce, at a modest cost, a much more comprehensive set of criteria than would have been possible by individual efforts. It has also ended the confusion caused by the use of different criteria by each provincial government. In Germany, a joint task force was established to develop water quality criteria and to establish water quality objectives. This task force consists of scientists and water managers appointed by the Federal Government and the *Länder* authorities responsible for water management.

In some countries attempts have been made to apply water quality criteria elaborated in other countries (see Box 2.1). In such cases, it is necessary to establish that the original criteria were developed for similar environmental conditions and that at least some of the species on which toxicity studies were carried out occur in relevant water bodies of the country considering adoption of other national criteria. On many occasions, the application of water quality criteria from other countries requires additional ecotoxicological testing. An example of the adaptation of a traditional water pollution indicator is the use of a 3-day BOD in the tropics rather than the customary 5-day BOD developed for temperate countries.

2.3.2   Raw water used for drinking-water supply

These criteria describe water quality requirements imposed on inland waters intended for abstraction of drinking water and apply only to water which is treated prior to use. In developing countries, large sections of the population may be dependent on raw water for drinking purposes without any treatment whatsoever. Microbiological requirements as well as inorganic and organic substances of significance to human health are included.

Quality criteria for raw water generally follow drinking-water criteria and even strive to attain them, particularly when raw water is abstracted directly to drinking-water treatment works without prior storage. Drinking-water criteria define a quality of water that can be safely consumed by humans throughout their lifetime. Such criteria have been developed by international organisations and include the WHO *Guidelines for Drinking-water Quality* (WHO, 1984, 1993) and the EU *Council Directive of 15 July 1980 Relating to the Quality of Water Intended for Human Consumption* (80/778/EEC), which covers some 60 quality variables. These guidelines and directives are used by countries, as appropriate, in establishing enforceable national drinking-water quality standards.

Water quality criteria for raw water used for drinking-water treatment and supply usually depend on the potential of different methods of raw water treatment to reduce the concentration of water contaminants to the level set by drinking-water criteria. Drinking-water treatment can range from simple physical treatment and disinfection, to chemical treatment and disinfection, to intensive physical and chemical treatment. Many countries strive to ensure that the quality of raw water is such that it would only be necessary to use near-natural conditioning processes (such as bank filtration or low-speed sand filtration) and disinfection in order to meet drinking-water standards.

In member states of the European Union, national quality criteria for raw water used for drinking-water supply follow the EU *Council Directive of 16 June 1975 Concerning the Quality Required of Surface Water Intended for the Abstraction of Drinking Water in Member States* (75/440/EEC). This directive covers 46 criteria for water quality variables directly related to public health (microbiological characteristics, toxic compounds and other substances with a deleterious effect on human health), variables affecting the taste and odour of the water (e.g. phenols), variables with an indirect effect on water quality (e.g. colour, ammonium) and variables with general relevance to water quality (e.g. temperature). A number of these variables are now being revised.

2.3.3 Irrigation

Poor quality water may affect irrigated crops by causing accumulation of salts in the root zone, by causing loss of permeability of the soil due to excess sodium or calcium leaching, or by containing pathogens or contaminants which are directly toxic to plants or to those consuming them. Contaminants in irrigation water may accumulate in the soil and, after a period of years, render the soil unfit for agriculture. Even when the presence of pesticides or pathogenic organisms in irrigation water does not directly affect plant growth, it may potentially affect the acceptability of the agricultural product for sale or consumption. Criteria have been published by a number of countries as well as by the Food and Agriculture Organization of the United Nations (FAO). Some examples are given in Table 2.3. Quality criteria may also differ considerably from one country to another, due to different annual application rates of irrigation water.

Water quality criteria for irrigation water generally take into account, amongst other factors, such characteristics as crop tolerance to salinity, sodium concentration and phytotoxic trace elements. The effect of salinity on the osmotic pressure in the unsaturated soil zone is one of the most important water quality considerations because this has an influence on the availability of water for plant consumption. Sodium in irrigation waters can adversely affect soil structure and reduce the rate at which water moves into and through soils. Sodium is also a specific source of damage to fruits. Phytotoxic trace elements such as boron, heavy metals and pesticides may stunt the growth of plants or render the crop unfit for human consumption or other intended uses.

Table 2.3   Selected water quality criteria for irrigational waters (mg l^{-1})

Element	FAO	Canada	Nigeria
Aluminium	5.0	5.0	5.0
Arsenic	0.1	0.1	0.1
Cadmium	0.01	0.01	0.01
Chromium	0.1	0.1	0.1
Copper	0.2	0.2–1.0[1]	0.2–1.0[1]
Manganese	0.2	0.2	0.2
Nickel	0.2	0.2	0.2
Zinc	2.0	1.0–5.0[2]	0.0–5.0[2]

[1]   Range for sensitive and tolerant crops, respectively.
[2]   Range for soil pH > 6.5 and soil pH > 6.5, respectively.

Sources: FAO, 1985; CCREM, 1987; FEPA, 1991

As discussed in the chapters on wastewater as a resource (Chapter 4) and the case study on wastewater use in the Mezquital Valley, Mexico (Case Study 7), both treated and untreated wastewater is being used for the irrigation of crops. In these cases, the WHO *Health Guidelines for the Use of Wastewater in Agriculture and Aquaculture* (WHO, 1989) should be consulted to prevent adverse impacts on human health and the environment (Hespanhol, 1994).

2.3.4   Livestock watering

Livestock may be affected by poor quality water causing death, sickness or impaired growth. Variables of concern include nitrates, sulphates, total dissolved solids (salinity), a number of metals and organic micropollutants such as pesticides. In addition, blue-green algae and pathogens in water can present problems. Some substances, or their degradation products, present in water used for livestock may occasionally be transmitted to humans. The purpose of quality criteria for water used for livestock watering is, therefore, to protect both the livestock and the consumer.

Criteria for livestock watering usually take into account the type of livestock, the daily water requirements of each species, the chemicals added to the feed of the livestock to enhance the growth and to reduce the risk of disease, as well as information on the toxicity of specific substances to the different species. Some examples of criteria for livestock watering are given in Table 2.4.

Table 2.4   Selected water quality criteria for livestock watering (mg l^{-1})

Water quality variable	Canadian criteria	Nigerian criteria
Nitrate plus nitrite	100	100
Sulphates	1,000	1,000
Total dissolved solids	3,000	3,000
Blue-green algae	Avoid heavy growth of blue-green algae	Avoid heavy growth of blue-green algae
Pathogens and parasites	Water of high quality should be used	Water of high quality should be used (chlorinate, if necessary, sanitation and manure management must be emphasised to prevent contamination of water supply sources)

Sources: CCREM, 1987; FEPA, 1991; ICPR, 1991

2.3.5   Recreational use

Recreational water quality criteria are used to assess the safety of water to be used for swimming and other water-sport activities. The primary concern is to protect human health by preventing water pollution from faecal material or from contamination by micro-organisms that could cause gastro-intestinal illness, ear, eye or skin infections. Criteria are therefore usually set for indicators of faecal pollution, such as faecal coliforms and pathogens. There has been a considerable amount of research in recent years into the development of other indicators of microbiological pollution including viruses that could affect swimmers. As a rule, recreational water quality criteria are established by government health agencies.

The EU *Council Directive of 8 December 1975 Concerning the Quality of Bathing Water* (76/160/EEC)for example, established quality criteria containing both guideline values and maximum allowable values for microbiological parameters (total coliforms, faecal coliforms, faecal , streptococci, salmonella, entero viruses) together with some physico-chemical parameters such as pH, mineral oils and phenols. This Directive also prescribes that member states should individually establish criteria for eutrophication-related parameters, toxic heavy metals and organic micropollutants.

Recreational use of water is often given inadequate consideration. For example, in the United Nations Economic Commission for Latin America and the Caribbean (ECLAC) region, several tourist areas are effected to various degrees by water pollution, including such popular resorts as Guanabara Bay in Brazil, Vina del Mar in Chile and Cartagena in Colombia. Offensive smells, floating materials (particularly sewage solids) and certain other pollutants can create aesthetically repellent conditions for recreational uses of water and reduce its visual appeal. Even more important, elevated levels of bacteriological contamination and, to a lesser extent, other types of pollution can render water bodies unsuitable for recreational use. This is of particular concern in those countries of the region where tourism is an important source of foreign exchange and employment. In general, recreation is a much neglected use of water within the ECLAC region and is hardly considered in the process of water management despite the available information that suggests that pollution in recreational areas is a serious problem. This is of particular concern as the recreational use of water is very popular in the region and is also concentrated in water bodies closest to the large metropolitan areas. Many of these are increasingly contaminated by domestic sewage and industrial effluents (ECLAC, 1989).

2.3.6   Amenity use

Criteria have been established in some countries aimed at the protection of the aesthetic properties of water. These criteria are primarily orientated towards visual aspects. They are usually narrative in nature and may specify, for example, that waters must be free of floating oil or other immiscible liquids, floating debris, excessive turbidity, and objectionable odours. The criteria are mostly non-quantifiable because of the different sensory perception of individuals and because of the variability of local conditions.

2.3.7   Protection of aquatic life

Within aquatic ecosystems a complex interaction of physical and biochemical cycles exists. Anthropogenic stresses, particularly the introduction of chemicals into water, may adversely affect many species of aquatic flora and fauna that are dependent on both abiotic and biotic conditions. Water quality criteria for the protection of aquatic life may take into account only physico-chemical parameters which tend to define a water quality that protects and maintains aquatic life, ideally in all its forms and life stages, or they may consider the whole aquatic ecosystem.

Water quality parameters of concern are traditionally dissolved oxygen (because it may cause fish kills at low concentrations) as well as phosphates, ammonium and nitrate (because they may cause significant changes in community structure if released into aquatic ecosystems in excessive amounts). Heavy metals and many synthetic chemicals can also be ingested and absorbed by organisms and, if they are not metabolised or excreted, they may bioaccumulate in the tissues of the organisms. Some pollutants can also cause carcinogenic, reproductive and developmental effects.

Table 2.5   Water quality objectives for the River Rhine related to metals in suspended matter

Water quality variable	Quality objective (mg kg^{-1})
Cadmium	1.0
Chromium	100.0
Copper	50.0
Lead	100.0
Mercury	0.5
Nickel	50.0
Zinc	50.0

Source: ICPR, 1991

International Commission for the Protection of the Rhine against Pollution, for example, criteria related to metals in suspended matter have been converted into water quality objectives (Table 2.5). At present the quality objectives are mainly based on limit values developed for the spreading of sewage sludge on agricultural areas and taking into account, if available, information related to the adverse impacts of sewage sludge on soil organisms. At a later stage, the quality objectives will be revised in order to protect organisms living in or on sediment, as well as to protect the marine ecosystem (for situations where dredged sediment is disposed of at sea).

Recent experience in Germany and the Netherlands suggests that a far greater number of substances than previously considered are a potential threat to aquatic and terrestrial life. Consequently, present water quality criteria for sediment are now under revision.

2.4   Water quality objectives

A major advantage of the water quality objectives approach to water resources management is that it focuses on solving problems caused by conflicts between the various demands placed on water resources, particularly in relation to their ability to assimilate pollution. The water quality objectives approach is sensitive not just to the effects of an individual discharge, but to the combined effects of the whole range of different discharges into a water body. It enables an overall limit on levels of contaminants within a water body to be set according to the required uses of the water.

The advantage of the fixed emission approach (see Chapter 5) is that it treats industry equitably requiring the use of best available technology for treating hazardous, as well as a number of conventional, water pollutants wherever the industry is located. This is seen to be a major advantage for transboundary catchment areas where all riparian countries are required to meet the same standards and no country has an unfair trade advantage.

It is generally recognised that water quality objectives, the setting of emission limits on the basis of best available technology, and the use of best environmental practice should all form part of an integrated approach to the prevention, control and reduction of pollution in inland surface waters. In most cases, water quality objectives serve as a means of assessing pollution reduction measures. For example, if emission limits are set for a given water body on the basis of best available technology, toxic effects may, nevertheless, be experienced by aquatic communities under certain conditions. In addition, other sensitive water uses, such as drinking-water supplies, may be adversely affected. The water quality objectives help to evaluate, therefore, whether additional efforts are needed when water resources protection is based on using emission limits for point sources according to the best available technology or on best environmental practice for non-point sources.

Experience gained in some countries suggests that catchment planning plays an essential role in setting water quality objectives (see Box 2.2). It provides the context in which the demands of all water users can be balanced against water quality requirements. Catchment planning also provides the mechanism for assessing and controlling the overall loading of pollutants within whole river catchments and, ultimately, into the sea, irrespective of the uses to which those waters are put. The need for "catchment accountability" is becoming increasingly important in order to ensure that both national and international requirements to reduce pollutant loadings are properly planned and achieved.

The elaboration of water quality objectives and the selection of the final strategy for their achievement necessarily involves an analysis of the technical, financial and other implications associated with the desired improvements in water quality. The technical means available to reduce inputs of pollutants into waters have a direct bearing on the elaboration of water quality objectives by indicating the technical feasibility of attaining the threshold values set in the objectives. Economic factors are also taken into account because the attainment of a certain objective may require the allocation of considerable financial resources and may also have an impact on investment, employment and, inevitably, on prices paid by consumers.

The establishment of a time schedule for attaining water quality objectives is mainly influenced by the existing water quality, the urgency of control measures and the prevailing economic and social conditions. In some countries, a step-by-step approach to establish water quality objectives is applied.

Box 2.2   Examples of the setting of water quality objectives

Canada and the United States of America
Water quality objectives for watercourses may also take into account quality requirements of downstream lakes and reservoirs. For example, water quality objectives for nutrient concentrations in tributaries of the Great Lakes consider the quality requirements of the given watercourse, as well as of the lake system. Similarly, requirements for the protection of the marine environment, in particular of relatively small enclosed seas, need to be taken into consideration when setting water quality objectives for watercourses (as has been done, for example, in the setting of water quality objectives for the Canadian rivers flowing into the sea).

Germany
A methodology to establish water quality objectives for aquatic communities, fisheries, suspended particulate matter/sediment, drinking-water supply, irrigation, and recreation has been drawn up by a German task force (see section 2.3.1). This task force will further develop its methodology, for example, by comparing numerical values established according to its methodology with the results of the monitoring of 18 toxic and carcinogenic substances in surface waters. Once water quality objectives are established, they will be used by regional authorities as a basis for water resources planning. However, such water quality objectives will not be considered as generally obligatory but regional authorities will have to decide, case by case, which water uses are to be protected in a given water body and which water quality objectives are to be applied. Obligatory limit values will only be established in the course of the implementation of water management plans by competent water management authorities. The authorities will decide on the specific uses of a given water body that should be protected and the relevant water quality objective that should be used, taking into account the water uses that have been licensed for that water body.

Sources: McGirr *et al.*, 1991; UNECE, 1993

This gradual introduction is probably also the best approach for developing countries. For example, in order to establish a baseline for water pollution control measures, priority should be given to setting objectives for variables related to the oxygen regime and nutrients (e.g. dissolved oxygen, BOD, NH_3-N) because many rivers in the world suffer from pollution by organic matter (Meybeck *et al.*, 1989). Experience also suggests that establishing water quality objectives initially only for a limited number of variables can focus attention on key water quality attributes and lead to marked improvements in water quality in a cost-effective manner. It is of the utmost importance that the objectives are understandable to all parties involved in pollution control and are convertible into operational and cost-effective measures which can be addressed through targets to reduce pollution. It should also be possible to monitor, with existing networks and equipment, compliance with such objectives. Objectives that are either vague or too sophisticated should be avoided. The objectives should also have realistic time schedules.

Targets to improve water quality are usually set at two levels. The first represents the ultimate goal at which no adverse effects on the considered human uses of the water would occur and at which the functions of the aquatic ecosystems would be maintained and/or protected. This level corresponds, in most countries, with the most stringent water quality criterion among all of the considered water uses, with some modifications made to account for specific site conditions. A second level is also being defined that should be reached within a fixed period of time. This level is a result of a balance between what is desirable from an environmental point of view and what is feasible from an economic and technical point of view. This second level allows for a step-by-step approach that finally leads to the first level. Additionally, some countries recommend a phased approach, which starts with rivers and catchments of sensitive waters and is progressively extended to other water bodies during a second phase.

In many countries, water quality objectives are subject to regular revisions in order to adjust them, among other things, to the potential of pollution reduction offered by new technologies, to new scientific knowledge on water quality criteria, and to changes in water use.

Current approaches to the elaboration and setting of water quality objectives differ between countries. These approaches may be broadly grouped as follows:

- Establishment of water quality objectives for individual water bodies (including transboundary waters) or general water quality objectives applicable to all waters within a country.
- Establishment of water quality objectives on the basis of water quality classification schemes.

The first approach takes into account the site-specific characteristics of a given water body and its application requires the identification of all current and reasonable potential water uses. Designated uses of waters or "assets" to be protected may include: direct extraction for drinking-water supply, extraction into an impoundment prior to drinking-water supply, irrigation of crops, watering of livestock, bathing and water sports, amenities, fish and other aquatic organisms.

In adopting water quality objectives for a given water body, site-specific physical, chemical, hydrological and biological conditions are taken into consideration. Such conditions may be related to the overall chemical composition (hardness, pH, dissolved oxygen), physical characteristics (turbidity, temperature, mixing regime), type of aquatic species and biological community structure, and natural concentrations of certain substances (e.g. metals or nutrients). These site-specific factors may affect the exposure of aquatic organisms to some substances or the usability of water for human consumption, livestock watering, irrigation and recreation.

In some countries general water quality objectives are set for all surface waters in a country, irrespective of site-specific conditions. They may represent a compromise after balancing water quality requirements posed by individual water uses and economic, technological and other means available to meet these requirements at a national level. Another approach is to select water quality criteria established for the most sensitive uses (e.g. drinking-water supply or aquatic life) as general water quality objectives.

2.4.1 Water quality classification schemes

Many countries in the ECE and ESCAP regions have established water quality objectives for surface waters based on classification schemes (see Box 2.3). A number of these countries require, as a policy goal, the attainment of water quality classes I or II (which characterise out of a system of four or five quality classes, excellent or good water quality) over a period of time. In the UK, this approach has even led to statutory water quality objectives for England and Wales under the 1989 Water Act (NRA, 1991). Generally, before establishing quality objectives on the basis of classification systems, comprehensive water quality surveys have to be carried out.

The ECE has recently adopted a *Standard Statistical Classification of Surface Freshwater Quality for the Maintenance of Aquatic Life* (UNECE, 1994). The class limits are primarily derived from ecotoxicological considerations and based on the research work of the US EPA. As a general rule, the orientation of the classification system towards aquatic life implies that the class limits are more conservative than they would be if targeted at other water uses. In addition to variables that characterise the oxygen regime, eutrophication and acidification of waters, the system includes hazardous substances such as aluminium, arsenic, heavy metals, dieldrin, dichloro-diphenyltrichloroethane (DDT) and its metabolites, endrin, heptachlor, lindane, pentachlorophenol, polychlorinated biphenyls (PCBs) and free ammonia. It also includes gross α- and β-activity. Concentrations of

Box 2.3 Examples of water quality classification schemes

India
In India, five water quality classes have been designated (A–E) on the basis of the water quality requirements for a particular use:
Class A waters for use as drinking water source without conventional treatment but after disinfection.
Class B waters for use for organised outdoor bathing.
Class C waters for use as drinking water source with conventional treatment followed by disinfection.
Class D waters to maintain aquatic life (i.e. propagation of wildlife and fisheries).
Class E waters for use for irrigation, industrial cooling and controlled waste disposal.

The five classes have been used to set quality objectives for stretches of the Yamuna and Ganga rivers, and surveys have been carried out to compare the actual river-quality classification with that required to sustain the designated best use. Where a river has multiple uses, the quality objectives are set for the most stringent (best) use requirements. After comparing ambient water quality with the designated water quality objective, any deficiencies will require appropriate pollution control measures on the discharges, including discharges in upstream stretches. This system is also helpful for the planning and siting of industry. No industries are permitted to discharge any effluent in stretches of rivers classified in Class A.

A pollution control action plan was drawn up for the Ganga in 1984 and the Ganga Project Directorate was established under the Central Ganga Authority in 1985. This Directorate oversees pollution control and abatement (ESCAP, 1990). The table below shows the improvements in water quality classification that were achieved by 1987. The classification and zoning of 12 other major rivers has also been recently accomplished.

A comparison between water quality objectives for the Ganga and results of classifications in 1982 and 1987

Zone	River length (km)	Water quality objective class	Results of water quality classification 1982	1987	Critical primary water quality characteristics
Source to Rishikesh	250	A	B	B	Total coliform
Rishikesh to Kannauj	420	B	C	B	Total coliform, BOD
Kannauj to Trighat	730	B	D	B	Total coliform, BOD
Trighat to Kalyani	950	B	C	B	Total coliform
Kalyani to Diamond Harbour	100	B	D	B	Total coliform

Thailand
There are many forms of legislation on water quality control and management in Thailand including laws, acts, regulations and ministerial notifications established by various agencies, depending on their relative areas of responsibility. The objectives of setting water quality requirements and standards in Thailand are: to control and maintain water quality at a level that suits the activities of all concerned, to protect public health, and to conserve natural resources and the natural environment.

The Ministry of Agriculture and Cooperatives has established, for example, regulations concerning water quality for irrigation, wildlife and fisheries. The Office of the

Box 2.3   Continued

National Environmental Board (ONEB) is responsible for defining the water quality requirements of receiving waters, as well as for setting quality standards for fresh-waters, domestic effluents and effluents from agricultural point sources (e.g. pig farms and aquaculture). These standards are based on sets of water quality criteria. For example, in order to protect commercial fishing, ONEB has set the following allowable concentrations of pesticides in aquatic organisms: DDT 5.0 mg kg^{-1}, endrin 0.5 mg kg^{-1}, lindane 0.5 mg kg^{-1}, heptachlor 0.3 mg kg^{-1} and parathion 0.2 mg kg^{-1} (ESCAP, 1990).

The system of surface water resources classification and standards in Thailand is based on the idea that the concentrations of water quality parameters in Class I shall correspond to the natural concentrations. Variables characterising the oxygen and nutrient regimes, the status of coliform bacteria, phenols, heavy metals, pesticides and radioactivity are being considered.

Sources: ESCAP, 1990; Venugupal, 1994

United Kingdom
The Water Resources Act of 1991 enabled the UK Government to prescribe a system for classifying the quality of controlled waters according to specified requirements. These requirements (for any classification) consist of one or more of the following:

- General requirements as to the purposes for which the waters to which the classification is applied are to be suitable.
- Specific requirements as to the substances that are to be present, in or absent from, the water and as to the concentrations of substances which are, or are required to be, present in the water.
- Specific requirements as to other characteristics of those waters.

Future regulations will describe whether such requirements should be satisfied by reference to particular sampling procedures. Then, for the purpose of maintaining or improving the quality of controlled waters the Government may, by serving a notice on the National Rivers Authority (NRA), establish with reference to one or more of the classifications to be described as above, the water quality objectives for any waters and the date by which the objectives shall apply.

The purpose of the new system is to provide a firmer framework for deciding the policy that governs the determination of consent for discharges into each stretch of controlled waters and the means by which pollution from diffuse sources can be dealt with. The system will be extended to coastal waters, lakes and groundwater. It will pro-vide a basis for a requirement for steady improvement in quality in polluted waters.

The 1994 Surface Waters (River Ecosystem) (Classification) Regulations intro-duced a component of the scheme designed to make water quality targets statutory. The NRA has set water quality targets for all rivers and these are known as river quality objectives (RQO) and they establish a defined level of protection for aquatic life. They are used for planning the maintenance and improvement of river quality and to provide a basis for setting consent to discharge effluent into rivers, and guide decisions on the NRA's other actions to control and prevent pollution. Achieving the required RQO will help to sustain the use of rivers for recreation, fisheries and wildlife, and to protect the interest of abstractors. The water quality classification scheme used to set RQO plan-ning targets is known as the river ecosystem scheme. It provides a nationally consistent basis for setting RQO. The scheme comprises five classes which reflect the chemical quality requirements of communities of plants and animals occurring in the rivers. The standards defining these classes reflect differing degrees of pollution by organic matter and other common pollutants.

Sources: NRA, 1991, 1994; UNECE, 1993

hazardous substances in Class I and Class II should be below current detection limits. In Class III, their presence can be detected but the concentrations should be below chronic and acute values. For Class IV, concentrations may exceed the chronic values occasionally but should not lead to chronically toxic conditions, either with respect to concentration, duration or frequency (Table 2.6).

The system has been applied to a number of internal and transboundary waters within the region, and is expected to constitute a basis for setting water quality objectives at border sections of transboundary waters under the *Convention on the Protection and Use of Transboundary Watercourses and International Lakes* (UNECE, 1992). The system is expected to be supplemented by water quality objectives for specific hazardous substances as well as by a system of biologically-based water quality objectives.

2.4.2   Transboundary waters

To date, there are only a few examples of transboundary waters for which water quality objectives have been established. Examples include the Great Lakes and some transboundary rivers in North America (St Croix, St John, St Lawrence, River Poplar, River Rainy, Red River of the North) and the River Rhine in Europe (Tables 2.5 and 2.7 and Box 2.4). Following the provisions of the *Convention on the Protection and Use of Transboundary Watercourses and International Lakes* (UNECE, 1992), water quality objectives are being developed for some other transboundary surface waters in Europe, including the rivers Danube, Elbe and Oder and their tributaries. In the ESCAP region, countries riparian to the Mekong river are jointly developing water quality objectives for the main river and other watercourses in the catchment area.

2.4.3   The ecosystem approach

The application of the ecosystem approach in water management has led to the development of objectives for safeguarding the functional integrity of aquatic ecosystems. The functional integrity of aquatic ecosystems is characterised by a number of physical, chemical, hydrological, and biological factors and their interaction.

Ecosystem objectives attempt to describe a desired condition for a given ecosystem through a set of variables, taking into account the ecological characteristics and uses of the water. Ecosystem objectives may specify the level or condition of certain biological properties that could serve as indicators of the overall condition or "health" of the aquatic ecosystem. Ecosystem objectives are used in combination with water quality objectives, and objectives relating to hydrological conditions.

Table 2.6 ECE standard statistical classification of surface freshwater quality for the maintenance of aquatic life

Variables	Class I	Class II	Class III	Class IV	Class V
Oxygen regime					
DO (%)					
epilimnion (stratified waters)	90–110	70–90 or 110–120	50–70 or 120–130	30–50 or 130–150	< 30 or > 150
hypolimnion (stratified waters)	90–70	70–50	50–30	30–10	< 10
unstratified waters	90–70	70–50 or 110–120	50–30 or 120–130	30–10 or 130–150	< 10 or > 150
DO (mg l^{-1})	> 7	7–6	6–4	4–3	< 3
COD-Mn (mg O$_2$ l^{-1})	< 3	3–10	10–20	20–30	> 30
COD-Cr (mg O$_2$ l^{-1})	–	–	–	–	–
Eutrophication					
Total P (µg l^{-1}) [1]	< 10 (< 15)	10–25 (15–40)	25–50 (40–75)	50–125 (75–190)	> 125 (> 190)
Total N (µg l^{-1}) [1]	< 300	300–750	750–1,500	1,500–2,500	> 2,500
Chlorophyll a (µg l^{-1}) [1]	< 2.5 (< 4)	2.5–10 (4–15)	10–30 (15–45)	30–110 (45–165)	> 110 (> 165)
Acidification					
pH [2]	9.0–6.5	6.5–6.3	6.3–6.0	6.0–5.3	< 5.3
Alkalinity (mg CaCO$_3$ l^{-1})	> 200	200–100	100–20	20–10	< 10
Metals					
Aluminium (µg l^{-1}; pH 6.5)	< 1.6	1.6–3.2	3.2–5	5–75	> 75
Arsenic (µg l^{-1}) [3]	< 10	10–100	100–190	190–360	> 360
Cadmium (µg l^{-1}) [4]	< 0.07	0.07–0.53	0.53–1.1	1.1–3.9	> 3.9
Chromium (µg l^{-1}) [3]	< 1	1–6	6–11	11–16	> 16

Continued

Table 2.6 Continued

Variables	Class I	Class II	Class III	Class IV	Class V
Metals					
Copper (µg l^{-1}) [4]	< 2	2–7	7–12	12–18	> 18
Lead (µg l^{-1}) [4]	< 0.1	0.1–1.6	1.6–3.2	3.2–82	> 82
Mercury (µg l^{-1}) [4]	< 0.003	0.003–0.007	0.007–0.012	0.012–2.4	> 2.4
Nickel (µg l^{-1}) [4]	< 15	15–87	87–160	160–1,400	> 1,400
Zinc (µg l^{-1}) [4]	< 45	45–77	77–110	110–120	> 120
Chlorinated micropollutants and other hazardous substances					
Dieldrin (µg l^{-1})	na	na	< 0.0019	0.0019–2.5	> 2.5
DDT and metabolites (µg l^{-1})	na	na	< 0.001	0.001–1.1	> 1.1
Endrin (µg l^{-1})	na	na	< 0.0023	0.0023–0.18	> 0.18
Heptachlor (µg l^{-1})	na	na	< 0.0038	0.0038–0.52	> 0.52
Lindane (µg l^{-1})	na	na	< 0.08	0.08–2.0	> 2.0
Pentachlorophenol (µg l^{-1})	na	na	< 13	13–20	> 20
PCBs (µg l^{-1})	na	na	< 0.014	0.014–2.0	> 2.0
Free ammonia (NH$_3$)	na	na	–	–	–
Radioactivity					
Gross-alpha activity (mBq l^{-1})	< 50	50–100	100–500	500–2,500	> 2,500
Gross-beta activity (mBq l^{-1})	< 200	200–500	500–1,000	1,000–2,500	> 2,500

Measures falling on the boundary between two classes are to be classified in the lower class.

na Not applicable

– No value set at present

1 Data in brackets refer to flowing waters.

2 Values > 9.0 are disregarded in the classification of acidification.

3 Applicable for hardness from about 0.5 to 8 meq l^{-1}. Arsenic V and chromium III to be converted to arsenic III and chromium VI, respectively.

4 Applicable for hardness from about 0.5 to 8 meq l^{-1}.

Source: UNECE, 1994

Table 2.7   Water quality objectives for the River Rhine related to organic
substances

Water quality variable	Water quality objective ($\mu g \ l^{-1}$)	Basis for elaboration[1]
Tetrachloromethane	1.0	Drw+aqL
Trichloromethane	0.6	aqL
Aldrin, Dieldrin, Endrin, Isodrin	0.0001 (per substance)	aq+terrL
Endosulfan	0.003	aqL
Hexachlorobenzene	0.0005	aqL
Hexachlorobutadien	0.001	aqL
PCB 28, 52, 101,180, 138, 153	0.001 (per substance)	aqL
1-Chloro-4-nitro-Benzen	1.0	Drw
1-Chloro-2-nitro-Benzen	1.0	Drw+aqL
Trichlorobenzene	0.1	aqL
Pentachlorophenol	0.001	aq+terrL
Trichloroethen	1.0	Drw
Tetrachloroethen	1.0	Drw
3,4-Dichloroanilin	0.1	aqL
2-Chloroanilin	0.1	Drw+aqL
3-Chloroanilin	0.1	Drw
4-Chloroanilin	0.01	aqL
Parathion(-ethyl)	0.0002	aqL
Parathion(-methyl)	0.01	aqL
Benzene	0.1	aqL
1, 1, 1-Trichloroethane	1.0	Drw
1, 2-Dichloroethane	1.0	aqL
Azinphos-methyl	0.001	aqL
Bentazon	0.1	Drw
Simazine	0.1	Drw+aqL
Atrazine	0.1	Drw+aqL
Dichlorvos	0.001	aqL
2-Chlorotoluol	1.0	Drw
4-Chlorotoluol	1.0	Drw
Tributyl tin-substances	0.001	aqL
Triphenyl tin-substances	0.001	aqL
Trifluralin	0.1	aqL
Fenthion	0.01	aqL

[1]   Water quality objectives have been set on the basis of water quality criteria for drinking-water supply (Drw), drinking-water supply and aquatic life (Drw+aqL) and/or aquatic life (aqL), as well as on the basis of toxicity testing on selected species of aquatic and terrestrial life (aq+terrL).

Source: ICPR, 1991

Box 2.4 An example of water quality objectives for transboundary rivers:
the Rhine

Water quality objectives established for the River Rhine are based on the four
major elements of the Rhine Action Programme aimed at:
- Improving the ecosystem of the river in such a way that sensitive species
  which were once indigenous in the Rhine will return.
- Guaranteeing the future production of drinking water from the Rhine.
- Reducing the pollution of the water by hazardous substances to such a
  level that sediment can be used on land or dumped at sea without causing
  harm.
- Protecting the North Sea against the negative effects of the Rhine water.
  At present, water quality objectives for the River Rhine cover 50 priority
substances, such as heavy metals, organic micropollutants as well as ammonium and phosphorus discharged from industries, municipalities or
agriculture. The list of these substances was established on the basis of catchment inventories of point and diffuse sources of discharges of substances into
the Rhine. The established water quality objectives should be complied with
by the year 2000.

Source: ICPR, 1994

Ecosystem objectives are expressed by a set of species, referred to as the
target variables. The target variables as a whole are usually a cross-section of
the aquatic ecosystem that provides a fairly representative picture of ecosys-
tem conditions and include, for example:
- Species from all types of aquatic habitats.
- Species from the benthos, water column, water surface and shores.
- Species from high and low parts of the food web.
- Plants and animals.
- Sessile, migratory and non-migratory species.

In order to ensure, for example, the functional integrity of Lake Ontario,
specific ecosystem objectives were developed that enabled the waters of the
lake to support diverse, healthy, reproducing and self-sustaining communities in a dynamic equilibrium. Human health considerations were also taken
into account in this process, because the lake should be usable for drinking water
and recreation, as well as for the safe human consumption of fish and wildlife.

Determining whether the functioning integrity of the ecosystem is
achieved requires a set of measurable and quantitative indicators. Extensive
studies were undertaken to select appropriate biological indicators that would
supplement conventional physical and chemical measurements of water

quality. Comprehensive criteria were elaborated by the Aquatic Ecosystems Objectives Committee (established within the framework of the 1978 Great Lakes Water Quality Agreement) to judge the suitability of candidate organisms to serve as indicators of the quality of the ecosystem.

Based on these criteria, a number of organisms were considered suitable indicators for the Great Lakes. For oligotrophic systems of Lake Superior, the lake trout *Salvelinus namaycush* (the top aquatic predator) and the amphipod *Pontoporeia hoyi* (the major benthic macro-invertebrate of a cold-water community) were selected. For mesotrophic systems, the walleye *Stizostedion vitreum*, which has many characteristics in common with the lake trout, has recently been chosen, together with the mayfly *Hexagenia limbata* which was considered as representative of a diverse benthic community because of its requirements for clean, well-oxygenated sediment. Work is under way to select mammalian, avian and reptilian species.

The absence or presence of Atlantic salmon is used as an indicator of the functional integrity of the Rhine riverine ecosystem and of the quality of its water. Other indicator species and groups of species are also being observed. A method of ecological and biological assessment known as AMOEBA, the Dutch acronym for "a general method of ecosystem description and assessment", was developed in the Netherlands (ten Brink *et al.*, 1990). As indicators for the Rhine ecosystems, for example, some 30 species have been selected. For each species, the abundance for the period 1900–30 (a pragmatic selection to represent an unaffected situation) was estimated and compared with that of the present day, thus showing the deviation from the quasi-natural situation. Other aquatic ecosystems have also been characterised by choosing about 30 species which can be regarded as representative for their specific ecosystem.

2.4.4   Implementation and monitoring compliance

Usually, a two-step approach is applied for achieving compliance with water quality objectives. The urgency of control measures, for example, has a direct bearing on the time schedule for attaining water quality objectives for specific hazardous substances. For examples, the immediate and substantial reduction of emissions of three organic substances (carbon tetrachloride, DDT and pentachlorophenol) was stipulated by the EU *Council Directive 86/280/EEC of 12 June 1986 on Limit Values and Quality Objectives for Discharges of Certain Dangerous Substances Included in List I of the Annex to Directive 76/464/EEC.* Water quality objectives for these substances had to be complied with after a period of one and a half years (as of 1 January 1988). In some countries and for other hazardous substances, a time period of

5–10 years has been set to attain water quality objectives by the substantial reduction of emissions from point sources. Some countries, notably those participating in the Rhine Action Programme, have chosen the year 2000 as the deadline for attaining water quality objectives. Phasing out the use of certain substances, reducing nutrient discharges and changing agricultural practices usually requires a longer time period and the need to comply with relevant water quality objectives should take this fact into consideration.

Water quality objectives may be subject to revision and to adjustment in order to take account of potential reductions in pollution offered by new technology, of new scientific knowledge on water quality criteria and of changes in water use. Practical experience suggests, however, that dischargers should not be asked to review their practices on the basis of newly elaborated water quality objectives too often, or too soon after establishing practices designed to comply with earlier water quality objectives. In the UK, for example, the 1991 Water Act allows for the revision of water quality objectives although such a review can only take place at intervals of at least five years, or if the NRA requests such a review following consultation with water users and other appropriate bodies.

Adaptation of monitoring programmes, surveillance systems and laboratory practices are necessary in the implementation of water quality objectives. Two problems deserve special mention in this respect: the detection limit of laboratory equipment, and agreement on a criterion for the attainment of water quality objectives. Experience in many countries shows that laboratory techniques should have a detection limit that is preferably, one order of magnitude lower than the water quality objective for the substance in question. In the case of hazardous substances, this may require sophisticated laboratory equipment and specially trained personnel and may lead to high costs for laboratory analyses.

Usually, water quality criteria used as a basis for elaborating water quality objectives already have a built-in margin of safety so that, for the most part, a certain number of monitoring data may exceed the established water quality objective and forewarn of a certain risk, without requiring immediate action. In most cases, this advance warning ensures that action can be taken before real damage occurs. For hazardous substances some countries consider that the water quality objective has been attained if at least 90 per cent of all measurements (within a period of three years) comply with the water quality objective, or if the mean value of the concentration of the substance is less than, or equal to, half the concentration value of the water quality objective. Another approach requires the use of the mean concentration of a substance as an evaluation criterion. This approach is followed, for example, by the EU

Council Directive 86/280/EEC. In some countries, the median value for phosphorus is taken as a criterion for assessing the attainment of its water quality objective.

2.5 Conclusions and recommendations

Many chemical substances emitted into the environment from anthropogenic sources pose a threat to the functioning of aquatic ecosystems and to the use of water for various purposes. The need for strengthened measures to prevent and to control the release of these substances into the aquatic environment has led many countries to develop and to implement water management policies and strategies based on, amongst others, water quality criteria and objectives. To provide further guidance for the elaboration of water quality criteria and water quality objectives for inland surface waters, and to strengthen international co-operation the following recommendations have been put forward (UNECE, 1993):

- The precautionary principle should be applied when selecting water quality parameters and establishing water quality criteria to protect and maintain individual uses of waters.
- In setting water quality criteria, particular attention should be paid to safeguarding sources of drinking-water supply. In addition, the aim should be to protect the integrity of aquatic ecosystems and to incorporate specific requirements for sensitive and specially protected waters and their associated environment, such as wetland areas and the surrounding areas of surface waters which serve as sources of food and as habitats for various species of flora and fauna.
- Water-management authorities in consultation with industries, municipalities, farmers' associations, the general public and others should agree on the water uses in a catchment area that are to be protected. Use categories, such as drinking-water supply, irrigation, livestock watering, fisheries, leisure activities, amenities, maintenance of aquatic life and the protection of the integrity of aquatic ecosystems, should be considered wherever applicable.
- Water-management authorities should be required to take appropriate advice from health authorities in order to ensure that water quality objectives are appropriate for protecting human health.
- In setting water quality objectives for a given water body, both the water quality requirements for uses of the relevant water body, as well as downstream uses, should be taken into account. In transboundary waters, water quality objectives should take into account water quality requirements in the relevant catchment area. As far as possible, water quality requirements for water uses in the whole catchment area should be considered.

- Under no circumstances should the setting of water quality objectives (or modification thereof to account for site-specific factors) lead to the deterioration of existing water quality.
- Water quality objectives for multipurpose uses of water should be set at a level that provides for the protection of the most sensitive use of a water body. Among all identified water uses, the most stringent water quality criterion for a given water quality variables should be adopted as a water quality objective.
- Established water quality objectives should be considered as the ultimate goal or target value indicating a negligible risk of adverse effects on use of the water and on the ecological functions of waters.
- The setting of water quality objectives should be accompanied by the development of a time schedule for compliance with the objectives that takes into account action which is technically and financially feasible and legally implementable. Where necessary, a step-by-step approach should be taken to attain water quality objectives, making allowance for the available technical and financial means for pollution prevention, control and reduction, as well as the urgency of control measures.
- The setting of emission limits on the basis of best available technology, the use of best environmental practices and the use of water quality objectives as integrated instruments of prevention, control and reduction of water pollution, should be applied in an action-oriented way. Action plans covering point and diffuse pollution sources should be designed, that permit a step-by-step approach to water pollution control which are both technically and financially feasible.
- Both the water quality objectives and the timetable for compliance should be subject to revision at appropriate time intervals in order to adjust them to new scientific knowledge on water quality criteria, to changes in water use in the catchment area, and to achievements in pollution control from point and non-point sources.
- The public should be kept informed about water quality objectives that have been established and about measures taken to attain these objectives.

2.6   References

Alabaster, J.S. and Lloyd, R. 1982 *Water Quality Criteria for Freshwater Fish.* 2nd edition. Published on behalf of Food and Agriculture Organization of the United Nations by Butterworth, London, 361 pp.

ten Brink, B.J.E., Hosper, S.H. and Colijn, F. 1990 *A Quantitative Method for Description and Assessment of Ecosystems: the AMOEBA Approach.* ECE Seminar on Ecosystems Approach to Water Management, Oslo, May 1991.

ENVWA/SEM.5/R.33, United Nations Economic Commission for Europe, United Nations, Geneva.

CCREM 1987 *Canadian Water Quality Guidelines*. Prepared by the Task Force on Water Quality Guidelines of the Canadian Council of Resource and Environment Ministers, Ottawa.

Chiaudani, G. and Premazzi, G. 1988 *Water Quality Criteria in Environmental Management*. Report EUR 11638 EN, Commission of the European Communities, Luxembourg.

Dick, R.I. 1975 *Water Quality Criteria, Goals and Standards*. Second WHO Regional Seminar on Environmental Pollution: Water Pollution, Manila, WPR/W.POLL/3, WHO Regional Office for the Western Pacific, Manila.

ECLAC 1989 *The Water Resources of Latin America and the Caribbean: Water Pollution*. LC/L.499, United Nations Economic Commission for Latin America and the Caribbean, United Nations, Santiago de Chile.

Enderlein, R.E. 1995 Protecting Europe's water resources: policy issues. *Wat. Sci. Tech.*, **31**(8), 1–8.

Enderlein, R.E. 1996 Protection and sustainable use of waters: agricultural policy requirements in Europe. *HRVAT. VODE*, **4**(15), 69–76.

EPA 1976 *Quality Criteria for Water*. EPA-440/9-76-023, United States Environmental Protection Agency, Washington, D.C.

EPA 1985 *Ambient Water Quality Criteria for Ammonia*. EPA-440/5-85-001, United States Environmental Protection Agency, Washington, D.C.

EPA 1986 *Ambient Water Quality Criteria for Dissolved Oxygen*. EPA 440/5-86-003, United States Environmental Protection Agency, Washington, D.C.

ESCAP 1990 *Water Quality Monitoring in the Asian and Pacific Region*. Water Resources Series No. 67, United Nations Economic and Social Commission for Asia and the Pacific, United Nations, New York.

FAO 1985 *Water Quality for Agriculture*. Irrigation and Drainage Paper No. 29, Rev. 1. Food and Agriculture Organization of the United Nations, Rome.

FEPA 1991 *Proposed National Water Quality Standards*. Federal Environmental Protection Agency, Nigeria.

Hespanhol, I. 1994 WHO Guidelines and National Standards for Reuse and Water Quality. *Wat. Res.*, **28**(1), 119–124.

ICPR 1991 *Konzept zur Ausfüllung des Punktes A.2 des APR über Zielvorgaben. Lenzburg, den 2. Juli 1991 (Methodology to implement item A.2 of the Rhine Action Programme related to water quality objectives, prepared at Lenzbourg on 2 July 1991)*. PLEN 3/91, International Commission for the Protection of the Rhine against Pollution, Koblenz, Germany.

ICPR 1994 Unpublished contribution of the secretariat of the International Commission for the Protection of the Rhine against Pollution, Koblenz (Germany), to the ECE project on policies and strategies to protect transboundary waters. United Nations Economic Commission for Europe, Geneva.

ICWE 1992 *The Dublin Statement and Report of the Conference, Development Issues for the 21st Century.* International Conference on Water and the Environment. 26–31 January 1992, Dublin, Ireland.

McGirr, D., Gottschalk, Ch. and Lindholm, O. 1991 Unpublished contributions of the Governmentally designated rapporteurs from Canada, Germany and Norway for the ECE project on water quality criteria and objectives. United Nations Economic Commission for Europe, Geneva.

Meybeck, M., Chapman, D. and Helmer, R. 1989 *Global Freshwater Quality. A First Assessment.* Published on behalf of WHO and UNEP by Blackwell Reference, Oxford, 306 pp.

NRA 1991 *Proposals for Statutory Water Quality Objectives.* Report of the National Rivers Authority, England and Wales, Water Quality Series No. 5., HMSO, London.

NRA 1994 *Water Quality Objectives. Procedures used by the National Rivers Authority for the Purpose of the Surface Waters (River Ecosystem) (Classification) Regulation 1994.* National Rivers Authority, England and Wales, Bristol.

Pham Thi Dung, 1994 Residue pesticides monitoring in the Mekong basin. In: *Mekong Water Quality Monitoring and Assessment Expert Meeting.* Bangkok, 29–30 November 1993. Report prepared by the Mekong Secretariat, MKG/R 94002, Mekong Secretariat, Bankok.

UNCED 1992 *Agenda 21, Chapter 18. Protection of the Quality and Supply of Freshwater Resources: Application of Integrated Approaches to the Development, Management and Use of Water Resources.* United Nations Conference on Environment and Development, Rio de Janeiro, 14 June 1992.

UNECE 1992 *Convention on the Protection and Use of Transboundary Watercourses and International Lakes*, Helsinki, 17 March 1992, United Nations Economic Commission for Europe, United Nations, New York and Geneva.

UNECE 1993 *Protection of Water Resources and Aquatic Ecosystems.* Water Series, No. 1. ECE/ENVWA/31, United Nations Economic Commission for Europe, United Nations, New York.

UNECE 1994 Standard Statistical Classification of Surface Freshwater Quality for the Maintenance of Aquatic Life. In: *Readings in International*

Environment Statistics, United Nations Economic Commission for Europe, United Nations, New York and Geneva.

UNECE 1995 *Protection and Sustainable Use of Waters: Recommendations to ECE Governments.* Water Series, No. 2. ECE/CEP/10, United Nations Economic Commission for Europe, United Nations, New York and Geneva.

UNESCO/WHO 1978 *Water Quality Surveys. A Guide for the Collection and Interpretation of Water Quality Data.* Studies and Reports in Hydrology, No. 23, United Nations Educational Scientific and Cultural Organization, Paris, 350 pp.

United Nations, 1994 *Consolidated List of Products Whose Consumption and/or Sale Have Been Banned, Withdrawn, Severely Restricted or Not Approved by Governments.* Fifth issue, ST/ESA/239, United Nations, New York.

Venugopal, T. 1994 Water and air quality monitoring programme in India: An overview for GEMS/Water. Unpublished report of the Central Pollution Control Board, Ministry of Environment and Forests of India, New Delhi.

WHO 1984 *Guidelines for Drinking-Water Quality, Volume 2, Health Criteria and Other Supporting Information.* World Health Organization, Geneva.

WHO 1989 *Health Guidelines for the Use of Wastewater in Agriculture and Aquaculture. Report of a Scientific Group Meeting.* Technical Report Series, No. 778, World Health Organization, Geneva.

WHO 1993 *Guidelines for Drinking-Water Quality, Volume 1, Recommendations.* 2nd edition, World Health Organization, Geneva.

Chapter 3[*]

TECHNOLOGY SELECTION

3.1   Integrating waste and water management

Economic growth in most of the world has been vigorous, especially in the so-called newly industrialising countries. Nearly all new development activity creates stress on the "pollution carrying capacity" of the environment. Many hydrological systems in developing regions are, or are getting close to, being stressed beyond repair. Industrial pollution, uncontrolled domestic discharges from urban areas, diffuse pollution from agriculture and livestock rearing, and various alterations in land use or hydro-infrastructure may all contribute to non-sustainable use of water resources, eventually leading to negative impacts on the economic development of many countries or even continents. Lowering of groundwater tables (e.g. Middle East, Mexico), irreversible pollution of surface water and associated changes in public and environmental health are typical manifestations of this kind of development.

Technology, particularly in terms of performance and available wastewater treatment options, has developed in parallel with economic growth. However, technology cannot be expected to solve each pollution problem. Typically, a wastewater treatment plant transfers 1 m^3 of wastewater into 1–2 litres of concentrated sludge. Wastewater treatment systems are generally capital-intensive and require expensive, specialised operators. Therefore, before selecting and investing in wastewater treatment technology it is always preferable to investigate whether pollution can be minimised or prevented. For any pollution control initiative an analysis of cost-effectiveness needs to be made and compared with all conceivable alternatives. This chapter aims to provide guidance in the technology selection process for urban planners and decision makers. From a planning perspective, a number of questions need to be addressed before any choice is made:

- *Is wastewater treatment a priority in protecting public or environmental health?* Near Wuhan, China, an activated sludge plant for municipal

*   *This chapter was prepared by S. Veenstra, G.J. Alaerts and M. Bijlsma*

sewage was not financed by the World Bank because the huge Yangtse River was able to absorb the present waste load. The loan was used for energy conservation, air pollution mitigation measures (boilers, furnaces) and for industrial waste(water) management. In Wakayama, Japan, drainage was given a higher priority than sewerage because many urban areas were prone to periodic flooding. The human waste is collected by vacuum trucks and processed into dry fertiliser pellets. Public health is safeguarded just as effectively but the huge investment that would have been required for sewerage (two to three times the cost of the present approach) has been saved.

- *Can pollution be minimised by recovery technologies or public awareness?* South Korea planned expansion of sewage treatment in Seoul and Pusan based on a linear growth of present tap water consumption (from 120 l cap^{-1} d^{-1} to beyond 250 l cap^{-1} d^{-1}). Eventually, this extrapolation was found to be too costly. Funds were allocated for promoting water saving within households; this allowed the eventual design of sewers and treatment plants to be scaled down by half.

- *Is treatment most feasible at centralised or decentralised facilities?* Centralised treatment is often devoted to the removal of common pollutants only and does not aim to remove specific individual waste components. However, economies of scale render centralised treatment cheap whereas decentralised treatment of separate waste streams can be more specialised but economies of scale are lost. By enforcing land-use and zoning regulations, or by separating or pre-treating industrial discharges before they enter the municipal sewer, the overall treatment becomes substantially more effective.

- *Can the intrinsic value of resources in domestic sewage be recovered by reuse?* Wastewater is a poorly valued resource. In many arid regions of the world, domestic and industrial sewage only has to be "conditioned" and then it can be used in irrigation, in industries as cooling and process water, or in aqua- or pisciculture (see Chapter 4). Treatment costs are considerably reduced, pollution is minimised, and economic activity and labour are generated. Unfortunately, many of these potential alternatives are still poorly researched and insufficiently demonstrated as the most feasible.

Ultimately, for each pollution problem one strategy and technology are more appropriate in terms of technical acceptability, economic affordability and social attractiveness. This applies to developing, as well as to industrialising, countries. In developing countries, where capital is scarce and poorly-skilled workers are abundant, solutions to wastewater treatment should preferably be low-technology orientated. This commonly means that the technology chosen is less mechanised and has a lower degree of automatic process control, and that

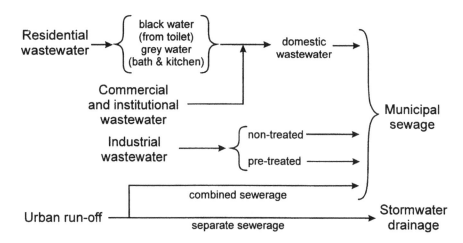

Figure 3.1 Origin and flows of wastewater in an urban environment

construction, operation and maintenance aim to involve locally available personnel rather than imported mechanised components. Such technologies are rather land and labour intensive, but capital and hardware extensive. However, the final selection of treatment technology may be governed by the origin of the wastewater and the treatment objectives (see Figure 3.2).

3.2 Wastewater origin, composition and significance

3.2.1 Wastewater flows

Municipal wastewater is typically generated from domestic and industrial sources and may include urban run-off (Figure 3.1). Domestic wastewater is generated from residential and commercial areas, including institutional and recreational facilities. In the rural setting, industrial effluents and stormwater collection systems are less common (although polluting industries sometimes find the rural environment attractive for uncontrolled discharge of their wastes). In rural areas the wastewater problems are usually associated with pathogen-carrying faecal matter. Industrial wastewater commonly originates in designated development zones or, as in many developing countries, from numerous small-scale industries within residential areas.

In combined sewerage, diffuse urban pollution arises primarily from street run-off and from the overflow of "combined" sewers during heavy rainfall; in the rural context it arises mainly from run-off from agricultural fields and carries pesticides, fertiliser and suspended matter, as well as manure from livestock.

Table 3.1   Typical domestic water supply and wastewater production in industrial, developing and (semi-)arid regions ($l\ cap^{-1}\ d^{-1}$)

Water supply service	Industrial regions	Developing regions	(Semi-)arid regions
Handpump or well	na	< 50	< 25
Public standpost	na	50–80	20–40
House connection	100–150	50–125	40–80
Multiple connection	150–250	100–250	80–120
Average wastewater flow	85–200	65–125	35–75

na   Not applicable

Within the household, tap water is used for a variety of purposes, such as washing, bathing, cooking and the transport/flushing of wastes. Wastewater from the toilet is termed "black" and the wastewater from the kitchen and bathroom is termed "grey". They can be disposed of separately or they can be combined. Generally, the wealthier a community, the more waste is disposed by water-flushing off-site. Such wastewater disposal may become a public problem for downstream areas.

Domestic wastewater generation is commonly expressed in litres per capita per day ($l\ cap^{-1}\ d^{-1}$) or as a percentage of the specific water consumption rate. Domestic water consumption, and hence wastewater production, typically depends on water supply service level, climate and water availability (Table 3.1). In moderate climates and in industrialising countries, 75 per cent of consumed tap water typically ends up as sewage. In more arid regions this proportion may be less than 50 per cent due to high evaporation and seepage losses and typical domestic water-use practices.

Industrial water demand and wastewater production are sector-specific. Industries may require large volumes of water for cooling (power plants, steel mills, distillation industries), processing (breweries, pulp and paper mills), cleaning (textile mills, abattoirs), transporting products (beet and sugar mills) and flushing wastes. Depending on the industrial process, the concentration and composition of the waste flows can vary significantly. In particular, industrial wastewater may have a wide variety of micro-contaminants which add to the complexity of wastewater treatment. The combined treatment of many contaminants may result in reduced efficiency and high treatment unit costs (US$ m^{-3}).

Hourly, daily, weekly and seasonal flow and load fluctuations in industries (expressed as $m^3\ s^{-1}$ or $m^3\ d^{-1}$ and as $kg\ s^{-1}$ or $kg\ d^{-1}$ of contaminant, respectively) can be quite considerable, depending on in-plant procedures such as production shifts and workplace cleaning. As a consequence, treatment

Table 3.2   Major classes of municipal wastewater contaminants and their significance and origin

Contaminant	Significance	Origin
Settleable solids (sand, grit)	Settleable solids may create sludge deposits and anaerobic conditions in sewers, treatment facilities or open water	Domestic, run-off
Organic matter (BOD); Kjeldahl-nitrogen	Biological degradation consumes oxygen and may disturb the oxygen balance of surface water; if the oxygen in the water is exhausted anaerobic conditions, odour formation, fish kills and ecological imbalance will occur	Domestic, industrial
Pathogenic micro-organisms	Severe public health risks through transmission of communicable water borne diseases such as cholera	Domestic
Nutrients (N and P)	High levels of nitrogen and phosphorus in surface water will create excessive algal growth (eutrophication). Dying algae contribute to organic matter (see above)	Domestic, rural run-off, industrial
Micro-pollutants (heavy metals, organic compounds)	Non-biodegradable compounds may be toxic, carcinogenic or mutagenic at very low concentrations (to plants, animals, humans). Some may bioaccumulate in food chains, e.g. chromium (VI), cadmium, lead, most pesticides and herbicides, and PCBs	Industrial, rural run-off (pesticides)
Total dissolved solids (salts)	High levels may restrict wastewater use for agricultural irrigation or aquaculture	Industrial, (salt water intrusion)

Source: Metcalf and Eddy Inc., 1991

plants are confronted with varying loading rates which may reduce the removal efficiency of the processes. Removal of hazardous or slowly-biodegradable contaminants requires a constant loading and operation of the treatment plant in order to ensure process and performance stability. To accommodate possible fluctuations, equalisation or buffer tanks are provided to even out peak flows. Fluctuations in domestic sewage flow are usually repetitive, typically with two peak flows (morning and evening), with the minimum flow at night.

3.2.2   Wastewater composition

Wastewater can be characterised by its main contaminants (Table 3.2) which may have negative impacts on the aqueous environment in which they are discharged. At the same time, treatment systems are often specific, i.e. they are meant to remove one class of contaminants and so their overall performance deteriorates in the presence of other contaminants, such as from industrial effluents. In particular, oil, heavy metals, ammonia, sulphide and

Table 3.3 Variation in the composition of domestic wastewater

Contaminant	Specific production $(g\ cap^{-1}\ d^{-1})^2$	Concentration[1] $(mg\ l^{-1})^2$
Total dissolved solids	100–150	400–2,500
Total suspended solids	40–80	160–1,350
BOD	30–60	120–1,000
COD	70–150	280–2,500
Kjeldahl-nitrogen (as N)	8–12	30–200
Total phosphorus (as P)	1–3	4–50
Faecal coliform (No. per 100 ml)	10^6–10^9	4×10^6–1.7×10^7

BOD Biochemical oxygen demand
COD Chemical oxygen demand

[1] Assuming water consumption rate of 60–$250\ l\ cap^{-1}\ d^{-1}$
[2] Except for faecal coliforms

toxic constituents may damage sewers (e.g. by corrosion) and reduce treatment plant performance. Therefore, municipalities may set additional criteria for accepting industrial waste flows into their sewers.

Contaminated sewage may be rendered unfit for any productive use. Several in-factory treatment technologies allow selective removal of contaminants and their recovery to a high degree and purity. Such recovery may cover part of the investment if it is applied to concentrated waste streams. For example, in textile mills pigments and caustic solution can be recovered by ultra-filtration and evaporation, while chromium (VI) can be recovered by chemical precipitation in leather tanneries. In other situations, sewage can be made suitable for irrigation or for reuse in industry.

Domestic waste production per capita is fairly constant but the concentration of the contaminants varies with the amount of tap water consumed (Table 3.3). For example, municipal sewage in Sana'a, Yemen (water consumption of $80\ l\ cap^{-1}\ d^{-1}$), is four times more concentrated in terms of chemical oxygen demand (COD) and total suspended solids (TSS) than in Latin American cities (water consumption is around $300\ l\ cap^{-1}\ d^{-1}$). In addition, seepage or infiltration of groundwater may occur because the sewerage system may not be watertight. Similarly, many sewers in urban areas collect overflows from septic tanks which affects the sewage quality. Depending on local conditions and habits (such as level of nutrition, staple food composition and kitchen habits) typical waste parameters may need adjustment to these local conditions. Sewage composition may also be fundamentally altered if industrial discharges are allowed into the municipal sewerage system.

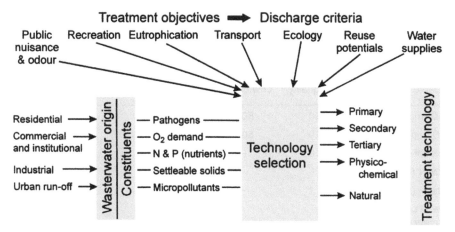

Figure 3.2 Treatment technology selection in relation to the origin of the wastewater, its constituents and formulated treatment objectives as derived from set discharge criteria

3.3   Wastewater management

3.3.1   Treatment objectives

Technology selection eventually depends upon wastewater characteristics and on the treatment objectives as translated into desired effluent quality. The latter depends on the expected use of the receiving waters. Effluent quality control is typically aimed at public health protection (for recreation, irrigation, water supply), preservation of the oxygen content in the water, prevention of eutrophication, prevention of sedimentation, preventing toxic compounds from entering the water and food chains, and promotion of water reuse (Figure 3.2). These water uses are translated into emission standards or, in many countries, water quality "classes" which describe the desired quality of the receiving water body (see also Chapter 2). Emission or effluent standards can be set which may take into account the technical and financial feasibility of wastewater treatment. In this way a treatment technology, or any other action, can be taken to remove or prevent the discharge of the contaminants of concern. Standards or guidelines may differ between countries. Table 3.4 gives some typical discharge standards applied in many industrialised and developing countries, in relation to the expected quality or use of the receiving waters.

3.3.2   Sanitation solutions for domestic sewage

The increasing world population tends to concentrate in urban communities. In densely populated areas the sanitary collection, treatment and disposal of wastewater flows are essential to control the transmission of waterborne

Table 3.4   Typical treated effluent standards as a function of the intended use of the receiving waters

Variable	Discharge in surface water		Discharge in water sensitive to eutrophication	Effluent use in irrigation and aquaculture
	High quality	Low quality		
BOD (mg l^{-1})	20	50	10	100[1]
TSS (mg l^{-1})	20	50	10	< 50[1]
Kjeldahl-N (mg l^{-1})	10	–	5	–
Total N (mg l^{-1})	–	–	10	–
Total P (mg l^{-1})	1	–	0.1	–
Faecal coliform (No. per 100 ml)	–	–	–	< 1,000
Nematode eggs per litre	–	–	–	< 1
SAR	–	–	–	< 5
TDS (salts) (mg l^{-1})	–	–	–	< 500[2]

–	No standards set		[1]	Agronomic norm
BOD	Biochemical oxygen demand		[2]	No resriction on crop selection
TSS	Total suspended solids			
SAR	Sodium adsorption ratio			Sources: Ayers and Westcot, 1985; WHO, 1989
TDS	Total dissolved solids			

diseases. They are also essential for the prevention of non-reversible degradation of the urban environment itself and of the aquatic systems that support the hydrological cycle, as well as for the protection of food production and biodiversity in the region surrounding the urban area. For rural populations, which still account for 75 per cent of the total population in developing countries (WHO, 1992), concern for public health is the main justification for investing in water and sanitation improvement. In both settings, the selected technologies should be environmentally sustainable, appropriate to the local conditions, acceptable to the users, and affordable to those who have to pay for them. Simple solutions that are easily replicable, that allow further upgrading with subsequent development, and that can be operated and maintained by the local community, are often considered the most appropriate and cost-effective.

The first issue to be addressed is whether sanitary treatment and disposal should be provided on-site (at the level of a household or apartment block) or whether collection and centralised, off-site treatment is more appropriate. Irrespective of whether the setting is urban or rural, the main deciding criteria are population density (people per hectare) and generated wastewater flow (m^3 ha^{-1} d^{-1}) (Figure 3.3). Population density

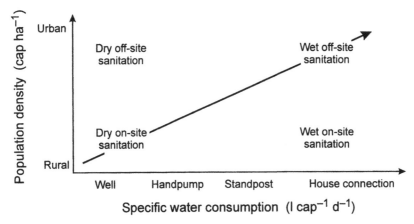

Figure 3.3 Classification of basic sanitation strategies. The trend of development is from dry on-site to wet off-site sanitation (After Veenstra, 1996)

determines the availability of land for on-site sanitation and strongly affects the unit cost per household. Dry and wet sanitation systems can be distinguished by whether water is required for flushing the solids and conveying them through a sewerage system. The present trend for increasing tap water consumption (1 cap^{-1} d^{-1}) together with increasing urban population densities, is creating a continuing interest in off-site sanitation as the main future strategy for wastewater collection, treatment and disposal.

In wealthier urban situations, off-site solutions are often more appropriate because the population density does not allow for percolation of large quantities of wastewater into the soil. In addition, the associated risk of groundwater pollution reported in many cities in Africa and the Middle East is prohibitive for on-site sanitation. Frequently, towns and city districts cannot afford such capital-intensive solutions due to the lower population density per hectare and the resultant high unit costs involved. Depending on the local physical and socio-economic circumstances, on-site sanitation may be feasible, although if this is not satisfactory, intermediate technologies are available such as small bore sewerage. The latter approach combines on-site collection of sewage in a septic tank followed by off-site disposal of the settled effluent by small-bore sewers. The settled solids accumulate in the septic tank and are periodically removed (desludged). The advantage of this system is that the unit cost of small bore sewerage is much lower (Sinnatamby *et al.*, 1986).

3.3.3 Level of wastewater treatment

To achieve water quality targets an extensive infrastructure needs to be developed and maintained. In order to get industries and domestic polluters to

Table 3.5   The phased expansion and upgrading of wastewater treatment plants in industrialised countries to meet ever stricter effluent standards

Decade	Treatment objective	Treatment	Operations included
1950–60	Suspended/coarse solids removal	Primary	Screening, removal of grit, sedimentation
1970	Organic matter degradation	Secondary	Biological oxidation of organic matter
1980	Nutrient reduction (eutrophication)	Tertiary	Reduction of total N and total P
1990	Micro-pollutant removal	Advanced	Physicochemical removal of micro-pollutants

pay for the huge cost of such infrastructure, legislation has to be set up based on the principle of "The Polluter Pays". Treatment objectives and priorities in industrialised countries have been gradually tightened over the past decades. This resulted in the so-called first, second and third generation of treatment plants (Table 3.5). This step-by-step approach allowed for determination of the "optimum" (desired) effluent quality and how it can be reached by wastewater treatment, on the basis of full scale experience. As a consequence, existing wastewater treatment plants have been continually expanding and upgrading; primary treatment plants were extended with a secondary step, while secondary treatment plants are now being completed with tertiary treatment phases.

In general, the number of available treatment technologies, and their combinations, is nearly unlimited. Each pollution problem calls for its specific, optimal solution involving a series of unit operations and processes (Table 3.6) put together in a flow diagram.

Primary treatment generally consists of physical processes involving mechanical screening, grit removal and sedimentation which aim at removal of oil and fats, settleable suspended and floating solids; simultaneously at least 30 per cent of biochemical oxygen demand (BOD) and 25 per cent of Kjeldahl-N and total P are removed. Faecal coliform numbers are reduced by one or two orders of magnitude only, whereas five to six orders of magnitude are required to make it fit for agricultural reuse.

Secondary treatment mainly converts biodegradable organic matter (thereby reducing BOD) and Kjeldahl-N to carbon dioxide, water and nitrates by means of microbiological processes. These aerobic processes require oxygen which is usually supplied by intensive mechanical aeration. For sewage with relatively elevated temperatures anaerobic processes can also be

Table 3.6   Classification of common wastewater treatment processes according to their level of advancement

Primary	Secondary	Tertiary	Advanced
Bar or bow screen	Activated sludge	Nitrification	Chemical treatment
Grit removal	Extended aeration	Denitrification	Reverse osmosis
Primary sedimentation	Aerated lagoon	Chemical precipitation	Electrodialysis
Comminution	Trickling filter	Disinfection	Carbon adsorption
Oil/fat removal	Rotating bio-discs	(Direct) filtration	Selective ion exchange
Flow equalisation	Anaerobic treatment/UASB	Chemical oxidation	Hyperfiltration
pH neutralisation	Anaerobic filter	Biological P removal	Oxidation
Imhoff tank	Stabilisation ponds	Constructed wetlands	Detoxification
	Constructed wetlands	Aquaculture	
	Aquaculture		

UASB   Upflow Anaerobic Sludge Blanket

applied. Here the organic matter is converted into a mixture of methane and carbon dioxide (biogas).

In primary and secondary treatment, sludges are produced with a volume of less than 0.5 per cent of the wastewater flow. Heavy metals and other micro-pollutants tend to accumulate in the sludge because they often adsorb onto suspended particles. Nowadays, the problems associated with wastewater treatment in industrialised countries have shifted gradually from the wastewater treatment itself towards treatment and disposal of the generated sludges.

Non-mechanised wastewater treatment by stabilisation ponds, constructed wetlands or aquaculture using macrophytes can, to a large extent, provide adequate secondary and tertiary treatment. As the biological processes are not intensified by mechanical equipment, large land areas are required to provide sufficient retention time to allow for a high degree of contaminant removal.

Tertiary treatment is designed to remove the nutrients, total N (comprising Kjeldahl-N, nitrate and nitrite) and total P (comprising particulate and soluble phosphorus) from the secondary effluents. Additional suspended solids removal and BOD reduction is achieved by these processes. The objective of tertiary treatment is mainly to reduce the potential occurrence of eutrophication in sensitive, surface water bodies.

Advanced treatment processes are normally applied to industrial wastewater only, for removal of specific contaminants. Advanced treatment is commonly preceded by physicochemical coagulation and flocculation. Where a high quality effluent may be required for reclamation of groundwater by recharge or for discharge to recreational waters, advanced treatment steps may also be added to the conventional treatment plant.

Table 3.7 reviews the degree to which contaminants are removed by treatment processes or operations. Most treatment processes are only truly efficient in the removal of a small number of pollutants.

3.3.4   Best available technology

In taking precautionary or preventive end-of-pipe treatment measures, authorities may by statute require the polluter, notably industry, to rely on the best available technology (BAT), the best available technology not entailing excessive costs (BATNEEC), the best environmental practices (BEP) and the best practical environmental option (BPEO) (see also Chapter 5).

The best available technology is generally accessible technology, which is the most effective in preventing or minimising pollution emissions. It can also refer to the most recent treatment technology available. Assessing whether a certain technology is the best available requires comparative technical assessment of the different treatment processes, their facilities and their methods of operation which have been recently and successfully applied for a prolonged period of time, at full scale.

The BATNEEC adds an explicit cost/benefit analysis to the notion of best available technology. "Not entailing excessive cost" implies that the financial cost should not be excessive in relation to the financial capability of the industrial sector concerned, and to the discharge reductions or environmental protection envisaged.

The best environmental practices and the best practicable environmental options have a wider scope. The BPEO requires identification of the least environmentally damaging method for the discharge of pollutants, whereas a requirement for the use of treatment processes must be based upon BATNEEC. Best practical environmental option policies also require that the treatment measures avoid transferring pollution or pollutants, from one medium to another (from water into sludge for example). Thus BPEO takes into account the cross-media impacts of the technology selected to control pollution.

3.3.5   Selection criteria

The general criteria for technology selection comprise:
- *Average, or typical, efficiency and performance of the technology.* This is usually the criterion considered to be best in comparative studies. The possibility that the technology might remove other contaminants than those which were the prime target should also be considered an advantage. Similarly, the pathways and fate of the removed pollutants after treatment

should be analysed, especially with regard to the disposal options for the sludges in which the micro-pollutants tend to concentrate.

- *Reliability of the technology.* The process should, preferably, be stable and resilient against shock loading, i.e. it should be able to continue operation and to produce an acceptable effluent under unusual conditions. Therefore, the system must accommodate the normal inflow variations, as well as infrequent, yet expected, more extreme conditions. This pertains to the wastewater characteristics (e.g. occasional illegal discharges, variations in flow and concentrations, high or low temperatures) as well as to the operational conditions (e.g. power failure, pump failure, poor maintenance). During the design phase, "what if" scenarios should be considered. Once disturbed, the process should be fairly easy to repair and to restart.

- *Institutional manageability.* In developing countries few governmental agencies are adequately equipped for wastewater management. In order to plan, design, construct, operate and maintain treatment plants, appropriate technical and managerial expertise must be present. This could require the availability of a substantial number of engineers with postgraduate education in wastewater engineering, access to a local network of research for scientific support and problem solving, access to good quality laboratories, and experience in management and cost recovery. In addition, all technologies (including those thought "simple") require devoted and experienced operators and technicians who must be generated through extensive education and training.

- *Financial sustainability.* The lower the financial costs, the more attractive the technology. However, even a low cost option may not be financially sustainable, because this is determined by the true availability of funds provided by the polluter. In the case of domestic sanitation, the people must be willing and able to cover at least the operation and maintenance cost of the total expenses. The ultimate goal should be full cost recovery although, initially, this may need special financing schemes, such as cross-subsidisation, revolving funds, and phased investment programmes.

- *Application in reuse schemes.* Resource recovery contributes to environmental as well as to financial sustainability. It can include agricultural irrigation, aqua- and pisciculture, industrial cooling and process water re-use, or low-quality applications such as toilet flushing. The use of generated sludges can only be considered as crop fertilisers or for reclamation if the micro-pollutant concentration is not prohibitive, or the health risks are not acceptable.

- *Regulatory determinants.* Increasingly, regulations with respect to the desired water quality of the receiving water are determined by what is considered to be technically and financially feasible. The regulatory agency

Table 3.7   Percentage efficiency for potential contaminant removal of different processes and operations used in wastewater treatment and reclamation

Variable or contaminant	Primary treatment	Activated sludge (AS)	Nitrification	Denitrification	Trickling filter	RBC	Coag.–Floc.–Sedim.[1]	Filtration after AS	Carbon adsorption
BOD	25–50	> 50	> 50	25	> 50	> 50	> 50	25–50	> 50
COD	25–50	> 50	> 50	25	> 50		> 50	25–50	25–50
TSS	> 50	> 50	> 50	25	> 50	> 50	>50	> 50	> 50
NH$_3$-N	25	> 50	> 50	25–50		> 50	25	25–50	25–50
NO$_3$-N					> 50			25–50	25
Phosphorus	25	25–50	> 50	> 50			> 50	> 50	> 50
Alkalinity		25–50					25–50	> 50	
Oil and grease	> 50	> 50	> 50				25–50		25–50
Total coliform		> 50	> 50		25		> 50		> 50
TDS									
Arsenic	25–50	25–50	25–50				25–50	> 50	25
Barium		25–50	25				25–50	25	
Cadmium	25–50	> 50	> 50		25	25–50	> 50	25–50	25
Chromium	25–50	> 50	> 50		25	> 50	> 50	25–50	25–50
Copper	25–50	> 50	> 50		> 50	> 50	> 50	25	25–50
Fluoride							25–50		25
Iron	25–50	> 50	> 50		25–50	> 50	> 50	> 50	> 50
Lead	> 50	> 50	> 50		25–50	> 50	> 50	25	25–50
Manganese	25	25–50	25–50		25		25–50	> 50	25–50
Mercury	25	25	25		25	> 50	25	25–50	25
Selenium	25	25	25				25	> 50	25
Silver	> 50	> 50	> 50		25–50		> 50		25–50
Zinc	25–50	25–50	> 50		> 50	> 50	> 50		> 50
Colour	25	25–50	25–50		25		> 50	25–50	> 50
Foaming agents	25–50	> 50	> 50		> 50		25–50		> 50
Turbidity	25–50	> 50	> 50	25	25–50		> 50	> 50	> 50
TOC	25–50	> 50	> 50	25	25–50		> 50	25–50	> 50

The percentage relates to the influent concentration. Where no percentage efficiency is indicated no data are available, the results are inconclusive or there is an increase.

[1] Coagulation–Floculation–Sedimentation
RBC   Rotating Biological Contactor (bio-disc)
BOD   Biochemical oxygen demand
COD   Chemical oxygen demand

Table 3.7   Continued

Variable or contaminant	Ammonia stripping	Selective ion exchange	Break point chlorination	Reverse osmosis	Overland flow	Irrigation	Infiltration–percolation	Chlorination	Ozone
BOD		25–50		> 50	> 50	> 50	> 50		25
COD	25	25–50		> 50	> 50	> 50	> 50		> 50
TSS		> 50		> 50	> 50	> 50	> 50		
NH₃-N	> 50	> 50	> 50	> 50	> 50	> 50	> 50		
NO₃-N					25–50				
Phosphorus				> 50	> 50	> 50	> 50		
Alkalinity							25–50		
Oil and grease					> 50	> 50	> 50		
Total coliform			> 50		> 50	> 50	> 50	> 50	> 50
TDS				> 50					
Arsenic									
Barium									
Cadmium							25		
Chromium									
Copper							> 50		
Fluoride							25–50		
Iron									
Lead							25–50		
Manganese				> 50					
Mercury									
Selenium									
Silver									
Zinc							> 50		
Colour				> 50	> 50	> 50	> 50		> 50
Foaming agents				> 50	> 50	> 50	> 50		25
Turbidity				> 50	> 50	> 50	> 50		
TOC	25	25		> 50	> 50	> 50	> 50		> 50

TSS   Total suspended solids
TDS   Total dissolved solids
TOC   Total organic carbon

Source: Metcalf and Eddy, 1991

then imposes the use of specified, up-to-date technology (BAT or BATNEEC) upon domestic or industrial dischargers, rather than prescribing the required discharge standards.

3.4   Pollution prevention and minimisation

Although end-of-pipe approaches have reduced the direct release of some pollutants into surface water, limitations have been encountered. For example, end-of-pipe treatment transfers contaminants from the water phase into a sludge or gaseous phase. After disposal of the sludge, migration from the disposed sludge into the soil and groundwater may occur. Over the past years, there has been growing awareness that many end-of-pipe solutions have not been as effective in improving the aquatic environment as was expected. As a result, the approach is now shifting from "waste management" to "pollution prevention and waste minimisation", which is also referred to as "cleaner production".

Pollution prevention and waste minimisation covers an array of technical and non-technical measures aiming at the prevention of the generation of waste and pollutants. It is the conceptual approach to industrial production that demands that all phases of the product life cycle should be addressed with the objective of preventing or minimising short- and long-term risks to humans and the environment. This includes the product design phase, the selection, production and preparation of raw materials, the production and assembly of final products, and the management of all used products at the end of their useful life. This approach will result in the generation of smaller quantities of waste reducing end-of-pipe treatment and emission control technologies. Losses of material and resources with the sewage are minimised and, therefore, the raw material is used efficiently in the production process, generally resulting in substantial financial savings to the factory.

In the past, pollution prevention and minimisation were an indirect, although beneficial, result of the implementation of water conservation measures. Water demand management aimed to conserve scarce water by reducing its consumption rates. This was an important and relevant issue in the industrial, domestic and agricultural sector because of the rapid growth in water demand in densely populated regions of the world.

With regard to the generation of wastewater, pollution prevention and minimisation technologies are mainly implemented in the industrial sector (Box 3.1). Minimisation of wastewater from domestic sources is possible to a limited extent only and is mainly achieved by the introduction of water-saving equipment for showers, toilet flushing and gardening. In the Netherlands a new concept has been developed for residential areas where the grey water fraction is used for toilet flushing after treatment by a constructed

---

Box 3.1   Examples of successful waste minimisation in industry

Example 1
Tanning is a chemical process which converts putrescible hides and skins into stable leather. Vegetable, mineral and other tanning agents may be used (either separately or in combination) to produce leather with different qualities and quantities. Trivalent chromium is the major tanning agent, producing a modern, thin, light leather. Limits have been set for the discharge of the chromium. Cleaner production technology was used to recover the trivalent chromium ion from the spent liquors and to reuse it in the tanning process, thereby reducing the necessary end-of-pipe treatment cost to remove chromium from the wastewater.

Tanning of hides is carried out with basic chromium sulphate, $Cr(OH)SO_4$. The chromium recovery process consists of collecting and treating the spent tanning solution after its use, instead of simply wasting it. The spent liquor is sieved to remove particles and fibres. Through the addition of magnesium oxide, the valuable chromium precipitates as a hydroxide sludge. By the addition of concentrated sulphuric acid, this sludge dissolves and yields the chromium salt ($Cr(OH)SO_4$) solution that can be reused. Whereas in a conventional tanning process 20–40 per cent of the used chrome is lost in the wastewater, in this waste minimisation process 95–98 per cent of the waste chromium can be recycled.

This recovery technique was first developed and applied in a Greek tannery. The increased yearly operating costs of about US$ 30,000 were more then compensated for by the yearly chromium savings of about US$ 74,000. The capital investment of US$ 40,000 was returned in only 11 months.

Example 2
Sulphur dyes are a preferred range of dyes in the textile industry, but cause a significant wastewater problem. Sulphur dyes are water-insoluble compounds that first have to be converted into a water-soluble form and then into a reduced form having an affinity for the fibre to be dyed. The traditional method of converting the original dye to the affinity form is treatment with an aqueous solution of sodium sulphide. The use of sodium sulphide results in high sulphide levels in the textile plant wastewater which exceed the discharge criteria. Therefore, end-of-pipe treatment technology is necessary.

To avoid capital expenditure for wastewater treatment, a study was undertaken in India of available methods of sulphur black colour dyeing and into alternatives for sodium sulphide. An alternative chemical for sodium sulphide was found in the form of hydrol, a by-product of the maize starch industry. Only minor adaptations in the textile dyeing process were necessary. The introduction of hydrol did not involve any capital expenditure and sulphide levels in the mill's wastewater were reduced from 30 ppm to less than 2 ppm. The savings resulting from not having to install additional end-of-pipe treatment to reduce sulphide level in the wastewater were about US$ 20,000 in investment and US$ 3,000 a year in running costs.

---

wetland (Figure 3.4). In the agricultural sector, measures are directed primarily at water conservation through the application of, for example, water-saving irrigation techniques.

Waste minimisation involves not only technology but also planning, good housekeeping, and implementation of environmentally sound management practices. Many obstacles prevent the introduction of these new concepts in

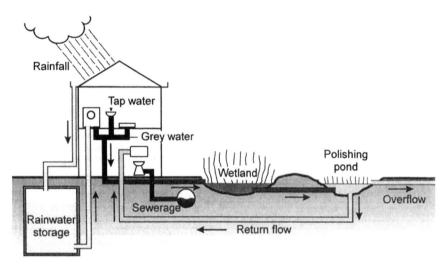

Figure 3.4 Potential reuse of grey water for toilet flushing after treatment by a constructed wetland (Based on van Dinther, 1995)

existing or even in new facilities, such as insufficient awareness of the environmental effects of the production process, lack of understanding of the true costs of waste management, no access to technical advice, insufficient knowledge of the implementation of new technologies, lack of financial resources and, last but not least, social resistance to change.

In the past, the requirements of most regulatory agencies have centred on treatment and control of industrial liquid wastes prior to discharge into municipal sewers or surface waters. As a result, over the last 20 years the number of industries emitting pollutants directly into aquatic environments reduced substantially. However, most of the implemented environmental protection measures consisted of end-of-pipe treatment technologies, with the "end" located either inside the factory or industrial zone, or at the entry of the municipal sewage treatment plant. As a consequence the industry pays for its share in the cost of sewer maintenance and treatment operation. In both cases, the industry should be charged for the treatment and management effort that has to take place outside the factory, in particular in the municipal treatment works. This charge should be made up of the true, overall treatment cost. By this principle, industries are specifically encouraged:

- To prevent waste production by interfering in the production process.
- To reduce the occurrence of hydraulic or organic peak loads that may render a municipal treatment system more expensive or vulnerable.
- To treat their waste flows to meet discharge requirements, to prevent damage to the municipal sewer or to realise cost savings for municipal treatment.

Table 3.8 Typical regulations for industrial wastewater discharge into a public sewer system in the United Kingdom, Hungary and The Netherlands

Variable	UK	Hungary	Netherlands
pH	6–10	6.5–10	6.5–10
Temperature (°C)	< 40	nrs	< 30
Suspended solids (mg l^{-1})	< 400	nrs	$-^1$
Heavy metals (mg l^{-1})	<10	specific	$-^1$
Cadmium (mg l^{-1})	< 100	< 10,000	$-^1$
Total cyanide (mg l^{-1})	< 2	< 1	$-^1$
Sulphate (mg l^{-1})	< 1,000	< 400	< 300
Oil and grease (mg l^{-1})	< 100	< 60	$-^1$

nrs No regulations set

[1] No coarse, explosive or inflammable solids are allowed. Contaminants that might interfere with biological treatment should be in concentrations that do not differ from domestic sewage

Sources: UN ECE, 1984; Appleyard, 1992

Table 3.8 provides examples of discharge criteria into municipal sewers. A method to calculate pollution charges into sewers or the environment is provided in Box 3.2.

3.5 Sewage conveyance

3.5.1 Storm water drainage

In many developing countries, stormwater drainage should be part of wastewater management because large sewage flows are carried into open storm water drains or because stormwater may enter treatment works with combined sewerage. In industrialised countries, stormwater drainage receives great attention because it may be polluted by sediments, oils and heavy metals which may upset the subsequent secondary and tertiary treatment steps.

In urbanised areas, the local infiltration capacity of the soil is not sufficient usually to absorb peak discharges of storm water. Large flows often have to be transported in short periods (20–100 minutes) over long distances (500–5,000 m). Drainage cost is determined, to a large extent, by the actual flow rate of the moment and, therefore, retention in reservoirs to dampen peak flows allows the use of smaller conduits, thereby reducing drainage cost per surface area. In tropical countries, peak flow reduction by infiltration may not be feasible because the peak flows can by far exceed the local infiltration capacities.

Box 3.2   Calculation of pollution charges based on "population
equivalents"

Calculation of the financial charges for industrial pollution in the Netherlands
is based on standard population equivalents (pe):

$$\text{pe load of industrial discharge} = \frac{Q \times [COD + 4.57 TKN]}{136}$$

where   Q   =   wastewater flow rate ($m^3\, d^{-1}$)
COD=   24 h-flow proportional COD concentration (mg COD l^{-1})
TKN =   24 h-flow proportional Kjeldahl-N concentration (mg N l^{-1})
136 =   waste load of one domestic polluter (136 g O_2-consuming
substances per day) and by definition set at one
population equivalent.

Heavy metal discharges are charged separately:
- Each 100 g Hg or Cd per day are equivalent to I pe.
- Each 1 kg of total other metal per day (As, Cr, Cu, Pb, Ni, Ag, Zn) is
equivalent to 1 pe.

An annual charge of US$ 25–50 (1994) is levied per population equivalent by
the local Water Pollution Control Board; the charge is region specific and
relates to the Board's overall annual expenses.

3.5.2   Separate and combined sewerage

In separate conveyance systems, storm water and sewage are conveyed in
separate drains and sanitary sewers, respectively. Combined sewerage systems
carry sewage and storm water in the same conduit. Sanitary and combined
sewers are closed in order to reduce public health risks. Separate systems
require investment in, and operation and maintenance of, two networks.
However, they allow the design of the sanitary sewer and the treatment plant to
account for low peak flows. In addition, a more constant and concentrated
sewage is fed to the treatment plant which favours reliable and consistent
process performance. Therefore, even in countries with moderate climate
where the rainfall pattern would favour combined sewerage (rainfall well
distributed over the year and with limited peak flows) newly developed residential areas are provided, increasingly, with separate sewerage. Combined
sewerage is generally less suitable for developing countries because:
- Sewerage and treatment are comparatively expensive, especially in
regions with high rain intensity during short periods of the year.
- It requires simultaneous investment for drainage, sewerage and treatment.

▪ There is commonly a lack of erosion control in unpaved areas.

Combined sewerage is most appropriate for more industrialised regions with a phased urban development, with an even rainfall distribution pattern over the year and with soil erosion control by road surface paving. The advantage of combined sewerage is that the first part of the run-off surge, which tends to be heavily polluted, is treated along with the sewage. The sewage treatment plants have to be designed to accommodate, typically, two to five times the average dry weather flow rate, which raises the cost and adds to the complexity of process control. The disadvantage of the combined sewer is that extreme peak flows cannot be handled and overflows are discharged to surface water, which gets contaminated with diluted sewage. These overflows can create serious local water quality problems.

Sanitary sewers are feasible only in densely populated areas because the unit cost per household decreases. Although most street sewers carry only small amounts of sewage, the construction cost is high because they require a minimum depth in order to protect them against traffic loads (minimum soil cover of 1 m), a minimum slope to ensure resuspension and hydraulic flushing of sediment to the end of the sewer, and a minimum diameter to prevent blockage by faecal matter and other solids (preferably 25 cm diameter). The required flushing velocity (a minimum of 0.6 m s^{-1} at least once a day) occurs when tap water consumption rates in the drainage area are in excess of 60 l cap^{-1} d^{-1}.

To reduce costs, sewers may use smaller diameters, may be installed at less depth and may apply a milder gradient. However, these measures require entrapment of settleable solids in a septic tank prior to discharge into the sewer. Such small-bore sewers are only cost-effective if they are maintained by the local community. This demands a high level of sustained community participation. Small-bore sewers may, ultimately, discharge into a municipal sanitary sewer or a treatment plant. Alternatively, in flat areas with unstable soils and low population density, small-bore pressure or vacuum sewers can be applied, but these are not considered a "low-cost" option.

Successful examples of low-cost small-bore sewerage are reported from Brazil, Colombia, Egypt, Pakistan and Australia. At population densities in excess of 200 persons per hectare, these small-bore sewer systems tend to become more cost effective than on-site sanitation. Companhia de Saneamento Basico do Estado de São Paulo (SABESP, São Paulo, Brazil) estimates the average construction cost (1988) for small towns to be US$ 150–300 per capita for conventional sewerage and US$ 80–150 per capita for simplified, small-bore sewerage (Bakalian, 1994). It is common in developing countries for most plot owners not to desludge their septic tank or cess pit regularly or adequately. Examples from Indonesia and India show that overflowing septic tanks are sometimes illegally connected to public open drains or sewers, and that during desludging operations often only the liquid is removed leaving the solids in the septic tank. Therefore, the implementation of small-bore sewerage requires substantial investment in community involvement to avoid the major failure of this technology.

3.6   Costs, operation and maintenance

Investment costs notably cover the cost of the land, groundwork, electro-mechanical equipment and construction. Recurring costs relate mainly to the paying back of loans (interest and principal), and to the costs for personnel, energy and other utilities, stores, laboratories, repair and sludge disposal. Both types of cost may vary considerably from country to country, as well as in time. Any financial feasibility analysis requires the use of a discount factor. This factor depends on inflation and interest rates and is also subject to substantial fluctuations. Therefore, comparing different technologies is always difficult and requires extensive expert analysis. Nevertheless, Figure 3.5 offers typical comparative cost levels (for industrialised countries) for primary, secondary and tertiary treatment of domestic wastewater. Table 3.9 provides a comparison of the unit construction costs for on-site and off-site sanitation for different world regions.

Operation and maintenance (O&M) is an essential part of wastewater management and affects technology selection. Many wastewater treatment projects fail or perform poorly after construction because of inadequate O&M. On an annual basis, the O&M expenditures of treatment and sewage collection are typically in the same order of magnitude as the depreciation on the capital investment. Operation and maintenance requires:

- Careful exhaustive planning.
- Qualified and trained staff devoted to its assignment.
- An extensive and operational system providing spare parts and O&M utilities.
- A maintenance and repair schedule, crew and facility.
- A management atmosphere that aims at ensuring a reliable service with a minimum of interruptions.
- A substantial annual budget that is uniquely devoted to O&M and service improvement.

Maintenance policy can be corrective, i.e. repair or action is undertaken when breakdown is noticed, but this leads to service interruption and hence dissatisfied customers. Ideally, maintenance is preventive, i.e. replacement of mechanical parts is carried out at the end of their expected life time. This

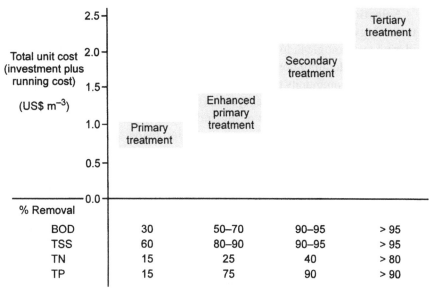

Figure 3.5 Typical total unit costs for wastewater treatment based on experience gained in Western Europe and the USA (After Somlyody, 1993)

Table 3.9   Typical unit construction cost (US$ cap^{-1}) for domestic wastewater disposal in different world regions (median values of national averages)

Region	Urban sewer connection	Rural on-site sanitation
Africa	120	22
Americas	120	25
South-East Asia	152	11
Eastern Mediterranean	360	73
Western Pacific	600	39

Source: WHO, 1992

allows optimal budgeting and maintenance schedules that have minimal impact on service quality. Clearly, O&M requirements are important factors when selecting a technology; process design should provide for optimal, but low cost, O&M.

The most common reasons for O&M failure are inadequate budgets due to poor cost recovery, poor planning of servicing and repair activities and weak spare parts management, and inadequately trained operational staff.

3.7   Selection of technology

The technology selection process results from a multi-criteria optimisation considering technological, logistic, environmental, financial and institutional factors within a planning horizon of 10–20 years. Key factors are:

- The size of the community to be served (including the industrial equivalents).
- The characteristics of the sewer system (combined, separate, small-bore).
- The sources of wastewater (domestic, industrial, stormwater, infiltration).
- The future opportunities to minimise pollution loads.
- The discharge standards for treated effluents.
- The availability of local skills for design, construction and O&M.
- Environmental conditions such as land availability, geography and climate.

Considerations for industrial technology selection tend to be relatively straightforward because the factors interfering in selection are primarily related to anticipated performance and extension potential. Both of these are associated directly with cost.

3.7.1   On-site sanitation technologies

For domestic wastewater the suitability of various sanitation technologies must be related appropriately to the type of community, i.e. rural, small town or urban (Table 3.10). Typically, in low-income rural and (peri-)urban areas, on-site sanitation systems are most appropriate because:

- They are low-cost (due to the absence of sewerage requirements).
- They allow construction, repair and operation by the local community or plot owner.
- They reduce, effectively, the most pressing public health problems.

Moreover, water consumption levels often are too low to justify conventional sewerage.

With on-site sanitation, black toilet water is disposed in pit latrines, soak-aways or septic tanks (Figure 3.6) and the effluent infiltrates into the soil or overflows into a drainage system. Grey water can infiltrate directly, or can flow into drainage channels or gullies, because its suspended solids and pathogen contents are low. The solids that accumulate in the pit or tank (approximately $40 \ \mathrm{l \ cap^{-1} \ a^{-1}}$) have to be removed periodically or a new pit has to be dug (dual-pit latrine). Depending on the system, the sludge may or may not be well stabilised. At the minimum solids retention time of six months the sludge may be considered to be pathogen-free and it can be used in agriculture as fertiliser or as a soil conditioner. Digestion of the full sludge content for several months can be carried out if a second, parallel pit is used while the first is digesting.

Table 3.10 Typical sanitation options for rural areas, small townships and urban residential areas

	Rural area	Township	Urban area
Community size	< 10,000 pe	10,000–50,000 pe	> 50,000 pe
Density (persons per hectare)	< 100	> 100 – < 200	> 200
Water supply service	Well, handpump	Public standpost	House connection
Water consumption	< 50 l cap^{-1} d^{-1}	50–100 l cap^{-1} d^{-1}	> 100 l cap^{-1} d^{-1}
Sewage production	< 5 m^3 ha^{-1} d^{-1}	5–20 m^3 ha^{-1} d^{-1}	> 20 m^3 ha^{-1} d^{-1}
Treatment options	Dry on-site sanitation by VIP or composting latrines	Dry and wet on-site sanitation; small-bore sewerage may be feasible depending on population density and soil conditions	Centre: Sewerage plus off-site treatment. Peri-urban: wet on-site sanitation with small-bore sewerage and septage handling

VIP   Ventilated Improved Pit latrine

The accumulating waste (septage) in septic tanks must be regularly collected and disposed of. After drying and dewatering in lagoons or on drying beds it can be disposed at a landfill site, or it can be co-composted with domestic refuse. Reuse in agriculture is only feasible following adequate pathogen removal and provided the septage is not contaminated with heavy metals. Alternatively, the septage can be disposed of in a sewage treatment plant, or it can be stabilised and rendered pathogen-free by adding lime (until the pH > 10) or by extended aeration. The latter two methods, however, are expensive.

3.7.2 On-site versus off-site options
In densely populated urban areas the generation of wastewater may exceed the local infiltration capacity. In addition, the risk of groundwater pollution and soil destabilisation often necessitates off-site sewerage. At hydraulic loading rates greater than 50 mm d^{-1} and less than 2 m unsaturated groundwater flow, nitrate and, in a later stage, faecal coliform contamination may occur (Lewis et al., 1980).

The unit cost for off-site sanitation decreases significantly with increasing population density, but sewering an entire city often proves to be very expensive. In cities where urban planning is uncoordinated, implementation of a balanced mix of on-site and off-site sanitation is most cost-effective. For example, in Latin America the population density at which small-bore sewerage becomes competitive with on-site sanitation is approximately 200

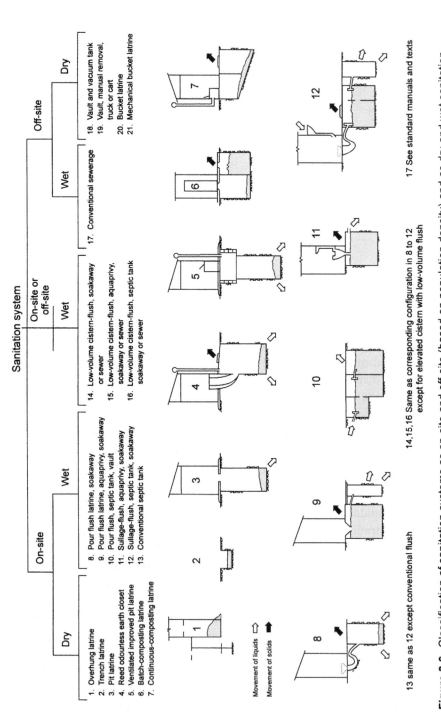

Figure 3.6 Classification of sanitation systems as on-site and off-site (based on population density) and as dry and wet sanitation (based on water supply) (After Kalbermatten *et al.*, 1980)

Sanitation system

On-site

Dry

1. Overhung latrine
2. Trench latrine
3. Pit latrine
4. Reed odourless earth closet
5. Ventilated improved pit latrine
6. Batch-composting latrine
7. Continuous-composting latrine

Wet

8. Pour flush latrine, soakaway
9. Pour flush latrine, aquaprivy, soakaway
10. Pour flush, septic tank, vault
11. Sullage-flush, aquaprivy, soakaway
12. Sullage-flush, septic tank, soakaway
13. Conventional septic tank

On-site or off-site

Wet

14. Low-volume cistern-flush, soakaway or sewer
15. Low-volume cistern-flush, aquaprivy, soakaway or sewer
16. Low-volume cistern-flush, septic tank soakaway or sewer

Off-site

Wet

17. Conventional sewerage

Dry

18. Vault and vacuum tank
19. Vault, manual removal, truck or cart
20. Bucket latrine
21. Mechanical bucket latrine

Movement of liquids ⇨
Movement of solids ➡

13 same as 12 except conventional flush

14,15,16 Same as corresponding configuration in 8 to 12 except for elevated cistern with low-volume flush

17 See standard manuals and texts

persons per hectare (Sinnatamby *et al.*, 1986). The deciding factor in these cost calculations is the cost of the collection and conveyance system.

Box 3.3 provides guidance for preliminary decision-making with respect to on- or off-site sanitation. In situations where there is a high wastewater production per hectare per day, sewerage is needed to transport either the liquids alone (in the case of small-bore sewerage) or the liquid plus suspended solids (in the case of conventional sewerage). Additional decisive parameters are whether shallow wells used for water supplies need to be protected, the population density, the soil permeability and the unit cost. To minimise groundwater contamination, a typical surface loading rate of $10 \text{ m}^3 \text{ ha}^{-1} \text{ d}^{-1}$ is recommended (Lewis *et al.*, 1980), provided that prevailing groundwater tables ensure at least 2 m unsaturated flow in a vertical direction.

When the wastewater production rate is in excess of $10 \text{ m}^3 \text{ ha}^{-1} \text{ d}^{-1}$, conventional sanitary sewerage may be feasible for managing municipal sewage, with or without the inclusion of storm water. Studies indicate that at 200–300 persons per hectare, gravity sewerage becomes economically feasible in developing countries; in industrialised countries the equivalent population density is about 50 persons per hectare.

If groundwater protection is not required, the infiltration rate may exceed $10 \text{ m}^3 \text{ ha}^{-1} \text{ d}^{-1}$, provided the soil permeability and stability allow it. If soil permeability is low, off-site sanitation needs consideration. Depending on the socio-economic environment and the degree of community involvement that can be generated, small-bore sewerage may be feasible. In such cases additional stormwater drainage facilities must be provided.

In addition to technical, logistic and financial criteria, reliable management by a local village-based entity or local government is essential for sustainable functioning of the system. Most off-site treatment technologies benefit from economies of scale although anaerobic technologies tend to scale down easily to township or local level without the unit cost rising seriously. This makes anaerobic technologies suitable for inclusion in urban sanitation at community level (Alaerts *et al.*, 1990). This "community on-site" option can stimulate more disciplined operation and desludging when compared with the often poor performance of individual units. At the same time, it retains the advantage that it can be managed by a local committee and semi-skilled caretakers.

3.7.3   Off-site centralised treatment technologies

There is a large variety of off-site treatment technologies. The selection of the most appropriate technology is determined, first of all, by the composition of the wastewater flow arriving at the treatment plant and also by the discharge

Box 3.3   Preliminary assessment for on-site sanitation, intermediate small-bore sewerage or conventional off-site sewerage for domestic or municipal wastewater disposal

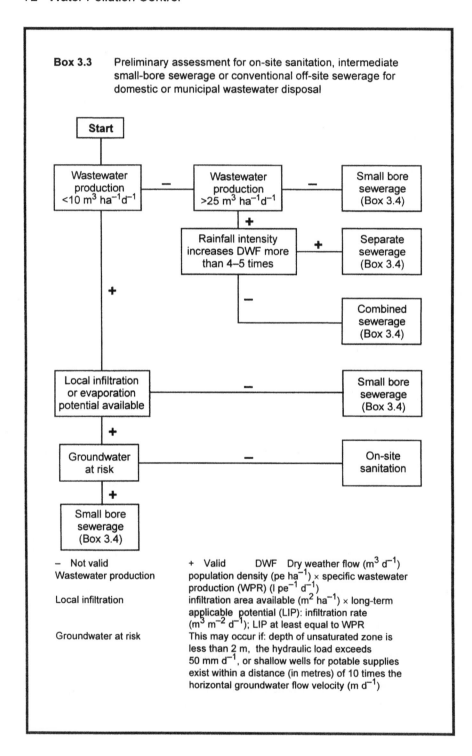

requirements. Questions for assessing the expected composition and behaviour of the sewage to be treated include:

- To what extent is industrial wastewater included?
- Will sewerage be separate, combined or small-bore?
- Is groundwater expected to infiltrate into the sewer?
- Are septic tanks removing settleable solids prior to discharge into the conveyance system?
- What is the specific water and food consumption pattern?
- What is the quality of the drinking water?

Each off-site treatment plant is composed of unit processes and operations that enable the effluent quality to meet the criteria set by the regulatory agency. Therefore, when selecting a technology the first step is to develop a complete flow diagram where all unit processes and operations are put together in a logical fashion. Off-site treatment systems are generally composed of primary treatment, usually followed by a secondary stage and, in some instances, a tertiary or advanced treatment stage. Table 3.7 summarises the potential performance of common technologies that can be applied in wastewater treatment.

Primary treatment
In most treatment plants mechanical primary treatment precedes biological and/or physicochemical treatment and is used to remove sand, grit, fibres, floating objects and other coarse objects before they can obstruct subsequent treatment stages. In particular, the grit and sand conveyed through combined sewers may settle out, block channels and occupy reactor space. Additional facilities may be designed to equalise peak flows. Approximately 50–75 per cent of suspended matter, 30–50 per cent of BOD and 15–25 per cent of Kjeldahl-N and total P are removed at moderate cost by means of settling. Settling tanks that include facilities for extended sludge or solids retention may facilitate the stabilisation of sludge and are, therefore, convenient for small communities.

Physicochemical processes may be incorporated in the primary treatment stage in order to further enhance removal efficiencies, to adjust (neutralise) the pH, or to remove any toxic or inhibitory compounds that may affect the functioning of the subsequent treatment steps. Flocculation with aluminium or iron salts is often used. Such enhanced primary treatment is comparatively cheap in terms of capital investment but the running costs are high due to the chemicals that are required and the additional sludge produced. This approach is attractive when it is necessary to expand the plant capacity due to a temporary (e.g. seasonal) overload.

Secondary treatment

The most common technology used for secondary treatment of wastewater relies on (micro)biological conversion of oxygen consuming substances such as organic matter, represented as BOD or COD, and Kjeldahl-N. The technologies can be classified mainly as aerobic or anaerobic depending on whether oxygen is required for their performance, or as mechanised or non-mechanised depending on the intensity of the mechanised input required. Table 3.11 provides a matrix classification of available (micro)biological treatment technologies. Further detailed information is available in Metcalf and Eddy (1991) and Arceivala (1986).

The choice between aerobic and anaerobic technologies has to consider mainly the added complexity of the oxygen supply that is need for aerobic technologies. The supply of large amounts of oxygen by a surface aeration or bubble dispersion system adds to the capital cost of the aeration equipment substantially, as well as to the running cost because the annual energy consumption is rather high (it can reach 30 kWh per population equivalent (pe)).

The choice between mechanised or non-mechanised technologies centres on the locally or nationally available technology infrastructure which may ensure a regular supply of skilled labour, local manufacturing, operational and repair potential for used equipment, and the reliability of supplies (e.g. power, chemicals, spare parts). Additional key considerations are land requirements and the potential for biomass resource recovery. In general, non-mechanised technologies rely on substantially longer retention time to achieve a high degree of contaminant removal whereas mechanised systems use equipment to accelerate the conversion process. If land costs are in excess of US$ 20 per square metre, non-mechanised systems lose their competitive cost advantage over mechanised systems. Resource recovery may be possible if, for example, the algal or macrophyte biomass generated is marketable, generating revenue and employment opportunities. For example, constructed wetlands using *Cyperus papyrus* may generate about 40–50 tonnes of standing biomass per hectare a year which can be used in handicraft or other artisanal activities.

For non-biodegradable (mainly industrial) wastewaters physicochemical alternatives have been developed that rely on the physicochemical removal of contaminants by chemical coagulation and flocculation. The generated sludges are typically heavily contaminated and have no potential for reuse other than for landfill.

Overall, the selection process for the most appropriate secondary technology may have to be decided using multi-criteria analysis. In addition to the overall unit costs, the environmental, aesthetic and health risks involved, the

Table 3.11   Classification of secondary treatment technology

Conversion method	Mechanised technology	Non-mechanised technology
Aerobic	Activated sludge	Facultative stabilisation ponds
	Trickling filter	Maturation ponds
	Rotating bio-contactor	Aquaculture (e.g. algal, duck weed or fish ponds)
		Constructed wetlands
Anaerobic	Upflow anaerobic sludge bed (UASB)	Anaerobic ponds
	Anaerobic (upflow) filter	

quality standards to be met, the skilled staff and land requirements, and the reliability of the potential for recovery by the technology, all have to be evaluated to give a total score that indicates the feasibility of each technology for a particular country or location (Handa *et al.*, 1990).

Physicochemical treatment. Physicochemical technologies can achieve significant BOD, P and suspended solids reduction, although it is generally not the preferred option for domestic sewage because removal rates for organic matter are rather poor (Table 3.12). It is often used for industrial wastewater treatment to remove specific contaminants or to reduce the bulk pollutant load to the municipal sewer. Physicochemical treatment can also be combined with primary treatment to enhance removal processes and to reduce the load on the subsequent secondary treatment stage. For wastewater with a high organic matter content, like domestic sewage, (micro)biological methods are commonly preferred because they have lower operational costs and achieve a higher reduction of BOD.

The skills required to operate chemical dosing equipment, and the difficulty in ensuring a reliable supply of chemicals are often prohibitive for the selection of physicochemical technologies in developing countries where systems are more prone to malfunctioning. In particular, the fluctuating flow and composition of the incoming sewage makes frequent adjustments of the chemical dosing necessary. Biological treatment systems are more sturdy and ensure a constant effluent quality because they have a high internal buffering capacity for peak flows and loads.

Examples of physicochemical processes used in industrial applications include:

- Chemical oxidation with, for example, O_2, O_3 or Cl_2 (cyanide removal and oxidation of refractory organic compounds).

Table 3.12 Advantages and disadvantages of physicochemical treatment of domestic or municipal wastewater

Advantages	Disadvantages
Compact technology with low area needs	Chemical dosing is labour intensive due to fluctuating sewage load and composition
Good removal of micro-pollutants and P	Generation of chemical sludges
Fast start-up	High unit cost per m^3 of water treated
Insensitivity to toxic compounds	

- Chemical reduction (for example, H_2S assisted conversion of Cr (VI) into Cr (III)).
- Desorption (stripping) (NH_3 and odorous gas removal).
- Adsorption on activated carbon (removal of refractory organics and heavy metals).
- Ultra- and micro-filtration (separation of colloidal and dissolved compounds).

Anaerobic treatment. Aerobic treatment methods have traditionally dominated treatment of domestic and industrial wastewater. Since the 1970s, however, anaerobic treatment has become the preferred technology for concentrated organic wastewater from, for example, breweries, alcohol distilleries, fermentation industries, canning factories, pulp and paper mills (Hulshoff Pol and Lettinga, 1986). The principal characteristic of anaerobic processes is that degradation of the organic pollutants takes place in the absence of oxygen. The bacteria produce considerable quantities of methane gas. In addition, the process can proceed at exceptionally high hydraulic loading rates. Of the many process design alternatives, the Upflow Anaerobic Sludge Blanket (UASB) process is the most cost-effective in most types of industrial wastewater treatment (Figure 3.7). The reactor consists of an empty volume covered with a plate settler zone to catch and to recycle suspended matter escaping from the sludge blanket below. The water flows upwards through a blanket of suspended granules or flocs containing the active biomass. The methane and CO_2 bubbles are caught below the plate settlers and taken out of the reactor separately.

World-wide, over 400 anaerobic plants treat industrial wastewater, whereas operational experience on domestic sewage derives from approximately 10 full-scale UASB plants (size 20,000–200,000 pe) in Colombia, Brazil and India (Alaerts *et al.*, 1990; Draaijer *et al.*, 1992; Schellinkhout and Collazos, 1992; van Haandel and Lettinga, 1994). Whereas the aerobic

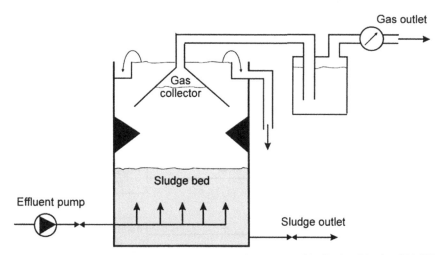

Figure 3.7 Schematic representation of the Upflow Anaerobic Sludge Blanket (UASB) reactor

process achieves 90–95 per cent removal of BOD, the anaerobic process achieves only 75–85 per cent necessitating, in most cases, post-treatment to meet effluent standards. Anaerobic treatment also provides minimal N and P removal but generates much less, and a better stabilised, sludge. Biogas recovery is only feasible on a large scale or in an industrial context. Many tropical developing countries would probably prefer anaerobic processes because of the numerous agro-industries and the (often) high domestic sewage temperatures.

The choice between aerobic and anaerobic treatment depends primarily on the wastewater characteristics (Box 3.4). If the average sewage temperature is above 20 °C (with a minimum of 18 °C over a maximum period of 2 months) and is highly biodegradable (COD:BOD ratio below 2.5) and concentrated (typically BOD > 1,000 mg l^{-1}), anaerobic treatment has clear economic advantages. If neither condition can be met, aerobic treatment is the only feasible option. If only one condition is met the choice is determined by additional considerations such as:

- Desired effluent quality: anaerobic technologies yield lower removal efficiencies. The presence of residual BOD, ammonium and, occasionally, sulphide in the effluent may require post-treatment.
- Sludge handling and disposal: anaerobic sludge production is less than half of that in aerobic treatment plants, and the sludge is already stabilised which facilitates further processing.
- Effluent use: anaerobic treatment retains more nutrients (N, P, K) and thus effluent have higher potentials for use in irrigation.

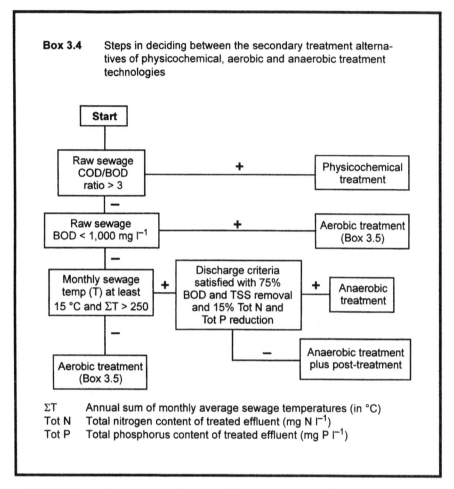

Box 3.4   Steps in deciding between the secondary treatment alternatives of physicochemical, aerobic and anaerobic treatment technologies

ΣT   Annual sum of monthly average sewage temperatures (in °C)
Tot N   Total nitrogen content of treated effluent (mg N l^{-1})
Tot P   Total phosphorus content of treated effluent (mg P l^{-1})

- Reliability of power supply: aerobic treatment performance is highly dependent on power input for aeration and mixing. Power failure may create rapid malfunctioning of aerobic plants while anaerobic systems are fairly resistant to periods of no power supply.
- Local potential for selling biogas.

When high effluent standards are to be met, and the cost of land is moderate to high, the combination of a UASB plant plus aerobic post-treatment is often decisively more cost-effective than conventional aerobic treatment.

Non-mechanised treatment. The availability of flat land is a decisive criterion in selecting between non-mechanised and mechanised technologies (Box 3.5). Land-intensive systems such as stabilisation ponds, aquaculture, pisciculture and constructed wetlands may be feasible only when flat land

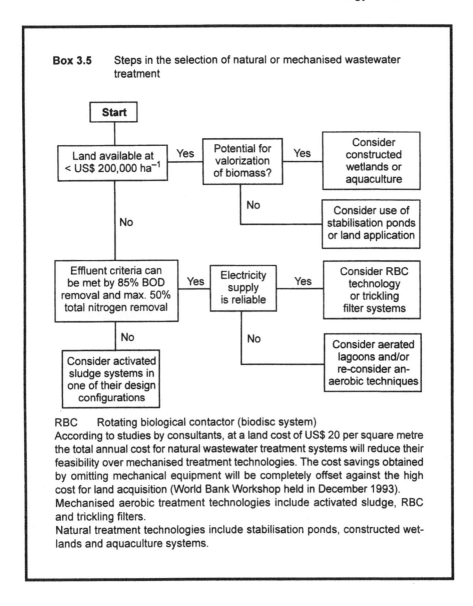

Box 3.5   Steps in the selection of natural or mechanised wastewater treatment

RBC   Rotating biological contactor (biodisc system)

According to studies by consultants, at a land cost of US$ 20 per square metre the total annual cost for natural wastewater treatment systems will reduce their feasibility over mechanised treatment technologies. The cost savings obtained by omitting mechanical equipment will be completely offset against the high cost for land acquisition (World Bank Workshop held in December 1993).

Mechanised aerobic treatment technologies include activated sludge, RBC and trickling filters.

Natural treatment technologies include stabilisation ponds, constructed wet-lands and aquaculture systems.

costs are below US$ 5 per square metre. Such systems typically require 5–10 m^2 per population equivalent and are not usually demanding with respect to O&M, provided the wastewater is of domestic origin. Land-intensive treatment may, particularly in developing countries, better fit a resource recovery scenario because the produced biomass can sometimes be harvested and used to generate income. Algae-based stabilisation ponds are

Table 3.13 Typical features of stabilisation ponds

Typical feature	Anaerobic pond	Facultative pond	Maturation pond
Objective	TSS removal	BOD removal	Nutrient and pathogen removal
Loading rate	0.1–0.3 kg BOD $m^{-3}\ d^{-1}$	100–350 kg BOD $ha^{-1}\ d^{-1}$	At least two ponds in series, each 5 days retention
Typical depth	2–5 m	1–2 m	1–1.5 m
Performance	TSS: 50–70 % BOD: 30–60 % Coliforms: 1 order of magnitude	TSS: increase BOD: 50-70 % Coliforms: 1–2 orders of magnitude	TSS: 20–30 % BOD: 20–50 % Coliforms: 3–4 orders of magnitude
Problems	Odour release	Algal TSS increase	Area requirement

TSS   Total suspended solids   BOD   Biochemical oxygen demand

in operation on all continents for sewage treatment or for additional treatment of partially treated effluent; although they sometimes suffer from sulphide or ammonium and from comparatively high suspended solids content in the effluent. Such ponds are characterised according to their purpose and dimensions (Table 3.13). Stabilisation ponds operate without forced retainment of the active biomass while the oxygen is provided from the photosynthesis of the algae present in the ponds and by re-aeration by the wind.

In aquaculture and constructed wetlands, macrophytes (plants) are grown to suppress algal growth by shielding the water column from light, by absorbing the nutrients and by assisting the oxygen transfer into the water. The floating plant duckweed (Lemnaceae), is particularly promising for aquaculture because it grows abundantly and can easily be harvested. In constructed wetlands, wastewater is made to flow either horizontally or vertically through the root zone of a permeable soil planted with vegetation. The plants, if regularly harvested, create a sink for the nutrients by their uptake and assimilation of N and P. Importantly, they also provide niches for bacteria that reduce BOD, and that enhance nitrification, denitrification and P-fixation. They also provide niches for predator organisms that contribute to pathogen removal. Such wetlands offer good prospects for small-scale operation in remote tropical areas, although this approach has not yet been demonstrated at full scale. Fish can also be grown in stabilisation ponds to control algal growth, although their consumption can present public health risks. Sewage-based pisciculture is applied on a small scale in China, Indonesia and other East Asian countries; large-scale applications can be found in Calcutta and Munich, amongst other places.

Aerobic mechanised treatment. If flat land is scarce or expensive, and if anaerobic technologies are not feasible, the remaining option is to use conventional, aerobic, mechanised technologies. Most wastewater treatment plants all over the world are presently of this type, although they tend to be less appropriate in low-cost environments. They can be divided according to their method of sludge retention, i.e. in fixed-biofilm or in suspended growth reactors with sludge recycling. In biofilm reactors, micro-organisms are immobilised because they are attached to an inert support (e.g. lava stones, plastic rings or bio-disc) and are in constant contact with the wastewater and with the air that flows through the open pores. In suspended growth systems, the micro-organisms and the wastewater are in constant contact through mechanical mixing, which also ensures aeration.

Biofilm reactors retain their biomass better than suspended growth reactors and can therefore handle hydraulic fluctuations and low BOD concentrations more efficiently. However, the operational control of biofilm reactors is fairly limited. By contrast, suspended growth reactors allow better control and generally produce a higher quality effluent.

Typical suspended growth systems are the activated sludge system and extended aeration; trickling filter and rotating bio-discs are both biofilm-based systems. These systems require less than 1 m^2 pe^{-1} but, depending on the situation, they consume somewhat more space than anaerobic technologies. The activated sludge system, in its various designs, is the most widely applied — offering operational flexibility, high reliability and resilience. An added advantage is that process control also offers the opportunity to have several processes integrated in the system such as carbon oxidation, nitrification, denitrification and biological P-removal. This is of great benefit in achieving high quality effluents that meet the European Union (EU) guidelines (Table 3.14). Although trickling filters are technically feasible and attractive because they are easy to operate and they consume less energy, they generally have a lower removal efficiency for BOD and TSS, they are sensitive to low temperatures and may be infested with flies and mosquitoes. Their N and P removal is too low to justify wide application in countries with stringent effluent quality standards (Table 3.15). Rotating bio-discs are not widely used because they have low operational flexibility, potential mechanical problems and, often, a complicated biofilm development.

A typical activated sludge process design that is becoming more popular in many industrialised countries is the oxidation ditch. The low sludge loading (kg BOD per kg of biomass per day) ensures, all in one reactor, BOD removal, advanced nitrification, substantial denitrification, biological P removal and modest generation of well-stabilised sludge. This even allows

Table 3.14   European Community guidelines for wastewater discharged to sensitive surface water bodies based on typical raw wastewater composition

Variable	Raw sewage composition	EU guideline	Percentage removal (%)
BOD_5 (mg l^{-1})	250	25	90
Total N (mg l^{-1})	48	10	80
Total P (mg l^{-1})	12	1	90

Source: CEC, 1991

Table 3.15   Comparative analysis of the performance of the trickling filter and the activated sludge process for secondary wastewater treatment

Parameter	Trickling filter	Activated sludge
BOD removal (%)[1]	80–90	90–98
Kjeldahl-N removal (%)	60–85	80–95
Total N removal (%)	20–45	65–90
Energy required (kWh cap^{-1} a^{-1})	10–15	20–30
O&M requirement	Medium	High
Pathogen removal	1–2 orders of magnitude	1–2 orders of magnitude

[1]   Not including BOD removal in primary treatment steps

the primary treatment to be skipped. The carousel is a modified version of the oxidation ditch with this enhanced capacity (Figure 3.8).

If pathogen removal is essential, only non-mechanised systems featuring hydraulic retention times of 20–30 days can provide satisfactory removal of faecal coliforms and nematode eggs to the standard required by the WHO guidelines (WHO, 1989). All mechanised treatment systems need additional chemical disinfection with chlorine or other oxidative chemicals, or with UV irradiation. This adds to the treatment cost and the operational complexity of the treatment technology and eventually may reduce the reliability of the treatment plant to provide "safe effluents".

3.8   Conclusions and recommendations

World-wide attitudes to sustainable water resources management for the future are being reconsidered. Conservation of water resources (with respect to quantity and quality) is being increasingly emphasised as the means to address the anticipated and increasing shortages of water resources of good

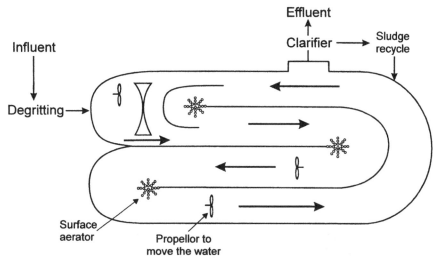

Figure 3.8 Novel carousel configuration of the oxidation ditch, activated sludge system for achieving a final effluent with low total N and P levels

quality in many parts of the world. This water is needed to meet ever increasing domestic, industrial and agricultural demands. Extrapolation of the increasing water consumption rates over the last ten years suggests that huge shortages will occur in many populated areas of the world, particularly in the arid and semi-arid world regions.

Solving sanitary problems of human and industrial waste flows in the future, especially those generated in urban environments, may not necessarily be feasible using water consuming technologies that rely on conventional sewerage, carrying and transporting the suspended waste material away from the place where it was generated. Water saving technologies, water recycling and reuse, will play an increasingly dominant role in the future and will draw attention away from pollution control policies to waste prevention and waste minimisation policies. Scenarios including the potential for recovery of valuable resources will be increasingly promoted as they become more feasible aspects of sustainable water resources management.

With urbanisation taking place world-wide, attention to water and sanitation will shift to the densely populated urban and peri-urban areas where new incentives are created for technology development. These incentives will be aimed at people with only marginal financial resources available and with water supply levels that are too low to justify conventional sewerage.

Separating wastewater flows (black and grey water, domestic and industrial, sewage and rainwater) and the development of technologies that aim to make these individual wastewater flows fit for reuse or recycling will, in the long run, contribute to sound water resources management. In addition, such approaches will reduce public health risks and environmental pollution, as well as the burden on the pollution carrying capacity of the environment.

Technology selection for waste flows may therefore have to take a broader perspective than purely meeting the present discharge standards formulated for the local situation. Anticipating the above trends might stimulate the use of an additional criterion in technology selection, i.e. sustainable use of scarce resources whether it be water, nutrients, energy or space.

3.9   References

Alaerts, G.J., Veenstra, S., Bentvelsen, M. and van Duijl, L.A. 1990 *Feasibility of Anaerobic Sewage Treatment in Sanitation Strategies in Developing Countries.* IHE Report No 20, International Institute for Infrastructural, Hydraulic and Environmental Engineering (IHE), Delft.

Appleyard, C. 1992 Industrial Wastewater Treatment. Lecture Notes for the International Post-Graduate Course in Sanitary Engineering, International Institute for Infrastructural, Hydraulic and Environmental Engineering (IHE), Delft.

Arceivala, S.J. 1986 *Wastewater Treatment for Pollution Control.* Tata Mc-Graw Hill Publ. Ltd, New Delhi.

Ayers, R.S. and Westcot, D.W. 1985 *Water Quality for Agriculture.* FAO Irrigation and Drainage Paper No. 29. United Nations Food and Agriculture Organization, Rome.

Bakalian, A. 1994 *Simplified Sewerage: Design Guidelines.* UNDP/World Bank Water and Sanitation Programme Report 7, World Bank, Washington D.C.

CEC 1991 Directive concerning urban wastewater treatment (91/271/EEC). Commission of the European Communities, *Off. J.* L135/40.

van Dinther, M. 1995 Greywater is good enough. De Volkskrant, April 22.

Draaijer, H., Maas, J.A.W., Schaapman, J.E. and Khan, A. 1992 Performance of the 5 MLD UASB reactor for sewage treatment at Kanpur, India. *Wat. Sci. Tech..,* 25(7), 123–132.

Eckenfelder, W.W., Patoczka, J.B. and Pulliam, G.W. 1988 Anaerobic versus aerobic treatment in the USA. In: E.R. Hall and P.N. Hobson [Eds] *Advances in Water Pollution Control.* 5th International IAWPRC Conference on Anaerobic Digestion, Bologna, International Association of Water Pollution Research and Control, London.

van Haandel, A.C. and Lettinga, G. 1994 *Anaerobic Sewage Treatment. A Practical Guide for Regions with a Hot Climate.* John Wiley & Sons, Chichester.

Handa, B.K. 1990 Ranking of technology options for municipal wastewater treatment. *Asian Env., 12*(3), 28–40.

Hulshoff Pol, L. and Lettinga, G. 1986 New technologies for anaerobic wastewater treatment. *Wat. Sci. Tech., 18*(12), 41–53.

Kalbermatten, J.M., Julius DeAnne, S., Mara, D.D. and Gunnerson, G.G. 1980 *Appropriate Technology for Water Supply and Sanitation.* Volume 2, World Bank, Washington, D.C.

Otis, R.J. and Mara, D.D. 1985 *The Design of Small Bore Sewers.* TAG Technical Note No. 14, Technology Advisory Group, World Bank, Washington D.C.

Lewis, W.J., Foster, S.S.D. and Drasar, B.S. 1980 *The Risk of Groundwater Pollution by On-site Sanitation in Developing Countries.* IRCWD Report No 01/82., International Reference Center for Waste Disposal, Duebendorf, Switzerland.

Metcalf and Eddy Inc. 1991 *Wastewater Engineering. Treatment Disposal and Reuse.* 3rd edition, Mc-Graw Hill Book Co, Singapore.

Schellinkhout, A. and Collazos, C.J. 1992 Full-scale application of the UASB technology for sewage treatment. *Wat. Sci. Tech., 25*(7), 159–166.

Sinnatamby, G., Mara, D. and McGarry, M. 1986 Shallow sewers offer hope to slums. *World Wat., 9*(1), 39–41.

Somlyody, L. 1993 Looking over the environmental legacy. *Wat. Qual. Int., 4*, 17–20.

UN ECE 1984 *Strategies, Technologies and Economics of Wastewater Management in ECE Countries.* Report E.84.II.E.18, UN European Commission for Europe, Geneva.

Veenstra, S. 1996 Environmental Sanitation. Lecture notes for the MSc course in Sanitary Engineering, International Institute for Infrastructural, Hydraulic and Environmental Engineering (IHE), Delft.

WHO 1989 *Health Guidelines for the Use of Wastewater in Agriculture and Aquaculture.* WHO Technical Report Series No 517, World Health Organization, Geneva.

WHO 1992 *The International Drinking Water and Sanitation Decade End of Decade Review* (as at December 1990). WHO/CW5/92.12, World Health Organization, Geneva.

World Bank 1994 *World Development Report 1994 — Infrastructure for Development.* Oxford University Press, Oxford, New York.

Chapter 4[*]

WASTEWATER AS A RESOURCE

4.1 Introduction

In many arid and semi-arid regions of the world water has become a limiting factor, particularly for agricultural and industrial development. Water resources planners are continually looking for additional sources of water to supplement the limited resources available to their region. Several countries of the Eastern Mediterranean region, for example, where precipitation is in the range of 100–200 mm a^{-1}, rely on a few perennial rivers and small underground aquifers that are usually located in mountainous regions. Drinking water is usually supplied through expensive desalination systems, and more than 50 per cent of the food demand is satisfied by importation.

In such situations, source substitution appears to be the most suitable alternative to satisfy less restrictive uses, thus allowing high quality waters to be used for domestic supply. In 1958, the United Nations Economic and Social Council provided a management policy to support this approach by stating that "*no higher quality water, unless there is a surplus of it, should be used for a purpose that can tolerate a lower grade*" (United Nations, 1958). Low quality waters such as wastewater, drainage waters and brackish waters should, whenever possible, be considered as alternative sources for less restrictive uses.

Agricultural use of water resources is of great importance due to the high volumes that are necessary. Irrigated agriculture will play a dominant role in the sustainability of crop production in years to come. By the year 2000, further reduction in the extent of exploitable water resources, together with competing claims for water for municipal and industrial use, will significantly reduce the availability of water for agriculture. The use of appropriate technologies for the development of alternative sources of water is, probably, the single most adequate approach for solving the global problem of water shortage, together with improvements in the efficiency of water use and with adequate control to reduce water consumption.

[*] *This chapter was prepared by I. Hespanhol*

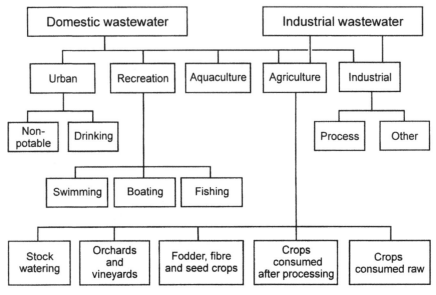

Figure 4.1 Types of wastewater use (After WHO, 1989)

4.2   Types of reuse

Water is a renewable resource within the hydrological cycle. The water recycled by natural systems provides a clean and safe resource which is then deteriorated by different levels of pollution depending on how, and to what extent, it is used. Once used, however, water can be reclaimed and used again for different beneficial uses. The quality of the once-used water and the specific type of reuse (or reuse objective) define the levels of subsequent treatment needed, as well as the associated treatment costs. The basic types of reuse are indicated in Figure 4.1 and described in more detail below (WHO, 1989).

4.2.1   Agriculture and aquaculture

On a world-wide basis wastewater is the most widely used low-quality water, particularly for agriculture and aquaculture. This rest of this chapter concentrates on this type of reuse because of the large volumes used, the associated health risks and the environmental concerns. Other types of reuse are only discussed briefly in the following sub-sections.

4.2.2   Urban

In urban areas, reclaimed wastewater has been used mainly for non-potable applications (Crook *et al.*, 1992) such as:

- Irrigation of public parks, recreation centres, athletic fields, school yards and playing fields, and edges and central reservations of highways.
- Irrigation of landscaped areas surrounding public, residential, commercial and industrial buildings.
- Irrigation of golf courses.
- Ornamental landscapes and decorative water features, such as fountains, reflecting pools and waterfalls.
- Fire protection.
- Toilet and urinal flushing in commercial and industrial buildings.

The disadvantages of urban non-potable reuse are usually related to the high costs involved in the construction of dual water-distribution networks, operational difficulties and the potential risk of cross-connection. Costs, however, should be balanced with the benefits of conserving potable water and eventually of postponing, or eliminating, the need for the development of additional sources of water supply.

Potable urban reuse can be performed directly or indirectly. Indirect potable reuse involves allowing the reclaimed water (or, in many instances, raw wastewater) to be retained and diluted in surface or groundwaters before it is collected and treated for human consumption. In many developing countries unplanned, indirect potable reuse is performed on a large scale, when cities are supplied from sources receiving substantial volumes of wastewater. Often, only conventional treatment (coagulation–flocculation–clarification, filtration and disinfection) is provided and therefore significant long-term health effects may be expected from organic and inorganic trace contaminants which remain in the water supplied.

Direct potable reuse takes place when the effluent from a wastewater reclamation plant is connected to a drinking-water distribution network. Treatment costs are very high because the water has to meet very stringent regulations which tend to be increasingly restrictive, both in terms of the number of variables to be monitored as well as in terms of tolerable contaminant limits.

Presently, only the city of Windhoek, Namibia is performing direct potable reuse during dry periods. The Goreangab Reclamation Plant constructed in 1968 is currently being enlarged to treat about 14,000 $m^3 d^{-1}$ by 1997 in order to further augment supplies to the city of Windhoek (Van Der Merwe *et al.*, 1994).

4.2.3   Industry

The most common uses of reclaimed water by industry are:

- Evaporative cooling water, particularly for power stations.
- Boiler-feed water.
- Process water.

- Irrigation of grounds surrounding the industrial plant.

The use of reclaimed wastewater by industry is a potentially large market in developed as well as in developing and rapidly industrialising countries. Industrial reuse is highly cost-effective for industries where the process does not require water of potable quality and where industries are located near urban centres where secondary effluent is readily available for reuse.

4.2.4   Recreation and landscape enhancement

The use of reclaimed wastewater for recreation and landscape enhancement ranges from small fountains and landscaped areas to full, water-based recreational sites for swimming, boating and fishing. As for other types of reuse, the quality of the reclaimed water for recreational uses should be determined by the degree of body contact estimated for each use. In large impoundments, however, where aesthetic appearance is considered important it may be necessary to control nutrients to avoid eutrophication.

4.3   Implementing or upgrading agricultural reuse systems

Land application of wastewater is an effective water pollution control measure and a feasible alternative for increasing resources in water-scarce areas. The major benefits of wastewater reuse schemes are economic, environmental and health-related. During the last two decades the use of wastewater for irrigation of crops has been substantially increased (Mara and Cairncross, 1989) due to:

- The increasing scarcity of alternative water resources for irrigation.
- The high costs of fertilisers.
- The assurances that health risks and soil damage are minimal, if the necessary precautions are taken.
- The high costs of advanced wastewater treatment plants needed for discharging effluents to water bodies.
- The socio-cultural acceptance of the practice.
- The recognition by water resource planners of the value of the practice.

Economic benefits can be gained by income generation and by an increase in productivity. Substantial increases in income will accrue in areas where cropping was previously limited to rainy seasons. A good example of economic recovery associated with the availability of wastewater for irrigation is the Mesquital Valley in Mexico (see Case Study VII) where agricultural income has increased from almost zero at the turn of the century when wastewater was made available to the region, to about 16 million Mexican Pesos per hectare in 1990 (CNA, 1993). The practice of excreta or wastewater fed aquaculture has also been a substantial source of income in many countries

Table 4.1   Increases in crop yields (tons ha^{-1} a^{-1}) arising from wastewater irrigation in Nagpur, India

Irrigation water	Wheat 8 yrs[1]	Moong beans 5 yrs[1]	Rice 7 yrs[1]	Potato 4 yrs[1]	Cotton 3 yrs[1]
Raw wastewater	3.34	0.90	2.97	23.11	2.56
Settled wastewater	3.45	0.87	2.94	20.78	2.30
Stabilisation pond effluent	3.45	0.78	2.98	22.31	2.41
Freshwater + NPK	2.70	0.72	2.03	17.16	1.70

[1] Years of harvest used to calculate average yield   Source: Shende, 1985

such as India, Bangladesh, Indonesia and Peru. The East Calcutta sewage fisheries in India, the largest wastewater use system involving aquaculture in the world (about 3,000 ha in 1987), produces 4–9 t ha^{-1} a^{-1} of fish, which is supplied to the local market (Edwards, 1992). Economic benefits of wastewater/excreta-fed aquaculture can also be found elsewhere (Bartone, 1985; Bartone *et al.*, 1990; Ikramullah, 1994).

Studies carried out in several countries have shown that crop yields can increase if wastewater irrigation is provided and properly managed. Table 4.1 shows the results of field experiments made in Nagpur, India, by the National Environmental Research Institute (NEERI), which investigated the effects of wastewater irrigation on crops (Shende, 1985).

Effluents from conventional wastewater treatment systems, with typical concentrations of 15 mg l^{-1} total N and 3 mg l^{-1} P, at the usual irrigation rate of about 2 m a^{-1}, provide application rates of N and P of 300 and 60 kg ha^{-1} a^{-1}, respectively. Such nutrient inputs can reduce, or even eliminate, the need for commercial fertilisers. The application of wastewater provides, in addition to nutrients, organic matter that acts as a soil conditioner, thereby increasing the capacity of the soil to store water. The increase in productivity is not the only benefit because more land can be irrigated, with the possibility of multiple planting seasons (Bartone and Arlosoroff, 1987).

Environmental benefits can also be gained from the use of wastewater. The factors that may lead to the improvement of the environment when wastewater is used rather than being disposed of in other ways are:

- Avoiding the discharge of wastewater into surface waters.
- Preserving groundwater resources in areas where over-use of these resources in agriculture are causing salt intrusion into the aquifers.
- The possibility of soil conservation by humus build-up and by the prevention of land erosion.

- The aesthetic improvement of urban conditions and recreational activities by means of irrigation and fertilisation of green spaces such as gardens, parks and sports facilities.

Despite these benefits, some potential negative environmental effects may arise in association with the use of wastewater. One negative impact is groundwater contamination. The main problem is associated with nitrate contamination of groundwaters that are used as a source of water supply. This may occur when a highly porous unsaturated layer above the aquifer allows the deeper percolation of nitrates in the wastewater. Provided there is a deep, homogeneous, unsaturated layer above the aquifer which is capable of retaining nitrate, there is little chance of contamination. The uptake of nitrogen by crops may reduce the possibility of nitrate contamination of groundwaters, but this depends on the rate of uptake by plants and the rate of wastewater application to the crops.

Build up of chemical contaminants in the soil is another potential negative effect. Depending on the characteristics of the wastewater, extended irrigation may lead to the build up of organic and inorganic toxic compounds and increases in salinity within the unsaturated layers. To avoid this possibility irrigation should only use wastewater of predominantly domestic origin. Adequate soil drainage is also of fundamental importance in minimising soil salinisation.

Extended irrigation may create habitats for the development of disease vectors, such as mosquitoes and snails. If this is likely, integrated vector control techniques should be applied to avoid the transmission of vector-borne diseases.

Indirect health-related benefits can occur because wastewater irrigation systems may contribute to increased food production and thus to improving health, quality of life and social conditions. However, potential negative health effects must be considered by public health authorities and by institutions managing wastewater reuse schemes because farm workers, the consumers of crops and, to some extent, nearby dwellers can be exposed to the risk of transmission of communicable diseases.

4.3.1 Policy and planning

The use of wastewater constitutes an important element of a water resources policy and strategy. Many nations, particularly those in the arid and semi-arid regions such as the Middle Eastern countries, have adopted (in principle) the use of treated wastewater as an important concept in their overall water resources policy and planning. A judicious wastewater use policy transforms wastewater from an environmental and health liability to an economic and environmentally sound resource (Kandiah, 1994a).

Governments must be prepared to establish and to control wastewater reuse within a broader framework of a national effluent use policy, which itself forms part of a national plan for water resources. Lines of responsibility and cost-allocation principles should be worked out between the various sectors involved, i.e. local authorities responsible for wastewater treatment and disposal, farmers who will benefit from effluent use schemes, and the state which is concerned with the provision of adequate water supplies, the protection of the environment and the promotion of public health. To ensure long-term sustainability, sufficient attention must be given to the social, institutional and organisational aspects of effluent use in agriculture and aquaculture.

The planning of wastewater-use programmes and projects requires a systematic approach. Box 4.1 gives a system framework to support the characterisation of basic conditions and the identification of possibilities and constraints to guide the planning phase of the project (Biswas, 1988).

Government policy on effluent use in agriculture has a deciding effect on the achievement of control measures through careful selection of the sites and the crops that may be irrigated with treated effluent. A decision to make treated effluent available to farmers for unrestricted irrigation removes the possibility of taking advantage of careful selection of sites, irrigation techniques and crops, and thereby of limiting the health risks and minimising the environmental impacts. However, if crop selection is not applied but a government allows unrestricted irrigation with effluent in specific controlled areas, public access to those areas can be prevented (and therefore some control is achieved). The greatest security against health risk and adverse environmental impact arises from limiting effluent use to restricted irrigation on controlled areas to which the public has no access.

It has been suggested that the procedures involved in preparing plans for effluent irrigation schemes are similar to those used in most forms of resource planning, i.e. in accordance with the main physical, social and economic dimensions summarised in Figure 4.2. The following key issues or tasks are likely to have a significant effect on the ultimate success of effluent irrigation schemes:

- The organisational and managerial provisions made to administer the resource, to select the effluent-use plan and to implement it.
- The importance attached to public health considerations and to the levels of risk taken.
- The choice of single-use or multiple-use strategies.
- The criteria adopted in evaluating alternative reuse proposals.
- The level of appreciation of the scope for establishing a forest resource.

Box 4.1   Framework for the analysis of wastewater irrigation projects

Nature of the problem
- How much wastewater will be produced and what will be the seasonal distribution?
- At what places will wastewater be produced?
- What will be the characteristics of wastewater that will be produced?
- What are feasible alternative disposal possibilities?

Legal feasibility
- What uses of wastewater are possible under national and/or state regulations if they exist?
- If no regulations exist, what uses seem feasible under WHO and FAO guidelines or irrigation?
- What are the prevailing water rights and how will these be affected by wastewater use?

Technical feasibility
- Is the quality of treated wastewater produced acceptable for restricted or unrestricted irrigation?
- How much land is available or required for wastewater irrigation?
- What are the soil characteristics of land to be irrigated?
- What are the present land use practices? Can these be changed?
- What types of crops can be grown?
- How do crop-water requirements match with seasonal availability of wastewater?
- What types of irrigation techniques can be used?
- If groundwater recharge is a consideration, are the hydrogeological characteristics of the study are suitable?
- What will be the impact of such recharge on groundwater quality?
- Are there additional health and environmental hazards that should be considered?

Political and social feasibility
- What have been the political reactions to past health and environmental hazards which may have been associated with wastewater reuse?
- What is the publics perception of wastewater reuse?
- What are the attitudes of influential people in areas where wastewater will be reused?
- What are the potential benefits of reuse to the community?
- What are the potential risks?

Economic feasibility
- What are the capital costs?
- What are the operation and maintenance costs?
- What is the economic rate of return?
- What are the cost of development of effluent-irrigated agriculture, e.g. cost of conveyance of wastewater to the irrigation site, and-levelling, installation or irrigation system, agricultural inputs, etc.?
- What are the benefits from the effluent-irrigated agricultural system?
- What is the benefit-cost ratio for the irrigation project?

Personnel feasibility
- Is adequate local labour and expertise available for adequate operation and maintenance of: wastewater treatment, irrigation and groundwater recharge works, agricultural facilities, and health and environmental control aspects?
- If not, what types of training programmes should be instituted?

Source: Biswas, 1988

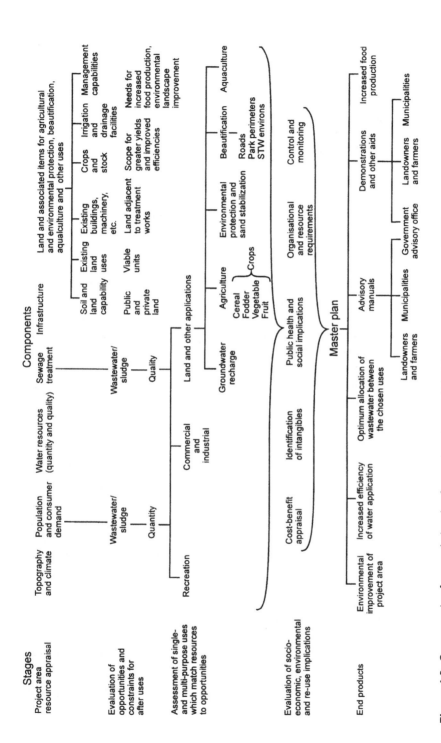

Figure 4.2 Components of general planning for wastewater use (After Cobham and Johnson, 1988)

Adopting a mix of effluent use strategies normally has the advantages of allowing greater flexibility, increased financial security and more efficient use of wastewater throughout the year, whereas a single-use strategy gives rise to seasonal surpluses of effluent for unproductive disposal.

4.3.2   Legal and regulatory issues

The use of wastewater, particularly for irrigation of crops, is associated with two main types of legal issues:

- Establishment of a legal status of wastewater and the delineation of a legal regime for its use. This may include the development of new, or the amendment of existing, legislation; creation of new institutions or the allocation of new powers to existing institutions; attributing roles of, and relationships between, national and local government in the sector; and public health, environmental and agricultural legislation such as standards and codes of practice for reuse.
- Securing tenure for the users, particularly in relation to rights of access to and ownership of waste, and including public regulation of its use. Legislation should also include land tenure, without which security of access to wastewater is worthless.

The delineation of a legal regime for wastewater management should address the following aspects (WHO, 1990):

- A definition of what is meant by wastewater.
- The ownership of wastewater.
- A system of licensing of wastewater use.
- Protection of other users of the water resources that may be adversely affected by the loss of return flows into the system arising from the use of wastewater.
- Restrictions for the protection of public and environmental health with respect to intended use of the wastewater, treatment conditions and final quality of wastewater, and conditions for the siting of wastewater treatment facilities.
- Cost allocation and pricing.
- Enforcement mechanisms.
- Disposal of the sludges which result from wastewater treatment processes.
- Institutional arrangements for the administration of relevant legislation.
- The interface of this legal regime with the general legal regime for the management of water resources, particularly the legislation for water and environmental pollution control and the legislation governing the provision of water supply and sewerage services to the public, including the relevant responsible institutions.

At the operational level, regulatory actions are applied and enforced through guidelines, standards and codes of practice (see Chapters 2 and 5).

Guidelines

One of the many functions of the World Health Organization (WHO) is to propose regulations and to make recommendations with respect to international health matters. Guidelines for the safe use of wastewater, produced as part of this function are intended to provide background and guidance to governments for risk management decisions related to the protection of public health and to the preservation of the environment.

It must be stressed that guidelines are not intended for absolute and direct application in every country. They are of advisory nature and are based on the state-of-the-art in scientific research and epidemiological findings. They are aimed at the establishment of a health basis and the health risks and, as such, they provide a common background against which national or regional standards can be derived (Hespanhol and Prost, 1994).

Agriculture. The Scientific Group on Health Guidelines for the Use of Wastewater in Agriculture and Aquaculture, held in Geneva in 1987 (WHO, 1989) established the basic criteria for health protection of the groups at risk from agricultural reuse systems and recommended the microbiological guidelines shown in Table 4.2. These criteria and guidelines were the result of a long preparatory process and the epidemiological evidence available at the time. They are related to the category of crops, the reuse conditions, the exposed groups and the appropriate wastewater treatment systems, in order to achieve microbiological quality.

Aquaculture. The use of wastewater or excreta to fertilise ponds for fish production has been associated with a number of infections caused by excreted pathogens, including invasion of fish muscle by bacteria and high pathogen concentrations in the digestive tract and the intra-peritoneal fluid of the fish. Limited experimental and field data on health effects of excreta or wastewater fertilised aquaculture are available and, therefore, the Scientific Group Meeting recommended the following tentative guidelines:
- A geometric mean of less than 10^3 faecal coliform per 100 ml for fish pond water, to ensure that bacterial invasion of fish muscle is prevented. The same guideline value should be maintained for pond water in which edible aquatic vegetables (macrophytes) are grown because in many areas they are eaten raw. This can be achieved by treating the wastewater supplied to the ponds to a concentration of 10^3–10^4 faecal coliforms per 100 ml

Table 4.2   Recommended microbiological guidelines for wastewater use in agriculture

Category	Reuse conditions	Exposed group	Intestinal nematodes[1] (No. of eggs per litre)[2]	Faecal coliforms (No. per 100 ml)[3]	Wastewater treatment expected to achieve microbiological quality
A	Irrigation of crops likely to be eaten uncooked, sports fields, public parks[4]	Workers, consumers, public	≤ 1	≤ 1,000	A series of stabilisation ponds designed to achieve the microbiological quality indicated, or equivalent treatment
B	Irrigation of cereal crops, industrial crops, fodder crops, pasture and trees[5]	Workers	≤ 1	na	Retention in stabilisation ponds for 8–10 days or equivalent helminth and faecal coliform removal
C	Localised irrigation of crops in category B if exposure of workers and public does not occur	None	na	na	Pre-treatment as required by irrigation technology, but no less than primary sedimentation

In specific cases, local epidemiological, socio-cultural and environmental factors should be taken into account, and these guidelines modified accordingly.

na   Not applicable

[1]   *Ascaris, Trichuris* and hookworms
[2]   During the irrigation period. Arithmetic mean
[3]   During the irrigation period. Geometric mean

[4]   A more stringent guideline (200 faecal coliforms per 100 ml) is appropriate for public lawns, such as hotel lawns, with which the public may have direct contact
[5]   In the case of fruit trees, irrigation should cease two weeks before fruit is picked, and no fruit should be picked off the ground. Sprinkler irrigation should not be used.

Source: WHO, 1989

(assuming that the pond will allow one order of magnitude dilution of the incoming wastewater).

- Total absence of trematode eggs, to prevent infection by helminths such as clonorchiasis, fascialopsiasis and schistosomiasis. This can be readily achieved by stabilisation pond treatment.
- High standards of hygiene during fish handling and gutting to prevent infection of fish muscle by the intra-peritoneal fluid of the fish.

The chemical quality of treated domestic effluents used for irrigation is also of particular importance. Several variables are relevant to agriculture in relation to the yield and quality of crops, the maintenance of soil productivity and the protection of the environment. These variables are total salt concentration, electrical conductivity, sodium adsorption ratio (SAR), toxic ions, trace elements and heavy metals. A thorough discussion of this subject is available in FAO (1985).

Standards and Codes of Practice. Standards are legal impositions enacted by means of laws, regulations or technical procedures. They are established by countries by adapting guidelines to their own national priorities and by taking into account their own technical, economical, social, cultural and political characteristics and constraints (see Chapter 5). They are established by competent national authorities by adopting a risk–benefit approach. This infers that the standards produced will consider not only health-related concerns but also a wide range of economic and social consequences. At any time, national standards can be changed or modified whenever new scientific evidence or new technologies become available, or in response to changes in national priorities or tendencies.

Standards are, in many countries, complemented by codes of practice which provide guidance for the construction, operation and maintenance and surveillance of wastewater use schemes. Codes of practice should be prepared according to local conditions, but the following basic elements are frequently included:
- Crops allowed under crop restriction policies.
- Wastewater treatment and effluent quality.
- Wastewater distribution network.
- Irrigation methods.
- Operation and maintenance.
- Human exposure control.
- Monitoring and surveillance.
- Reporting.
- Charges and fines.

4.3.3 Institutional arrangements
Wastewater-use projects at national level touch on the responsibilities of several ministries and government agencies. For adequate operation and minimisation of administrative conflicts, the following ministries should be involved from the planning phase onwards:
- Ministry of Agriculture and Fisheries: overall project planning; management of state-owned land; installation and operation of an irrigation infrastructure; agricultural and aquacultural extension, including training; and control of marketing.
- Ministry of Health: surveillance of effluent quality according to local standards; health protection and disease surveillance; responsibility for human

exposure control, such as vaccination, control of anaemia and diarrhoeal diseases (see section 4.4); and health education.

- Ministry of Water Resources: integration of wastewater use projects into overall water resources planning and management.
- Ministry of Public Works and Water Authorities: wastewater or excreta collection and treatment.
- Ministry of Finance/Economy/Planning: economic and financial appraisal of projects; and cost/benefit analysis, financing, criteria for subsidising, etc.

According to national arrangements, other ministries such as those concerned with environmental protection, land tenure, rural development, co-operatives and women's affairs may also be involved (Mara and Cairncross, 1989).

Countries starting activities involving wastewater use for the first time can benefit greatly from the establishment of an executive body, such as an inter-agency technical standing committee, which is under the aegis of a leading ministry (Agriculture or Water Resources) and which takes responsibility for sector development, planning and management. Alternatively, existing organisations may be given responsibility for the sector (or parts of it), for example a National Irrigation Board might be responsible for wastewater use in agriculture and a National Fisheries Board might be responsible for the aquacultural use of excreta and wastewater. Such organisations should then co-ordinate a committee of representatives from the different agencies having sectoral responsibilities. The basic responsibilities of inter-agency committees are:

- Developing a coherent national or regional policy for wastewater use and monitoring its implementation.
- Defining the division of responsibilities between the respective ministries and agencies involved and the arrangements for collaboration between them.
- Appraising proposed reuse schemes, particularly from the point of view of public health and environmental protection.
- Overseeing the promotion and enforcement of national legislation and codes of practice.
- Developing a rational staff development policy for the sector.

In countries with a regional or federal administration, such arrangements for inter-agency collaboration are even more important at regional or state level. Whereas the general framework of waste-use policy and standards may be defined at national level, the regional body will have to interpret and add to these, taking into account local conditions.

In Mexico, the National Water Commission (CNA), which is attached to the Ministry of Agriculture and Water Resources, administers the water resources of the country and, as such, is the institution in charge of the planning, administration and control of all wastewater use schemes at national level. Other governmental departments, such as the Ministry of Health, the Ministry of the Environment and the Ministry of Social Development, also participate according to specific interests within their own field of activity. At regional level, the State government is also integrated with the administration of local schemes. In the Mesquital Valley, for example, the State of Hidalgo collaborates with the local agency of CNA for the operation and maintenance of the irrigation districts as well as for monitoring, surveillance and enforcement actions. In the Mesquital Valley there is also a strong participation by the private sector, dealing with the administration of small irrigation units integrated into co-operative systems.

4.3.4   Economic and financial aspects

Economic appraisal of wastewater irrigation projects should be based on the incremental costs and benefits accrued from the practice. One procedure adopted in many projects is to adjust marginal benefits and costs to the current value at a real discount rate and to design the system carefully in order that the benefit/cost ratio is greater than 1. Another procedure consists of determining the internal rate of return of the project and confirming that it is competitive (Forero, 1993).

The financial evaluation can be done by comparison with one of the following hypothetical scenarios, each of which is configured with different benefits and costs:
- No agriculture at all.
- No irrigation at all (rain-fed agriculture).
- Irrigation with water from an alternative source without fertiliser application.
- Irrigation with water from an alternative source with fertiliser application.

Costs. The following costs must be considered in a wastewater irrigation project (Papadopoulos, 1990):
- Wastewater treatment costs, including land and site preparation, civil engineering works, system design, materials and equipment.
- Irrigation costs, including water handling, storage, conveyance and distribution.
- On-farm costs, associated with institutional build-up, including facilities and training, measures for public health protection, hygiene facilities for field workers, and use of lower value crops associated with specific wastewater application.

- Operation and maintenance costs, including additional energy consumption, labour, protective clothing for field workers, supplementary fertiliser if needed, management and overhead costs, and monitoring and testing.

It is of fundamental importance that only marginal costs are taken into account in the appraisal. For example, only the additional costs required to attain local effluent standards for reuse should be considered (if they are needed). Costs associated with treatment systems for environmental protection (which would be implemented anyway), should not be accounted in the economic evaluation of reuse systems. In the same way, irrigation and on-farm costs that should be considered are solely the supplementary costs accrued in association with the use of wastewater rather than any other conventional source of water.

Benefits. Direct benefits are relatively easy to evaluate. In agriculture or aquaculture systems they can be directly evaluated, for example in terms of the increase in crop production and yields, savings in fertiliser costs and saving in freshwater supply. By contrast, indirect benefits are complex and difficult to quantify properly. Among the many other benefits that attract decision-making officials who are able to foresee the health and environmental advantages of wastewater use in agriculture are:
- The improved nutritional status of poor populations through increased food availability.
- The increase in jobs and settlement opportunities.
- The development of new recreation areas.
- Reduced damage to the urban environment.
- Protection of groundwater resources from depletion.
- Protection of freshwater resources against pollution and their conservation.
- Erosion control, reduced desertification, etc.

The indirect benefits are "non-monetary issues" and, unfortunately, they are not taken into account when performing economical appraisals of projects involving wastewater use. However, the environmental enhancement provided by wastewater use, particularly in terms of preservation of water resources, improvement of the health status of poor populations in developing countries, the possibility of providing a substitute for freshwater in water-scarce areas, and the incentive provided for the construction of urban sewerage works, are extremely relevant. They are also sufficiently important to make the cost/benefit analysis purely subsidiary when taking a decision on the implementation of wastewater reuse systems, particularly in developing and rapidly industrialising countries.

Cost recovery. Adopting an adequate policy for the pricing of water is of fundamental importance in the sustainability of wastewater reuse systems. The incremental cost basis, which allocates only the marginal costs associated with reuse, seems to be a fair criteria for adoption in developing countries, where wastewater reuse is assumed to be a social benefit. A charge in the form of tariffs, or fees, based on the volumes of treated wastewater distributed, or in terms of hours of distribution, has been used in many countries. Where the volumes are very large and the distribution network covers a wide area, as in the Mesquital Valley in Mexico, the charges are made to farmers in relation to the individual areas being irrigated.

Subsidising reuse systems may be necessary in the early stages of system implementation, particularly when the associated costs are very large. This would avoid any discouragement to farmers arising from the permitted use of the treated wastewater. In order to determine the necessity of governmental support for the cost–recovery scheme it would be advisable to investigate the willingness and the ability of the farmers to pay for the services. The easiest way to collect fees is by imposing charges that are payable just after the harvest season.

4.3.5   Socio-cultural aspects
Public acceptance of the use of wastewater or excreta in agriculture and aquaculture is influenced by socio-cultural and religious factors. In the Americas, Africa and Europe, for example, there is a strong objection to the use of excreta as fertiliser, whereas in some areas of Asia, particularly in China, Japan and Java, the practice is performed regularly and regarded as economical and ecologically sound.

In most parts of the world, however, there is no cultural objection to the use of wastewater, particularly if it is treated. Wastewater use is well accepted where other sources of water are not readily available, or for economic reasons. Wastewater is used for the irrigation of crops in several Islamic countries provided that the impurities (*najassa*) are removed. This results, however, from economical need rather than cultural preference. According to Koranic edicts, the practice of reuse is accepted religiously provided impure water is transformed to pure water (*tahur*) by the following methods (Farooq and Ansari, 1983): self-purification, addition of pure water in sufficient quantity to dilute the impurities, or removal of the impurities by the passage of time or by physical effects.

Due to the wide variability in cultural beliefs, human behaviour and religious dogmas, acceptance or refusal of the practice of wastewater reuse within a specific culture is not always applicable everywhere. A complete assessment of local socio-cultural contexts and religious beliefs is always necessary as a preliminary step to implementing reuse projects (Cross, 1985).

4.3.6   Monitoring and evaluation

As mentioned before (see section 4.3.3), projects and programmes associated with the use of wastewater should be led and co-ordinated by inter-agency committees under the aegis of a leading ministry. This entity should also be in charge of monitoring and evaluation programmes and should have the legal powers to enforce compliance with local legislation.

Monitoring activities for wastewater use projects are of two different types. Process control monitoring is carried out to provide data to support the operation and optimisation of the system, in order to achieve successful project performance. It includes the monitoring of treatment plants, water distribution systems, water application equipment, environmental aspects (such as salinisation, drainage waters, water logging), agricultural aspects (such as productivity and yield) and health-related problems (such as the development of disease vectors and health problems associated with the use of wastewater). In addition to providing data for process control, this level of monitoring generates information for project revision and updating as well for further research and development. Responsibility for process control monitoring belongs to the operating agency (for example, a state agency or a municipal sewerage board) which is part of the inter-agency committee.

Compliance monitoring is required to meet regulatory requirements and should not be performed by the same agency in charge of process control monitoring. This responsibility should be extended to an enforcement agency that possesses legal powers to enforce compliance with quality standards, codes of practice and other pertinent legislation. The responsibility for compliance monitoring is usually granted to Ministries of Health because health problems are of prime importance for wastewater use systems (see section 4.4).

A successful monitoring programme should be cost effective (only essential data should be collected and analysed); it should provide adequate coverage (only representative sectors of the system should be covered); it must be reliable (representative sampling, accurate analysis with adequate analytical quality control, appropriate storing, handling and reporting of information); and it should be timely, in order to provide operators and decision-making officials with fresh and up-to-date information that allows the application of prompt remedial measures during critical situations.

4.3.7 Public awareness and participation

To achieve general acceptance of reuse schemes, it is of fundamental importance that active public involvement is obtained from the planning phase to the full implementation process. Public involvement starts with early contact with potential users, leading to the formation of an advisory committee and the holding of public workshops on potential reuse schemes. The continuous exchange of information between authorities and the public representatives ensures that the adoption of a specific water reuse programme will fulfil real user needs and generally-recognised community goals for health, safety, ecological concerns, programme cost, etc. (Crook *et al.*, 1992).

Acceptance of reuse systems depends on the degree to which the responsible agencies succeed in providing the concerned public with a clear understanding of the complete programme; the knowledge of the quality of the treated wastewater and how it is to be used; confidence in the local management of the public utilities and on the application of locally accepted technology; assurance that the reuse application being considered will involve minimal health risks and minimal detrimental effects to the environment; and assurance, particularly for agricultural uses, of the sustainability of supply and suitability of the reclaimed wastewater for the intended crops.

Figure 4.3 provides a flow chart for establishing programmes to involve the concerned community with all phases of wastewater use projects, from the planning phase to full implementation of the project, and Table 4.3 presents a series of tools to address, educate and inform the public at different levels of involvement.

4.4   Technical aspects of health protection

Health protection in wastewater use projects can be provided by the integrated application of four major measures: wastewater treatment, crop selection and restriction, wastewater irrigation techniques and human exposure control.

4.4.1 Wastewater treatment

Wastewater treatment systems were first developed in response to the adverse conditions caused by the discharge of raw effluents to water bodies. With this approach, treatment is aimed at the removal of biodegradable organic compounds, suspended and floatable material, nutrients and pathogens. However, the criteria for wastewater treatment intended for reuse in irrigation differ considerably. While it is intended that pathogens are removed to the maximum

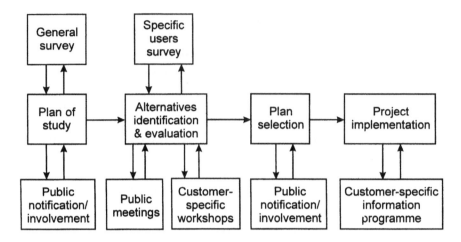

Figure 4.3   A flow chart illustrating a public participation programme (After Crook *et al.*, 1992)

Table 4.3   Removal of excreted bacteria and helminths by various wastewater treatment systems

Treatment process	Removal ($\log_{10}$ units) of			
	Bacteria	Helminths	Viruses	Cysts
Primary sedimentation				
Plain	0–1	0–2	0–1	0–1
Chemically assisted[1]	1–2	1–3 (G)	0–1	0–1
Activated sludge[2]	0–2	0–2	0–1	0–1
Biofiltration[2]	0–2	0–2	0–1	0–1
Aerated lagoon[3]	1–2	1–3 (G)	1–2	0–1
Oxidation ditch[2]	1–2	0–2	1–2	0–1
Disinfection[4]	2–6	0–1	0–4	0–3
Waste stabilisation ponds[5]	1–6 (G)	1–3 (G)	1–4	1–4
Effluent storage reservoirs[6]	1–6 (G)	1–3 (G)	1–4	1–4

G   With good design and proper operation the recommended guidelines are achievable
[1]   Further research is needed to confirm performance
[2]   Including secondary sedimentation
[3]   Including settling pond
[4]   Chlorination or ozonation

[5]   Performance depends on number of ponds in series and other environmental factors
[6]   Performance depends on retention time, which varies with demand

Source: Mara and Cairncross, 1989

Table 4.4   Reported effluent quality from stabilisation ponds with a retention time of 25 days

Location of ponds	No. of ponds in series	Effluent quality (fc/100 ml)[1]
Australia, Melbourne	8–11	100
Brazil, Extrabes	5	30
France, Cogolin	3	100
Jordan, Amman	9	30
Peru, Lima	5	100
Tunisia, Tunis	4	200

[1]   Faecal coliforms per 100 ml   Source: Bartone and Arlosoroff, 1987

extent possible, some of the biodegradable organic matter and most of the nutrients available in the raw wastewater need to be maintained.

Table 4.4 summarises the efficiency of wastewater treatment systems for the removal of pathogens, indicating where the proposed WHO guidelines for Category A (unrestricted irrigation) can be met. The following general comments provide technical support to guide the choice of adequate treatment systems for the use of wastewater in irrigation (Hespanhol, 1990).

Conventional primary and secondary treatments
Raw domestic wastewater contains between 10^7 and 10^9 faecal coliform per 100 ml. Conventional treatment systems, such as plain sedimentation, biofiltration, aerated lagoons and activated sludge, which are designed particularly for removal of organic matter, are not able to remove pathogens in order to produce an effluent that meets the WHO guideline for bacterial quality ($\leq$ 1,000 faecal coliform per 100 ml). In the same way, they are not generally effective in helminth removal. More research and adaptive work is required to improve the effectiveness of conventional systems in removing helminth eggs.

Waste stabilisation ponds
Ponding systems are the preferred technology to provide effluents for reuse in agriculture and aquaculture, particularly in warm climates and whenever land is available at reasonable cost (Mara, 1976; Arthur, 1983; Bartone, 1991). Ponding systems integrating anaerobic, facultative and maturation units, with an overall average retention time of 10–50 days (depending on temperature), can produce effluents that meet the WHO guidelines for both bacterial and helminth quality.

Table 4.5   Performance of five wastewater stabilisation ponds (mean temperature 26 °C) in Northeast Brazil

Sample	Retention time (days)	BOD$_5$ (mg l^{-1})	Suspended solids (mg l^{-1})	Faecal coliforms	Intestinal nematode eggs/litre
Raw wastewater		240	305	4.6×10^7	804
Effluent from					
Anaerobic pond	6.8	63	56	2.9×10^6	29
Facultative pond	5.5	45	74	3.2×10^5	1
Maturation pond No. 1	5.5	25	61	2.4×10^4	0
Maturation pond No. 2	5.5	19	43	450	0
Maturation pond No. 3	5.8	17	45	30	0

Sources: Mara *et al.*, 1983; Mara and Silva, 1986

Tables 4.5 and 4.6 illustrate the high confidence with which pond systems can meet the WHO guidelines and Table 4.6 also shows their excellent capacity for reducing BOD and suspended solids. The FAO Irrigation and Drainage Paper No. 47 *Wastewater Treatment in Agriculture* (FAO, 1985) also provides a good review of wastewater treatment systems which are recommended for wastewater use schemes. The following advantages are the reasons why stabilisation ponds are an adequate treatment system for the conditions prevailing in developing countries:

- Lower construction, operation and maintenance costs.
- No energy requirements.
- High ability to absorb organic and hydraulic loads.
- Ability to treat a wide variety of industrial and agricultural wastes.

Disinfection
Disinfection of wastewater through the application of chlorine has never been completely successful in practice, due to the high costs involved and the difficulty of maintaining an adequate, uniform and predictable level of disinfection efficiency. Effluents from well-operated conventional treatment systems, treated with 10–30 mg l^{-1} of chlorine and a contact time of 30–60 minutes, provide a good reduction of excreted bacteria, but have no capacity for removing helminth eggs and protozoa. As a well designed and operating stabilisation ponding system will provide an effluent with less than 1,000 faecal coliform per 100 ml and less than one egg of intestinal nematodes per litre, there is usually no need for disinfection of pond effluents intended for reuse.

Table 4.6 Evaluation of common irrigation methods in relation to the use of treated wastewater

Parameters of evaluation	Furrow irrigation	Border irrigation	Sprinkler irrigation	Drip irrigation
Foliar wetting and consequent leaf damage resulting in poor yield	No foliar injury as the crop is planted on the ridge	Some bottom leaves may be affected but the damage isnot serious enough to reduce yield	Severe leaf damage can occur resulting in significant yield loss	No foliar injury occurs under this method of irrigation
Salt accumulation in the root zone with repeated application	Salts tend to accumulate in the ridge which could harm the crop	Salts move vertically downwards and are not likely to accumulate in the root zone	Salt movement is downwards and root zone is not likely to accumulate salts	Salt movement is radial along the direction of water movement. A salt wedge is formed between drip points
Ability to maintain high soil water potential	Plants may be subject to stress between irrigations	Plants may be subject to water stress between irrigations	Not possible to maintain high soil water potential throughout the growing season	Possible to maintain high soil water potential throughout the growing season and minimise the effect of salinity
Suitability to handle brackish wastewater without significant yield loss	Fair to medium. With good management and drainage acceptable yields are possible	Fair to medium. Good irrigation and drainage practices can produce acceptable levels of yield	Poor to fair. Most crops suffer from leaf damage and yield is low	Excellent to good. Almost all crops can be grown with very little reduction in yield

Source: Kandiah, 1994b

Storage reservoirs

Water demand for irrigation occurs mainly in the dry season or during particular periods of the year. Wastewater intended for irrigation can, therefore, be stored in large, natural or specially constructed reservoirs, which provide further natural treatment, particularly in terms of bacteria and helminth removal. Such reservoirs have been used in Mexico and Israel (Shuval, *et al.*, 1986).

There are insufficient field data available to formulate an adequate design criterion for storage reservoirs, but pathogen removal depends on retention time and on the possibility of having the reservoir divided into compartments. The greater the retention time and the larger the number of compartments in series, the higher the efficiency of pathogen removal. A design recommendation, based particularly on data available from natural storage reservoirs operating in the Mesquital Valley, Mexico, is to provide a minimum hydraulic average retention time of 10 days, and to assume two orders of magnitude reduction in both faecal coliform and helminth eggs. Thus, the stored wastewater should contain no more than 10^2 eggs per litre and not more than 10^5 faecal coliform per 100 ml, in order that the WHO guidelines for unrestricted irrigation are attained.

Tertiary treatment

Tertiary or advanced treatment systems are used to improve the physico-chemical quality of biological secondary effluents. Several unit operations and unit processes, such as coagulation–flocculation–settling–sand filtration, nitrification and denitrification, carbon adsorption, ion exchange and electrodialysis, can be added to follow secondary treatment in order to obtain high quality effluents. None of these units are recommended for use in developing countries when treating wastewater for reuse, due to the high capital and operational costs involved and the need for highly skilled personnel for operation and maintenance.

If the objective is to improve effluents of biological plants (particularly in terms of bacteria and helminths), for the irrigation of crops or for aquaculture, a more appropriate option is to add one or two "polishing" ponds as a tertiary treatment. If land is not available for that purpose, horizontal or vertical-flow roughing filtration units (which have been used for pre-treatment of turbid waters prior to slow-sand filtration) may be considered. These units, which are low cost and occupy a relatively small area, have been shown to be very effective for the treatment of secondary effluents and remove a considerable proportion of intestinal nematodes. Detailed information on the design, operation and removal efficiencies of roughing filters can be found elsewhere (Wegelin, 1986; Wegelin *et al.*, 1991).

Sludge treatment
The excess sludge produced by biological treatment plants is valuable as a source of plant nutrient as well as a soil conditioner. It can also be used in agriculture or to fertilise aquaculture ponds. However, biological treatment processes concentrate organic and inorganic contaminants as well as pathogens in the excess sludge. Given the availability of nutrients and moisture, helminth eggs can survive and remain viable for periods close to one year. If adequate care is taken during the handling process, raw sludge can be applied to agricultural land in trenches and covered with a layer of earth. This should be done before the planting season starts and care should be taken that no tuberous plants, such as beets or potatoes, are planted along the trenches.

The following treatment methods can be applied to make sludges safe for use in agriculture or aquaculture:
- Storage, from 6–12 months, at ambient temperature in hot climates.
- Mesophyllic (around 35 °C) anaerobic digestion, which removes 90–95 per cent of total parasite eggs, but only 30–40 per cent of *Ascaris* eggs (Gunnerson and Stuckey, 1986).
- Thermophilic (around 55 °C) anaerobic digestion for about 13 days ensures total inactivation of all pathogens. Continuous reactors can allow pathogens to by-pass the removal process and therefore the digestion process should be performed under batch conditions (Strauss, 1985).
- Forced-aeration co-composting of sludge with domestic solid waste or some other organic bulking agent, such as wood chips, for 30 days at 55–60 °C followed by maturation for 2–4 months at ambient temperature, will produce a stable, pathogen-free compost (Obeng and Wright, 1987).

4.4.2   Crop selection
According to the WHO guidelines (see Table 4.2) wastewater of a high microbiological quality is needed for the irrigation of certain crops, particularly crops eaten uncooked. Nevertheless, a lower quality is acceptable for irrigation of certain types of crop and corresponding levels of exposure to the groups at risk, because lower quality waters will affect consumers and other exposed groups such as field workers and crop handlers. For example, crops which are normally cooked, such as potatoes, or industrial crops such as cotton and sisal, do not require a high quality wastewater for irrigation.

Crops can be grouped into two broad categories according to the group of persons likely to be exposed and the degree to which health protection measures are required:

Category A. Protection required for consumers, agricultural workers and the general public. This category includes crops likely to be eaten uncooked, spray-irrigated fruits, sports fields, public parks and lawns.

Category B. Protection required for agricultural workers only, because there would be no microbiological health risks associated with the consumption of the crops if they were irrigated with wastewater (there is no risk to consumers because crops in this category are not eaten raw, or they are processed before they reach the consumer). This category includes cereal crops, industrial crops, food crops for canning, fodder crops, pastures and trees. Some vegetable crops may be included in this category if they are not eaten raw (potatoes and peas), or if they grow well above the ground (chillies, tomatoes and green beans). In such cases it is necessary to ensure that the crop is not contaminated by sprinkler irrigation or by falling to the ground, and that contamination of kitchen utensils by such crops, before cooking, does not give rise to health risks.

The practice of crop restriction infers that crops that are allowed to be irrigated with wastewater are restricted to those specified under category B. This category protects consumers but additional protective measures are necessary for farm workers (see below).

Although it appears simple and straightforward, in practice it is very difficult to implement and to enforce crop restriction policies. A crop restriction policy is effective for health protection only if it is fully implemented and enforced. It requires a strong institutional framework and the capacity to monitor and to control compliance with the established crop restriction regulations. Farmers should be advised of the importance and necessity of the restriction policy and be assisted in developing a balanced mix of crops which makes full use of the available partially-treated wastewater. The likelihood of succeeding is greater where:

- A law-abiding society exists or the restriction policy is strongly enforced.
- A public body controls the allocation of wastewater under a strong central management.
- There is adequate demand for the crops allowed under the policy and they fetch a reasonable price.
- There is little market pressure in favour of crops in category A.

Crop restriction does not provide health protection in aquaculture schemes, because fish and macrophytes grown in wastewater or excreta-fertilised ponds are, in many places, eaten uncooked. An alternative and promising approach, already practised in many parts of the world, is to grow duckweed (*Lemna* sp.) in wastewater-fed ponds. The duckweed is then collected and dried, and fed to high-value fish grown in freshwater ponds. The same approach can be used to produce fishmeal for animal feed (or for fish food) by growing the fish to be used for the production of fishmeal in wastewater ponds.

4.4.3 Irrigation techniques

The different methods used by farmers to irrigate crops can be grouped under five headings (Kandiah, 1994b):

- Flood irrigation: water is applied over the entire field to infiltrate into the soil (e.g. wild flooding, contour flooding, borders, basins).
- Furrow irrigation: water is applied between ridges (e.g. level and graded furrows, contour furrows, corrugations). Water reaches the ridge (where the plant roots are concentrated) by capillary action.
- Sprinkler irrigation: water is applied in the form of a spray and reaches the soil in much the same way as rain (e.g. portable and solid set sprinklers, travelling sprinklers, spray guns, centre-pivot systems).
- Sub-surface irrigation: water is applied beneath the root zone in such a manner that it wets the root zone by capillary rise (e.g. subsurface canals, buried pipes).
- Localised irrigation: water is applied around each plant or group of plants so that only the root zone gets wet (e.g. drip irrigation, bubblers, micro-sprinklers).

The type of irrigation method selected depends on water supply conditions, climate, soil, the crops to be grown, the cost of irrigation methods and the ability of the farmer to manage the system.

There is considerable scope for reducing the negative effects of wastewater use in irrigation through the selection of appropriate irrigation methods. The choice of method is governed by the following technical factors:

- Type of crops to be irrigated.
- The wetting of foliage, fruits and aerial parts.
- The distribution of water, salts and contaminants in the soil.
- The ease with which high soil–water potential can be maintained.
- The efficiency of application.
- The potential to contaminate farm workers and the environment.

Table 4.7 analyses these factors in relation to four widely practised irrigation methods, namely border, furrow, sprinkler and drip irrigation.

A border (as well as a basin or any flood irrigation) system involves complete coverage of the soil surface with treated wastewater which is not normally an efficient method of irrigation. This system contaminates root

Table 4.7   Different levels of tools for public participation in the decision to reuse wastewater

Purpose	Tools
Education and information	Newspaper articles, radio and TV programmes, speeches and presentations, field trips, exhibits, information depositories, school programmes, films, brochures and newsletters, reports, letters, conferences
Review and reaction	Briefings, public meetings, public hearings, surveys and questionnaires, question and answer columns, advertised "hotlines" for telephone inquiries
Interaction dialogue	Workshops, special task forces, interviews, advisory boards, informal contacts, study group discussions, seminars

Source: Crook *et al.*, 1992

crops and vegetable crops growing near the ground and, more than any other method, exposes field workers to the pathogen content of wastewater. Thus, with respect to both health and water conservation, border irrigation with wastewater is not satisfactory.

Furrow irrigation does not wet the entire soil surface, and can reduce crop contamination, because plants are grown on ridges. Complete health protection cannot be guaranteed and the risk of contamination of farm workers is potentially medium to high, depending on the degree of automation of the process. If the treated wastewater is transported through pipes and delivered into individual furrows by means of gated pipes, the risk to irrigation workers is minimum. To avoid surface ponding of stagnant wastewater, which may induce the development of disease vectors, levelling of the land should be carried out carefully and appropriate land gradients should be provided.

Sprinkler, or spray, irrigation methods are generally more efficient in water use because greater uniformity of application can be achieved. However, such overhead irrigation methods can contaminate ground crops, fruit trees and farm workers. In addition, pathogens contained in the wastewater aerosol can be transported downwind and create a health hazard to nearby residents. Generally, mechanised or automated systems have relatively high capital costs and low labour costs compared with manually-operated sprinkler systems. Rough levelling of the land is necessary for sprinkler systems in order to prevent excessive head loss and to achieve uniformity of wetting. Sprinkler systems are more affected by the quality of the water than surface irrigation systems, primarily as a result of clogging of the orifices in the sprinkler heads but also due to sediment accumulation in pipes, valves and distribution systems. There is also the potential for leaf burn and phytotoxicity if the wastewater is saline and contains excessive toxic elements. Secondary treatment systems that meet the WHO microbiological guidelines have generally been found to produce an effluent suitable for distribution through sprinklers, provided that the wastewater is not too saline. Further precautionary measures, such as treatment with sand filters or micro-strainers and enlargement of the nozzle orifice to diameters not less than 5 mm, are often adopted.

Localised irrigation, particularly when the soil surface is covered with plastic sheeting or other mulch, uses effluent more efficiently. It produces higher crop yields and certainly provides the greatest degree of health protection to farm workers and consumers. However, trickle and drip irrigation systems are expensive and require a high quality of treated wastewater in order to prevent clogging of the orifices through which water is released into the soil. A relatively new technique called "bubbler irrigation", that was developed for localised irrigation of tree crops, avoids the needs for small orifices. This system requires, therefore, less treatment of the wastewater but needs careful setting for successful application.

When compared with other systems, the main advantages of trickle irrigation are:

- Increased crop growth and yield achieved by optimising the water, nutrients and air regimes in the root zone.
- High irrigation efficiency because there is no canopy interception, wind drift or conveyance losses, and minimal drainage loss.
- Minimal contact between farm workers and wastewater.
- Low energy requirements because the trickle system requires a water pressure of only 100–300 kPa (1–3 bar).
- Low labour requirements because the trickle system can be easily automated, even to allow combined irrigation and fertilisation.

In addition to the high capital costs of trickle irrigation systems, another limiting factor in their use is that they are mostly suited to the irrigation of crops planted in rows. Relocation of subsurface systems can be prohibitively expensive.

Special field management practices that may be required when wastewater irrigation is performed, include pre-planting irrigation, blending of wastewater with other water supplies, and alternating treated wastewater with other sources of supply.

The amount of wastewater to be applied depends on the rate of evapotranspiration from the plant surface, which is determined by climatic factors and can therefore be estimated with reasonable accuracy, using meteorological data. An extensive review of this subject is available in FAO (1984).

4.4.4   Human exposure control

The groups of people that are more susceptible to the potential risk from the use of wastewater in agriculture are agricultural field workers and their families, crop handlers, consumers of crops, meat and milk originating from wastewater irrigated fields, and those living near wastewater irrigated fields. The basic methods for eliminating or minimising exposure depend on the target groups. Agricultural field workers and crop handlers have higher potential risks mainly associated with parasitic infections. Protection can be achieved by:

- The use of appropriate footwear to reduce hookworm infection.
- The use of gloves (particularly crop handlers).
- Health education.
- Personal hygiene.
- Immunisation against typhoid fever and hepatitis A and B.
- Regular chemotherapy for intense nematode infections in children and the control of anaemia.
- Provision of adequate medical facilities to treat diarrhoeal diseases.

Protection of consumers can be achieved by:

- Cooking of vegetables and meat and boiling milk.
- High standards of personal and food hygiene.
- Health education campaigns.
- Meat inspection, where there is risk of tapeworm infections.
- Ceasing the application of wastes at least two weeks before cattle are allowed to graze (where there are risk of bovine cysticercosis).
- Ceasing the irrigation of fruit trees two weeks before the fruits are picked, and not allowing fruits to be picked up from the ground.
- Provision of information on the location of wastewater-irrigated fields together with the posting of warning notices along the edges of the fields.

There is no epidemiological evidence that aerosols from sprinklers cause significant risks of pathogen contamination to people living near wastewater irrigated fields. However, in order to allow a reasonable margin of safety and to minimise the nuisance caused by odours, a minimum distance of 100 m should be kept between sprinkler-irrigated fields and houses and roads.

4.4.5   Integrated measures for health protection

To planners and decision makers, wastewater treatment appears as a more straightforward and "visible" measure for health protection, second only to crop restriction. Both measures, however, are relatively difficult to implement fully. The first is limited by costs and operational problems and the second by lack of adequate markets for allowable crops or by legal and institutional constraints. The application of single, isolated measures will not, however, provide full protection to the groups at risk and may entail high costs of implementation and maintenance. Crop restriction, for example, if applied alone provides protection to consumers of crops but not to field workers.

To analyse the various measures in an integrated fashion aimed at the optimisation of a health protection scheme, a generalised model has been proposed (Mara and Cairncross, 1989; WHO, 1989). This model was conceived to help in decision making, by revealing the range of options for protecting agricultural workers and the crop-consuming public, and by allowing flexibility in responses to different situations. Each situation can be considered separately and the most appropriate option chosen after taking in account economic, cultural and technical factors.

The graphical conception of the model is shown in Figure 4.4. It was assumed that pathogens flow to the centre of the circle going through the five concentric rings representing wastewater or excreta, irrigated field or wastewater-fed fishpond, crops, field workers and consumers of crops. The thick black ring represents a barrier beyond which pathogens should not go if the health of the groups at risk is to be protected. The level of contamination of wastewater, field or crop, or the level of risk to consumers or workers, is indicated by the intensity of the shading. White areas in the three outer bands indicate zero or no significant level of contamination and, in the inner rings, they indicate a presumed absence of risk to human health, thereby indicating that the strategy will lead to the safe use of wastewater. If no protective measures are taken, both field workers and consumers will be at the highest risk of contamination. Assuming that a policy of crop restriction is enforced (regime A in Figure 4.4) consumers will be safe but workers will still be at high risk. Regime B assumes that application of wastewater is made through subsurface or localised irrigation, thereby avoiding crop contamination and, consequently, maintaining both workers and consumers virtually free of contamination.

If human exposure control is the single protective measure taken, both consumers and field workers will still be submitted to the same level of risk because such measures are rarely fully effective in practice. Regime D assumes partial treatment of wastewater through ponding (D-I) or conventional systems (D-II). Stabilisation ponds with an average retention time of 8–10 days are able to remove a significant proportion of helminth eggs, thus providing protection to field workers. However, the reduction of bacteria

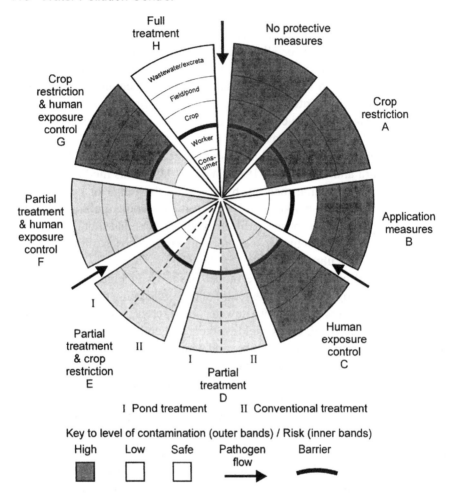

I Pond treatment   II Conventional treatment

Key to level of contamination (outer bands) / Risk (inner bands)

High   Low   Safe   Pathogen   Barrier
                        flow

Figure 4.4 A model illustrating the effect of control measures in reducing health risks from wastewater use (After Mara and Cairncross, 1989; WHO, 1989)

present is not sufficient to meet WHO guidelines and hence the risk to consumers remains high. Since conventional treatment systems are not efficient at helminth removal there will be some remaining risk for both consumers and field workers.

The regimes E, F and G are examples of the many possible associations of protective measures. Regime E integrates partial wastewater treatment with crop restriction, thus providing a large margin of protection to consumers of crops. However, full protection of field workers can be achieved only if the treatment is made through well-designed systems of stabilisation ponds. In regime F, human exposure control is integrated with partial treatment which may lead to complete protection of workers but some low level of risk remaining to consumers of the crops. The association of crop restriction with human exposure control (regime G) provides full protection to consumers but some risk remains to field workers. Finally, regime H provides full wastewater treatment allowing for complete protection to both field workers and consumers.

The feasibility and efficacy of any combination of protective measures will depend on several local factors which must be considered carefully before a final choice is made. Some factors to be considered are the availability of institutional, human and financial resources, the existing technological level (engineering and agronomic practices), socio-cultural aspects, and the prevalent pattern of excreta-related diseases.

4.5   Conclusions and recommendations

The incorporation of wastewater use planning into national water resource and agricultural planning is important, especially where water shortages exist. This is not only to protect sources of high quality waters but also to minimise wastewater treatment costs, safeguard public health and to obtain the maximum agricultural and aquacultural benefit from the nutrients that wastewater contains. Wastewater use may well help reduce costs, especially if it is envisaged before new treatment works are built, because the standards of effluents required for various types of use may result in costs lower than those for normal environmental protection. It also provides the possibility of recovering the resources invested in sewerage and represents a very efficient way of postponing investment of new resources in water supply (Laugeri, 1989).

The use of wastewater has been practised in many parts of the world for centuries. Whenever water of good quality is not available or is difficult to obtain, low quality waters such as brackish waters, wastewater or drainage waters are spontaneously used, particularly for agricultural or aquacultural purposes. Unfortunately, this form of unplanned and, in many instances unconscious, reuse is performed without any consideration of adequate health safeguards, environmentally sound practices or basic agronomic and on-farm principles.

Authorities, particularly the Ministries of Health and Agriculture, should investigate current wastewater reuse practices and take gradual steps for upgrading health and agronomic practices. This preliminary survey provides the basis for the clear definition of reuse priorities and the establishment of national strategies for reuse.

The implementation of an inter-sectoral institutional framework is the next step that should be taken. This entity should be able to deal with technological, health and environmental, economic and financial, and socio-cultural issues. It should also assign responsibilities and should create capacity for operation and maintenance of treatment, distribution and irrigation systems, as well as for monitoring, surveillance and the enforcement of effluent standards and codes of practice.

In countries with little or no experience on planned reuse, it is advisable to implement and to operate a pilot project. This experimental unit should include treatment, distribution and irrigation systems and provides the basis for the establishment of national standards and codes of practice which can then be fully adapted to local conditions and skills. Once the experimental phase has been completed, the system can be transformed into a demonstration and training project which could be able to disseminate the local experience to neighbouring countries.

4.6   References

Arthur, J.P. 1983 *Notes on the Design and Operation of Waste Stabilization Ponds in Warm Climates of Developing Countries*. Technical Paper No. 7, World Bank, Washington D.C.

Bartone, C.R. 1985 Reuse of wastewater at the San Juan de Miraflores stabilization ponds: public health, environmental, and socio-economic implications. *PAHO Bulletin*, **19**(2), 147–164.

Bartone, C.R. 1991 International perspective on water resources management and wastewater use - appropriate technologies. *Wat. Sci. Tech.*, **23**, 2039–2047.

Bartone, C.R. and Arlosoroff, S. 1987 Irrigation reuse of pond effluents in developing countries. *Wat. Sci. Tech.*, **19**(12), 289–297.

Bartone, C., Moscoso, J., Nava, H., 1990 Reuse of waste stabilization effluents for fishculture: productivity and sanitary quality results. In: Charles R. O'Melia [Ed.] *Environmental Engineering.* Proceedings of the 1990 Specialty Conference, Arlington, Virginia, 8–11 July 1990, American Society of Civil Engineers, New York, 673–680.

Bartone, C. Moscoso, J., Nava, H. And Mocetti, N. 1986 Aquaculture with treated wastewater: a status report on studies conducted in Lima, Peru. In: S.J. Cointreau [Ed.] *Applied Research and Technology. Technical Note No. 3, Integrated Resource Recovery Project.* UNDP/World Bank, Washington D.C.

Biswas, A.K. 1988 Role of wastewater reuse in water planning and management. In: A.K. Biswas and A. Arar [Eds] *Treatment and Reuse of Wastewater*, Butterworths, London, 3–15

CNA, 1993 *Information general de Los distritos de riego 03 Tula e 100, Alfajayucan, Gerencia Estatal, Pachuca, Hidalgo, Mexico.* Comision Nacional de Águas, Mexico City.

Cobham, R.O. and Johnson, P.R. 1988 The use of treated effluent for irrigation: case study from Kuwait. In: M.B. Pescod and A. Arar (Eds) *Treatment and Use of Sewage Effluent for Irrigation*. Butterworths, London, 289–305.

Crook, J., Ammermman, D.K., Okun, D.A. and Matthews, R.L. 1992 *Guidelines for Water Reuse*. Camp Dresser & McKee, Inc., Cambridge, Massachusetts.

Cross, P. 1985 Existing practices and beliefs in the utilization of human excreta. In: *Health Aspects of Nightsoil and Sludge use in Agriculture and Aquaculture*. Part I, IRCWD Report No. 04/85, International Reference Centre for Waste Disposal, Duebendorff, Switzerland.

Edwards, P. 1992 *Reuse of Human Excreta in Aquaculture - A Technical Review*. UNDP and World Bank Water and Sanitation Programme, World Bank, Washington, D.C.

FAO 1984 *Guidelines for Predicting Crop Water Requirements*. FAO Irrigation and Drainage Paper No. 24, Food and Agriculture Organization of the United Nations, Rome.

FAO 1985 *Water Quality for Agriculture*. FAO Irrigation and Drainage Paper No. 29, Rev.1, Food and Agriculture Organization of the United Nations, Rome.

Farroq, S. and Ansari, Z.I. 1983 Water reuse in muslim countries - an islamic perspective. *Environ. Manag.*, 7(2), 119–123.

Forero, R.S. 1993 Institutional, economic and sociocultural considerations. In: *WHO/FAO/UNCHS/UNEP Regional Workshop for the Americas on Health, Agriculture and Environmental Aspects of Wastewater Use*. Jiutepec, Morelos, Mexico, 8–12 November, 1993, Instituto Mexicano de Tecnologia de Agua (IMTA), Jiutepec, Mexico.

Gunnerson, C.G. and Stuckey, D.C. 1986 *Anaerobic Digestion, Principles and Practices for Biogas System*. World Bank Technical Paper No. 49, World Bank, Washington D.C.

Hespanhol, I. 1990 Health and technical aspects of the use of wastewater in agriculture and aquaculture. In: F. Rodrigues [Ed.] *Socioeconomic and Environmental Issues in Water Projects – Selected Readings*. Economic Development Institute of the World Bank/World Health Organization, Washington D.C., 157–190.

Hespanhol, I. and Prost, A. 1994 WHO guidelines and national standards for reuse and water quality. *Wat. Res.*, **28**(1), 119–124.

Ikramullah, M. 1994 Integrated duckweed-based aquaculture and rural enterprise promotion project. Paper presented at the WHO/FAO/UNCHS/UNEP Regional Workshop on Health, Agricultural and Environmental Aspects of Wastewater and Excreta Use, New Delhi, India, 2–6 May 1994.

Kandiah, A. 1994a The use of wastewater in the context of overall water resources planning and policy. Paper presented at the WHO/FAO/UNCHS/UNEP Workshop on Health, Agriculture and Environment Aspects of the Use of Wastewater, Harare, Zimbabwe, 31 October to 4 November, 1994, WHO, Geneva.

Kandiah, A. 1994b The use of wastewater in irrigation. Paper presented at the WHO/FAO/UNCHS/UNEP Workshop on Health, Agriculture and Environment Aspects of the Use of Wastewater, Harare, Zimbabwe, 31 October to 4 November, 1994, WHO, Geneva.

Laugeri, L. 1989 Economic aspects of wastewater reuse. Unpublished document. World Health Organization, Geneva.

Mara, D.D. 1976 *Sewage Treatment in Hot Climates*. John Wiley & Sons, Chichester.

Mara, D.D. and Cairncross, S. 1989 *Guidelines for the Safe Use of Wastewater and Excreta in Agriculture and Aquaculture*. World Health Organization/United Nations Environment Programme, Geneva.

Mara, D.D., Pearson, H.W. and Silva, S.A. 1983 Brazilian stabilization pond research suggests low cost urban applications. *World Wat.*, **6**(7), 20–24.

Mara, D.D. and Silva, S.A. 1986 Removal of intestinal nematode eggs in tropical waste stabilization ponds. *J. Trop. Med. and Hyg.*, **89**(2), 71–74.

Obeng, L.A. and Wright, F.W. 1987 *The Co-composting of Domestic Solid and Human Wastes*. World Bank Technical Paper No. 57, World Bank, Washington D.C.

Papadopoulos, I. 1990 *Wastewater Management for Agricultural Production and Environmental Protection in the Near East — A Manual*. Agricultural Research Institute, Nycosia, Cyprus.

Shende, G.B. 1985 Status of wastewater treatment and agricultural reuse with special reference to Indian experience and research and development needs. In: M.B. Pescod and A. Arar [Eds] *Proceedings of the FAO Regional Seminar on the Treatment and Use of Sewage Effluent for Irrigation.*. Nicosia, Cyprus, 7–9 October, Butterworths, London.

Shuval, H.I., Adin, A., Fattal, B., Rawitz, E. and Yekutiel, P. 1986 *Wastewater Irrigation in Developing Countries — Health Effects and Technical Solutions*. World Bank Technical Paper No. 51, World Bank, Washington D.C.

Strauss, M. 1985 Survival of excreted pathogens in excreta and faecal sludges. *IRCWD News*, **23**, 4–9, Duebendorff, Switzerland.

United Nations 1958 *Water for Industrial Use*. Economic and Social Council, Report E/3058ST/ECA/50, United Nations, New York.

van der Merwe, B., Peters, I. and Menge, J. 1994 Namibia case study. In: *Health, Agricultural and Environmental Aspects of Wastewater and Excreta Use.* Report of a joint WHO/FAO/UNEP/UNCHS Regional Workshop, Harare, Zimbabwe, 31 October to 4 November, 1994, WHO, Geneva.

Wegelin, M. 1986 *Horizontal-Flow Roughing Filtration (HRF) – A Design, Construction and Operation Manual.* IRCWD Report No. 06/86, International Reference Centre for Waste Disposal, Duebendorff, Switzerland.

Wegelin, *et al.* 1991 The decade of roughing filters — development of a rural water-treatment process for developing countries. *Aqua,* **40**(5), 304–316.

WHO 1989 *Health Guidelines for the Use of Wastewater in Agriculture and Aquaculture.* Technical Report Series No. 778, Report of a Scientific Group Meeting. World Health Organization, Geneva.

WHO 1990 *Legal issues in water resource allocation, wastewater use and water supply management.* Report of a Consultation of the FAO/WHO Working Group on Legal Aspects of Water Supply and Wastewater Management, Geneva 25–27 September 1990. World Health Organization, Geneva.

Chapter 5[*]

LEGAL AND REGULATORY INSTRUMENTS

5.1   Introduction

This chapter describes the legal and regulatory instruments that have been developed by a number of countries for the control of water pollution by governments or pollution control agencies.

In addition to the practical steps of treating liquid wastes by the construction of suitable treatment plants, there is a need to regulate the discharge of effluents and to control activities which may take place within a water catchment area and could contribute to water pollution. This chapter examines alternative approaches, ranging from the control of manufacture and use of dangerous or polluting materials (identified through the use of inventories and the use of risk assessment tools) to the development of standards which can be applied to effluent discharges. The use of water quality objectives and emission limit values as approaches to the development of standards for effluent control are described, as well as the use of process authorisations for pollution control as alternatives to simple, end-of-pipe controls for point source discharges. Waste minimisation and the use of cleaner technology can also contribute significantly to pollution reduction. Appropriate enforcement mechanisms are a prerequisite to successful pollution control. The difficulties of dealing with non-point source pollution, such as agricultural problems related to organic matter, nutrient enrichment and pesticide control are acknowledged, as is the problem of urban run-off from roads and pollution from storm water overflows. Finally, some means of tackling transboundary pollution problems are suggested.

It is important to stress that there are a large number of alternative approaches to pollution control through regulation and it is for policy makers to examine the facts in any particular situation and to decide which is likely to be the most successful method. Further advice is provided in Chapter 1. The

[*]   *This chapter was prepared by P.A. Chave*

regulatory instruments described here can be applied to all natural waters, i.e. inland surface freshwaters, groundwaters, estuaries and coastal waters.

It is important to realise that no one system of control is necessarily able to meet all the requirements of a particular situation. In practice, it is essential to use a combination of the available mechanisms, including legal, regulatory and financial regimes, to improve pollution control. Although this chapter discusses possible regulatory means, the development of financial systems of charging for pollution to encourage the adoption of good practices, or to provide incentives against over-production of potential pollutants and over-use of treatment facilities, must be considered alongside, or even in advance of, regulation. Such mechanisms can be especially useful where a large number of small industrial units are the cause of the pollution problems, as in many urban situations in developing countries.

A further issue to be taken into consideration is the amount of investment needed to meet any new regulations that come into force. Without suitable funding, regulations cannot be met and their practical usefulness is limited. This is an important policy area which must be examined by governments.

In most countries, controls on the discharge of substances which are liable to pollute natural waters have been limited to specific authorisations related to point source effluents discharging from pipes. Such sources are easily recognisable, and legal sanctions can be applied by subjecting the discharges to a licensing regime that includes conditions which the discharger is obliged to meet. The recognition that total pollution control is much more complex, possibly requiring potential polluters to spend a great deal of money to prevent pollution or to clean it up, has led to the emergence of a large number of alternative approaches. This chapter examines the regulatory regimes which can be applied to point sources and non-point sources of pollution, and includes examples of their use in a number of countries throughout the world.

5.2   Inventories for pollution control

In order to identify the need for pollution control measures, and to assist pollution control regulators in targeting the most significant problems (thereby making efficient use of scarce resources) and to assess the necessity for making changes to legislative provisions, a knowledge of the source and type of pollutant is necessary. Several countries have already realised the benefits of this approach and have developed requirements for surveys or inventories of pollution in their domestic legislation. However, these inventories usually consist of the amount of pollution actually observed and are little more than reports of the results of pollution surveys.

More recently, the benefits of targeted pollution control measures have been recognised and inventories have become, and are becoming, established (usually by statute) in countries throughout the world. They are of two distinct types: substances and polluting inputs.

5.2.1 Substance inventories

A number of substance-specific inventories have been established. In 1974 the United Nations Environment Programme (UNEP) decided to establish a register of chemicals and an associated network for the exchange of information. The resultant International Register of Potentially Toxic Chemicals (IRPTC) commenced in 1977 and is based in Geneva. The main aim of the IRPTC is to make data on chemicals readily available to those who need it. This is achieved by a query-response service aided by various computer databases and a library system. The main IRPTC chemical database (i.e. the "central file") has been available on the open market as a personal computer (PC) package since the end of 1994. The Register is aimed specifically at developing countries for which the acquisition of such data is often difficult.

In the USA, the Toxic Release Inventory (TRI) was established under the Emergency Planning and Community Right-to-Know Act of 1986. The TRI is a collection of information on releases of toxic chemicals into air, land and water across the nation. It is available through libraries and is an important resource for officials, as well as for the public, for discovering the presence and quantities of potential high-risk chemicals in specific localities. The inventory is compiled from information supplied by the potential polluters. Industries are required to report data if they have 10 or more employees, if they make or use designated chemicals in certain quantities, or if they conduct selected manufacturing operations.

Canada has established, through its Canadian Environmental Protection Act 1988, a Priority Substances List which ensures that when a substance is placed on the list it is subject to testing in order to establish the extent and the nature of the associated risk (if any). Public participation in the process is encouraged by allowing individual citizens to request that a substance be placed on the list.

The UK has also introduced a chemical release inventory under its Environmental Protection Act 1990. This applies to processes authorised as part of the Integrated Pollution Control provisions of the Act, i.e. largely processes which use or manufacture dangerous chemicals. The inventory aggregates, on an annual basis, the releases of pollutants into air, land and water, and is based on entries held in a statutory public register that describes the operation of the relevant industrial plant. The inventory also makes additional information available to the public.

In the European Union (EU) the Existing Substances Regulation (793/93/EEC) was adopted by the Council of Ministers on 23 March 1993 (EEC, 1992). This identifies, in an Annex, some 1,500 high tonnage chemicals which appear on the European Inventory of Existing Commercial Chemical Substances (EINECS) and which are produced or imported into the EU in quantities exceeding 1,000 t a^{-1}. Data on these substances must be reported to the European Commission (EC) which then derives, using common procedures, a priority list for comprehensive risk assessment. This list is used to ascertain the need for the adoption of improved control measures, including restrictions on supply and use and complete bans on the substances concerned. This is often achieved through the provisions of the Marketing and Use Directive (76/769/EEC) which harmonises member state's controls over the marketing and use of dangerous substances. The Directive sets a framework for banning or restricting substances and includes an Annex restricting some specific chemicals. Basic data are obtained from the EINECS database.

New chemicals have been reported to the EC since 1979 under the Dangerous Substances Directive (67/548/EEC), which is also known as the Classification, Packaging and Labelling Directive. This directive established a system for the hazard labelling of chemicals that indicates a range of human effects and physical properties, and also established a testing and notification protocol for chemicals placed on the market in quantities exceeding 1 t a^{-1}. The 12th Adaptation to technical progress (Directive 91/325/EEC), adopted in 1991, provides a methodology for classifying substances as *"dangerous for the environment"*. Thus both new and existing substances are required to be assessed for their environmental effects and steps must be taken to control their likely impact.

European Community Directive 76/464/EEC of 4 May 1976 on Pollution Caused by Certain Dangerous Substances Discharged into the Aquatic Environment of the Community (not to be confused with the Dangerous Substances Directive, 67/548/EEC, mentioned above) identifies two lists of dangerous substances. List I, the so-called "black list", contains dangerous substances which must be eliminated from the environment because of their harmful effects and List II contains substances that have a deleterious effect on the environment but which can be discharged under controls that reflect the particular circumstances of their location. Environmental quality standards (EQSs) for these substances are set by member states. As a result of this

Directive, inventories of sewage discharges, industrial sites, river sampling sites and sediment sites, together with details of standards which apply to discharges, are also held by member states wherever the discharges are affected by one or more of the 17 substances currently on List I.

Risks from potentially dangerous processes, rather than substances, are also taken into account by the application within member states of the Directive of Major Accident Hazards of Certain Industrial Processes (82/501/EEC). The Directive was agreed following the accidental release of dioxins as the result of an explosion at Seveso, Italy, in 1976 (Kletz, 1976). Annexes II and III set out threshold quantities of dangerous chemicals at, or above, which precautions or notification requirements must be met. This Directive has been modified several times and now includes the concept of dangers to the environment as expressed in the 6th and 7th Amendments of Directive 67/548/EEC referred to above. Member states have individually issued detailed guidance on implementation of this Directive.

5.2.2   Environmental risk assessment

The application of risk assessment techniques is a fundamental part of the procedures for classifying new substances as having potential environmental or health problems and, as a consequence, being worthy of legislative or administrative control. The rationale for applying techniques such as environmental risk assessment in the context of pollution control procedures relates to two separate issues. In the first place, the assessment enables the regulatory authorities to obtain early evidence of likely environmental damage in the event that the substance is allowed to enter the environment, either in a waste stream or as a result of its legitimate use. If the risk of harm is sufficient, the precautionary principle (see Chapter 1) can be applied and a justification made to control the use or manufacture of the substance, or even to ban its manufacture, in order to limit or prevent any possibility of pollution. Regulating a discharge then becomes of secondary importance. In the second place, if a predicted "no effect" concentration can be established, for example in the water environment, the regulator has a means of quantifying the amount of substance which can be released safely to the environment. Suitable environmental quality standards can then be established and authorisations for discharges which contain the substances can also be drawn up. Guidance is available in the UK from a government/industry working group established to devise an overall framework for environmental risk assessment (DOE, 1995). An example appropriate to the dangerous substance directive is outlined in Figure 5.1.

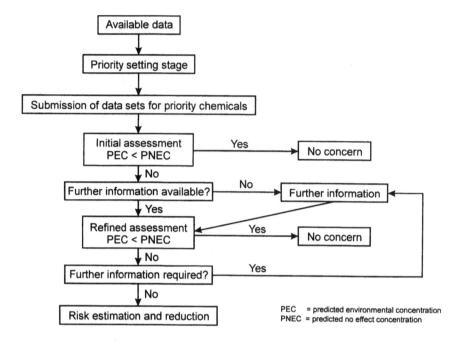

Figure 5.1 An overall framework for environmental risk assessment (Based on DOE, 1994)

The Organisation for Economic Co-operation and Development (OECD) is also active in this field and is engaged in a programme to identify and assess potentially hazardous chemicals. The OECD programme began in 1990 with selection criteria based on production volume. Initial assessments have been carried out on 35 chemicals.

There are many other international organisations engaged in compiling inventories related to polluting materials. These include the United Nations (UN), the Council of Europe, the North Sea Conference and the Paris Commission. The latter requires member states to compile an inventory of discharges from all industrial and sewage treatment works into estuaries and coastal waters, wherever one or more of the 36 dangerous substances listed in Annex 1A of the Commission's Declaration enter estuarine or coastal waters. The aim is to accomplish a pollution reduction programme. This particular inventory also includes data from all of the rivers used in the assessment of loads discharged to the North Sea, as required under the North Sea Declaration of 1987.

5.2.3 Pollution discharge inventories

The other major form of inventory which is of particular value to water pollution control is not related to the use or marketing of a specific substance, but is derived from knowledge of the quantity of pollutant discharging to a particular watercourse. The UK, for example, is applying a programme of catchment management planning to watercourses in which the catchment of each river (or, for larger rivers, individual tributaries) is examined in detail to discover what discharges are present that may affect water quality. The identified point sources are recorded using a geographic information system (GIS). The resultant catchment inventory forms the base data for planning reduction programmes or changes in control mechanisms used as a means of improving water quality. The information is held alongside other data relating to water abstractions, water quality survey reports and water quality objectives, and is used for planning purposes and for consultation with interested parties in the catchment area wherever new control proposals are needed to achieve an improvement in water quality. Such information is available to the public as a register.

A number of other countries, including developing countries, use the principle of catchment inventories for planning purposes. India, for example, has a documented system of inventories despite the difficulties of identifying the individual industries which contribute to emissions and in maintaining the database. In 1984 an inventory of larger water polluting industries was compiled, covering over 4,000 industrial sites of which half were installing suitable treatment works. No study has yet commenced on the smaller industrial sites, of which there are about 2 million. India is also examining its 14 river catchments and publishing reports on these. Such surveys, together with data collected from the Indian National Aquatic Resources Programme, assist in the planning of pollution control programmes.

5.3 Derivation of standards for point sources

Once a point source of pollution has been identified, two approaches are available for setting emission standards for its control. These are commonly referred to as the water quality objective approach or the limit value approach. In the former case the intrinsic capacity of the receiving watercourse to absorb and to degrade the pollutant is taken into account in the setting of standards, whereas in the latter case it is only the characteristics of the pollutant that are considered.

5.3.1 Water quality objectives

Most countries use the concept of water quality objectives for planning purposes but not all use them as a means of deciding effluent quality standards. In a recent publication, the United Nations Economic Commission for Europe (UNECE) recommended that water quality objectives should be set to encourage the promotion of ecosystems-based water management (UNECE, 1993). Water quality objectives are capable of addressing such diverse aspects as accumulation of toxins and eutrophication, in addition to taking account of the desired use of a particular watercourse.

General guidelines for the development of water quality objectives and their accompanying criteria are given in the 1992 UNECE Convention on the Protection and Use of Transboundary Watercourses and International Lakes (UNECE, 1994), but many countries have now developed their own approaches to such objectives.

Use-related water quality standards are becoming more common throughout the world. Such systems must first define the intended uses of the water body, for example as a source of drinking water, for particular industrial or agricultural use, as a recreational water, or that it possesses special characteristics which must be preserved. Often, although the water is not used for a particular purpose, it is considered necessary to maintain or improve its quality for general amenity purposes, or to ensure the survival of fish. These can also be considered as "uses". The identification of uses can be assisted by the participation of the local public in the decision-making process, because the ultimate definition of use of any watercourse can influence the activities that would be permitted in its vicinity.

In addition to the very specific environmental quality standards which are derived from a consideration of risk assessment and toxicity characteristics, the approach of developing use related water quality objectives leads to generalised standards reflecting the defined use.

A number of countries have developed water quality standards in which the basic concept is the protection of the natural environment. These are often based on an ecological approach and make the assumption that, provided the ecosystem is protected by the quality of the water thus defined, most other uses are also protected. Canada, for example, is developing a system based on an ecological classification to assist in identifying objectives, whereas the UK is introducing "use-related" objectives. In the UK, water quality objectives are a feature of the Water Resources Act 1991, and have statutory status. Regulations have recently been issued by the UK Government to institute the system and a guidance manual has also been published (NRA, 1994a).

It is significant that any system of water quality objectives leads directly to the derivation of quality standards for the waters themselves and that they reflect the chemical or biological requirements of the objectives. Such numerical values, or EQSs, may be used by the regulatory authorities whenever they are required to consider the impact of polluting activities.

One of the disadvantages of the water quality objective approach is the need for large quantities of real data relating to the water body concerned. In order to arrive at realistic "use" objectives, to determine what is capable of being achieved over a reasonable timescale and to set meaningful EQSs, information is required on:

- Current water quality.
- Natural variations in water quality over time.
- Inputs from industry and how these may vary given appropriate investment programmes.
- Knowledge of the likely effects of industrial effluents on the environment.

Nevertheless, provided reasonable data can be gathered, this approach is likely to ensure that investment in effluent treatment technology is well targeted and that the resultant water quality will meet a truly identified need.

5.3.2 Environmental quality standards

Environmental quality standards represent concentrations of substances which must not be exceeded if a specified use of the environment is to be maintained. Many standards in use today have been derived as a result of concerns over particular chemicals and often in association with events that have given rise to environmental or public health problems. As a result, a number of such standards are prescribed in national legislation.

It is practically impossible to define completely the water quality required for a particular use or for the general protection of aquatic life. Most countries, therefore, concentrate on a few key variables, together with other specific variables relating to known local or national problems. The common variables are biochemical oxygen demand (BOD), chemical oxygen demand (COD), dissolved oxygen, ammonia, nitrate, phosphate, suspended matter or turbidity, pH and temperature. The inclusion of metals, such as copper and zinc, allows the known toxicity of these elements to aquatic life to be taken into consideration.

In the EU, numerical standards are set out in the Annexes to a number of Directives. The following Directives contain EQSs which impact directly on the regulatory control of point source discharges (EEC, 1992):

- 76/160/EEC concerning the quality of bathing water.

- 77/795/EEC concerning the quality of fresh waters needing protection or improvement to support fish life.
- 86/280/EEC on limit values and quality objectives for the discharge of certain dangerous substances included in List 1 of the Annex to Directive 76/464/EEC.
- 75/440/EEC concerning the quality required of surface water intended for the abstraction of drinking water in the member states.

Where standards are prescribed by legislation, regulators have no option but to take account of them when a new discharge is proposed. There are many substances for which no statutory standards are available and in such cases individual nations must decide on the basis for control. The UK, for example, has developed a risk assessment type of protocol for determining appropriate environmental standards. The basic premise is that there is a certain acceptable concentration of each pollutant which does not produce unacceptable effects on the environment and its uses. The environment has, therefore, a certain capacity to accommodate pollutants and this capacity can be quantified. The protocol to determine this capacity examines toxicity data, the fate and behaviour of the pollutant, risk of accumulation in organisms and sediments, and existing concentrations in the environment. Inevitably, there is insufficient data available to answer all of the possible questions and therefore an extrapolation step is included in the protocol.

At a more general level, the OECD has also developed a risk assessment methodology for estimating the likely environmental impact of high production-volume chemicals. This uses the concept of Predicted No Effect Concentration (PNEC) and Predicted Environmental Concentration (PEC) to calculate the risk of harm to the environment from a particular chemical.

In the UK, the concept of Best Practical Environmental Option (BPEO) is becoming established. Together with use of Environmental Assessment Levels (EAL), it is used to calculate likely concentrations in the three environmental sectors (air, land and water) in order to establish the preferred disposal route for particular chemicals. For the purposes of assessing the BPEO, the EAL is the concentration or load of a pollutant above which harm is likely to occur in the environmental medium concerned at that location. Principles for establishing the BPEO for a particular site have been established that take into account the contribution of the plant effluent relative to the EAL for each medium, and which examine the environmental disposal routes to determine the one that is most environmentally beneficial. The procedure is site specific.

Once established, EQSs can contribute to the control of point source discharges. In this scenario the EQS values are used in a mass balance type of model which takes into account the relative volumes of effluent and receiving watercourse, dilution factors, and degradation factors where appropriate, to calculate the allowable concentration in the effluent which will permit the EQS to be met under all the likely conditions. This value is then used as either the maximum or 95 percentile limit value for the effluent. Limits determined using this approach are tailor-made for the conditions surrounding a particular outfall, and limits for similar industries would vary throughout a country.

5.3.3   Limit value or uniform emission standards

The limit value is an alternative approach used by many countries to set effluent quality standards. The principle of this approach is that all discharges of effluent must achieve the same minimum effluent standards as are laid down in regulations. Standards are usually related to a Best Available Technology (or Technique) (BAT) specification for the industry concerned. Sometimes the argument for this approach is that equity in the treatment of dischargers is more easily achieved and, as a result, barriers to trade are removed. This is because wherever the effluent discharges, and no matter what effect it has on the environment, the requirements for effluent quality will be the same. A counter argument to this is that unnecessary levels of treatment are imposed at many sites, leading to expenditure which could be better used elsewhere where, perhaps, real improvements are required.

The application of the limit value approach may be generalised or specific to a particular industrial sector. In Europe, the approach has varied from the setting of minimum standards for a particular industry (e.g. the titanium dioxide industry), to controls on particular chemicals irrespective of the industrial sector (e.g. controls on cadmium releases). The latest example of the limit value approach takes account of more general indicators of pollution, namely BOD, COD and suspended solids discharged from urban wastewater treatment plants. This is similar to the approach adopted by India where industry-wide standards are issued that set out the processes, effluent characteristics and methods of effluent disposal for specific industries, or for specific pollutants (e.g. mercury) or for industries within a specific geographical area.

The identification of the standards to be applied to effluents using the limit value approach often takes account of the state of development in the industry concerned, as well as the requirements to meet an environmental need expressed by an EQS. It is assumed that an industry will be able to take advantage of techniques available to minimise the level of contaminants in its effluent stream and this is allowed for when the emission limits are set. The principle of BAT is used widely throughout the world and it is a normal requirement for the principle to be applied whenever a permit to discharge is granted. In some countries the cost of installing BAT is also taken into account in the form of BATNEEC "best available technology not entailing excessive cost". When negotiating fixed emission limits, it is necessary to decide whether the regulatory system requires BAT or BATNEEC, and to consider the potential for individual firms to incorporate this within their processes.

The limit value approach requires much less water quality data than the alternative EQS system described previously, but there is sometimes a danger that spending large quantities of money on improvements to treatment plant and industrial processes may not lead to the desired water quality. In practice, especially for developing countries, a combined approach is needed to allow an examination of the needs of the environment, together with a system of prioritisation, so that the eventual objectives are identified. At the same time, and at an early stage in the process, investment is made in cleaning up those industries causing the greatest problems. The standards would ultimately be achieved over a number of years as financial resources become available. This approach has already been adopted in several industrialised countries and takes account of the phased improvement of industrial processes together with the long-term financial planning needed for the refurbishment of major industries.

5.4   Regulation of point sources

5.4.1   Permits

The common characteristic of point source discharges is that they are identifiable and that they can be monitored. Providing suitable legislation exists, they can usually be controlled. Most developed countries have had legislative provisions in place for many years that enable the authorising or licensing of potentially polluting operations.

There are two basic forms of control for point sources of pollution entering the water environment. These are end-of-pipe controls and process controls. In the former case the protection of the environment is accomplished by controlling only what is released from a discharge point, with little or no control of the processes which produce the effluents. In the latter, control starts at the beginning of the process and leads to minimisation of the effluent by using approaches such BAT in order to achieve the minimum impact on the environment from the process as a whole. Countries throughout the world use one or other or both of these techniques depending on the legislative system in place. In the UK, both systems have been used for many years.

End-of-pipe controls have been used mainly for discharges to water and process based controls for discharges to the atmosphere and for land disposal.

5.4.2   End-of-pipe controls

The UK Rivers (Prevention of Pollution) Act 1951 required authorisation for the discharge of sewage and other effluents into rivers, subject to their meeting quality standards. This was the first link between point source controls and EQSs, because the effluent standards were set so as to allow (subject to dilution criteria) the receiving water to maintain or improve its quality. The system has developed greatly since 1951. Water quality standards have been derived for a large number of substances and computer modelling has been used to assess the impact of industrial discharges on particular watercourses in order to ensure that the constituents of the discharge do not cause a breach of the EQS.

There is a general presumption that the discharge of polluting matter into water courses is illegal unless authorised. Section 85 of the UK Water Resources Act 1991 states that "*a person contravenes this section if he causes or knowingly permits any poisonous, noxious or polluting matter or any solid waste matter to enter any controlled waters etc.*". To avoid contravention of this law an authorisation, known as a consent, must be obtained. The consent is issued upon application to the water quality regulatory body (previously the National Rivers Authority and now the Environment Agency) which takes into account the above mentioned EQSs when it derives appropriate standards. The principal steps leading to the issue of a consent in England and Wales are given in Table 5.1. Full details of this procedure are described in a manual which is available to the public (NRA, 1994b).

End-of-pipe controls are used in a similar way in Canada where, for example, the Ontario Environmental Protection Act operates a general prohibition on discharging "*material of any kind into any well, lake, river, pond, stream, . . . in any place that may impair the quality of the water . . .*". The Ontario Water Resources Act also operates a similar prohibition. Such general prohibitions are modified by a licensing system which legalises certain discharges provided they are carried out in accordance with the Act. The New South Wales Environmental Offences and Penalties Act 1989, which governs the control of discharges in that part of Australia, has similar features in that discharges must be authorised and the existing quality of the receiving watercourse may influence the limits placed on the effluent. The same approach is also taken in the New Zealand Resource Management Act of 1991, under which it is illegal to discharge any contaminant into water or

Table 5.1   Principal steps for issuing consents in England and Wales

Step	Action	Commentary
1	Preliminary consultation	To assess likely problems before formal time period for issue begins
2	Formal application	Four month period in which to issue consent begins
3	Advertise	To enable public to comment/object
4	Consultation	With local authorities and others
5	Technical consideration	To decide conditions if consent is to be granted
6	Decision	Taking into account comments and objectives
7	Issue consent	Enter onto public register
8	Review	After two years, or before, by agreement with discharger

Source: Based on NRA, 1994

into, or onto, land in such a way that it may enter water. This also applies to trade premises.

A variation of this approach concerns the discharge of trade effluents into public sewers. In the UK, for example, the Drainage of Trade Premises Act 1937, now incorporated into the Water Industry Act 1991, gave industry the right to be connected to public sewers, subject to the consent of the sewerage agency. In most urban situations, trade effluent is rarely discharged direct to a watercourse. Instead, the local sewage works has the responsibility for treating the waste. Water pollution control is achieved by a consent for the discharge from the sewage works that takes into account the knowledge of the composition of any trade effluents discharged to the sewer. Therefore, although the sewerage undertaker issues a consent to discharge to the sewer, no active part is taken in controlling the industrial wastewater pre-treatment process.

Other countries have rather similar controls for substances discharged to the sewers. The Japanese system, for example, requires that industrial wastewater dischargers provide and operate their own pre-treatment plants to treat pollutants which pass through, or affect, municipal works or any sludge produced by them. National uniform standards are stipulated by government regulation for substances which are incompatible with municipal sewage treatment or are a threat to human health or the natural environment. In order to achieve legally enforceable EQSs, the Water Pollution Control Act of Japan sets national uniform standards for direct discharges to water (known as E standards) and defines mass limits for reduction programmes. More stringent standards can be set locally.

In the USA, point sources are regulated by means of the issue of National Pollutant Discharge Elimination System (NPDES) permits. Any person directly discharging a pollutant into any waters in the USA must obtain a NPDES permit. Individual states are authorised to issue and enforce those permits subject to Environment Protection Agency (EPA) oversight. Such permits must include State or Federal limits for pollutants and standards based on "best available demonstrated control technology" as set out in the Clean Water Act. The permits also specify interim compliance schedules, requirements for monitoring and collection and maintenance of effluent monitoring data. Data must be kept ready for agency inspection for at least three years and a manufacturing unit must also report emergencies within 24 hours. There is a special provision which does not allow the substitution of a permit by a less stringent one following review. A separate strategy for dealing with stormwater discharges will be developed. States take account of EQSs by setting receiving water quality standards according to the designated use of the water. If technology-based standards are insufficient to achieve these, additional limitations are prescribed in the permit.

5.4.3   Toxicity-based controls
Controls based on chemical variables are difficult to apply to complex and changing effluent streams. A number of countries, among them the USA, UK and the Republic of Ireland, have been developing toxicity-based controls which take account of the effects of the whole effluent on the receiving watercourse. In such cases data from acute and chronic toxicity tests on algae, invertebrates and fish are used to determine a No Observable Effect Concentration (NOEC). Using data from these tests an acceptable environmental concentration for the effluent is determined and a consent is written to achieve this level of effect. Such consents are independent of the exact constitution of the effluent at any time.

5.4.4   Process-based controls
The alternative to end-of-pipe controls is to authorise the process itself. In many countries there is a general move towards this type of approach, sometimes in conjunction with limited, or site-specific, effluent quality specification. The EC is promulgating a new Directive, the Integrated Pollution Prevention and Control (IPPC) Directive, which firmly endorses this approach. The principles of Integrated Pollution Control have been established in a number of countries for some time. The main objectives are:

- To prevent or minimise the release of toxic or dangerous substances and to render harmless any such substances which are released.
- To develop an approach to pollution control that considers discharges from industrial processes to all media in the context of the effect on the environment as a whole.

A fundamental factor influencing the use of process regulation is the ability to reduce the amount of waste that needs to be discharged to the environment. This is generally achieved by the application of the principle of BAT. "Best" means, in the context of pollution control, that which is most effective at preventing or minimising pollution, or rendering harmless any pollutant. "Available" is interpreted as being readily available to the operator without the need for development work before being able to use the technique. "Techniques" include both the plant in which the process is undertaken and the method of its operation. It is usual to include reference to working methods and management practices at a particular site. In the UK, and in some other countries, the costs of the operation are taken into account and BATNEEC is used (see section 5.3.3).

In the UK, Integrated Pollution Control with its emphasis on BAT is only applied to a limited range of industrial processes defined by the Environmental Protection (Prescribed Processes and Substances) Regulations 1991. These processes are those considered to be a major pollution threat to all environmental sectors. The regime considers the impact on all air, land and water and, in applying the principles of BATNEEC, it also assesses the Best Practical Environmental Option (BPEO) where the process involves a release of pollutants to more than one medium. The releases are not permitted to cause a breach in any statutory EQS and if they did the process would not be authorised. Her Majesty's Inspectorate of Pollution (HMIP), the UK regulator, has issued a series of guidance notes describing BATNEEC for those industries included in the list of prescribed processes (HMIP, 1991). The Inspectorate also assists in producing EuroBAT notes (in conjunction with the EC) for use in the EU.

In a survey of the use of BAT in 18 countries around the world, the OECD investigated the relationship between BAT and EQSs (OECD, 1994). In 14 of these countries, even when the EQS was met, no relaxation of the use of the BAT principle was permitted, usually on the grounds that BAT was necessary to ensure continual environmental improvement. In the other countries, site-specific decisions were generally taken. In most situations where the EQS was not met either stricter requirements were applied or the emission was not allowed. Where it was not possible to apply stricter criteria, a variety of other measures were used including compensatory payments to affected parties

(Finland), sectoral reduction plans (Switzerland), water charges per kilogram of pollutant discharged (Mexico) or, in some cases, closure of industrial units. The OECD survey also indicated that cost considerations generally played a significant role in the final decisions on authorising the processes where BAT was involved (OECD, 1994).

The proposed IPPC Directive of the EC permits relaxation of BAT provided that an increase in local, transboundary or global pollution is unlikely to occur wherever the EQS is not exceeded, but requires additional measures where the EQS is breached. This Directive provides a framework to which those countries already using integrated pollution control can adapt, and will harmonise the approach throughout Europe. It will apply to an extensive list of industries and will rely upon the local authority setting emission limits (based on BAT) that ensure that the EQSs derived by the Council of Ministers or the World Health Organization (WHO) will be met.

The EC Urban Wastewater Treatment Directive, which came into force in 1991, uses a mixed approach. It requires the use of "best technical means" in the design of sewerage systems, it sets minimum levels of treatment which depend on the population size served by the works as well as the characteristics of the receiving watercourse, and it also lists specific limit values which must be achieved by effluent from primary, secondary and tertiary treatment plants. The Directive only applies to municipal sewage treatment works and certain organic-based industries.

5.4.5   Public participation

An important part of the regulatory regime for the control of point sources of pollution (using the authorisation route) is the participation of the public and of the regulated sectors themselves. Such participation is essential to ensure that industry recognises the need for, and accepts the obligations placed upon it by, the regulatory regime, and that the public is satisfied that adequate control is being exercised. Information is freely available in many countries through "Freedom to Access of Environmental Information" type regulations. Particularly open arrangements exist in countries such as Sweden and Finland where all information, including internal communications, are available to the public.

In the UK the legislative provisions have been designed to allow a significant degree of public involvement in the decision-making process. All applications for authorisation under the Environmental Protection Act are referred to specific statutory consultees and must be advertised locally. The regulator has to consider any representations received. The Water Resources Act has similar advertising provisions. In addition, both Acts require details of applications, authorisations and consents and other relevant information to be kept available on a public register. In the USA a public hearing must be arranged if there is significant interest in a permit application, but in the UK this would be an unusual step.

It is very important that developing countries achieve a balanced view with respect to what is desired, what can be achieved and what can be afforded. The involvement of the public and industrialists is essential to enable regulators to understand the impact of any proposed measures prior to setting standards for water and effluents, and to ensure that any programmes for improvement are attainable within the financial and technical capabilities of the country concerned.

5.4.6   Waste minimisation and cleaner technology

While the principles of BAT give rise to the introduction of less polluting technologies through a regulated system, there are a number of initiatives in existence which are designed to encourage the use of clean technologies and better production systems. Waste minimisation is a technique which is being tested in a number of studies around the world in an attempt to reduce the amount of waste produced from industrial production units. The project methodology has been developed in the USA, the Netherlands (particularly studies carried out at 12 companies in Amsterdam and Rotterdam) and in a series of case studies on cleaner production carried out by UNEP. In the UK, a collaborative programme between a number of companies and regulators in the catchments of the Rivers Aire and Calder and along the River Mersey has enabled further development of the principles in the form of a demonstration project.

The key to waste minimisation is the adoption of a systematic approach to evaluating processes and quantifying the consumption of water, materials and energy. The six most important steps in the methodology are:

- *Commitment.* The need for the company to have policy commitment to waste minimisation including senior management support and clear objectives, targets and timescales.
- *Organisation for action.* Multidisciplinary teams should be set up covering all major aspects of the business.
- *Audit and review.* Examination and quantification of processes such as waste streams and consumption of materials.
- *Options for improvement.* These should be costed and prioritised.
- *Action.* Implementation of the programme of changes with targets and timescales.
- *Review and identification of further opportunities.*

The options for minimising environmental impacts include reduction at source through product or process changes, on-site recycling or material recovery, or off-site recycling.

An important question to be considered in this approach is whether it is necessary to use any particular material, or indeed whether the product itself is required. There are many instances where an alternative, less toxic or persistent substance could be used in the production process and a number of products have already been phased out completely in recent years where their pollution potential is greater than the benefits of their production and alternatives have been found. The "cradle-to grave" approach, in which all aspects of environmental impact are examined, from extraction of raw materials to the final disposal of the used product, is gaining in popularity.

5.4.7   Voluntary schemes

There are a growing number of environmental management systems which are voluntary and which may assist in the drive towards cleaner technology; some are enforced by government and some are international in extent, although most are applicable at the site level. Examples include the Environmental Management and Audit Scheme of the EC which became operable in April 1995 and the British Standards Institute BS 7750 Environmental Management System, both of which include arrangements for formal assessment and certification. Systems for environmental reporting are promoted by industrial groups such as the World Industry Council for the Environment (WICE) and the Public Environmental Reporting Initiative (PERI). All these schemes require the adoption of an environmental policy by the commercial organisation and that an environmental management system must be in place. They also require a statement about releases to the environment. Commitment to compliance with all regulatory requirements is an essential provision of all the schemes, and failure to achieve this would result in the company being removed from the accredited list. A common theme is the commitment to continual improvement by the company concerned. Environmental auditing of the company's operations and independent certification of the audit are also important features.

5.4.8   Enforcement mechanisms

A key issue in respect of point source discharge control is the ability of the regulator to take enforcement action against the discharger when the conditions of the authorisation are breached. Legal provisions vary widely from country to country, but in all cases the laws permit regulatory action. At the minimum level this involves using prosecution through the courts. Many countries, however, have mechanisms which allow for less severe action to be taken before recourse to criminal law sanctions. In some countries, for example Denmark, a system of agreements is in use in which an informal notice is sent by the regulator to the offender requiring certain work to be done to bring the discharge into compliance. This is not a legal measure. The same authority may issue a notice of violation, i.e. a formal notice indicating that a permit has been violated. In more extreme situations, a prohibition notice can be issued requiring the activity to stop and finally legal action can be taken. The authorities can, in addition, step in to solve the problem were necessary. A typical decision tree for the use of enforcement provisions, describing the Danish system, is shown in Figure 5.2.

In the UK, in addition to criminal sanctions, an oral or written warning can be issued to companies which are authorised under the integrated pollution control regime of the Environmental Protection Act. An authorisation can also be varied at any time by the enforcing authority, or the authority may require the discharger to submit a plan for improvement to a process or plant. For more serious breaches, there are three possibilities for action: an enforcement notice can be served, requiring action to resolve the problem, or a prohibition notice may be used to stop the process, or the enforcing authority can revoke an authorisation at any time. Failure to take the required action leads to court proceedings being taken. Where point source discharges are consented under the Water Resources Act, sanctions are limited to prosecution in the courts.

Some countries have available the possibility to use administrative acts to enforce their legislation in addition to criminal sanctions. Examples of this approach exist in The Netherlands, Germany and Belgium. Here, the criminal law is not invoked but fines, administrative orders or economic sanctions can be imposed either directly by the enforcing authority or through the public prosecutor. In such cases the courts are not involved and the use of court proceedings is reserved for more serious offences, or for situations where the administrative action itself is unsuccessful. In most countries recourse to civil law is also available where the polluting discharge has, for example, caused damage to a downstream user.

There is wide variation in the responsibility for taking action. In most countries a combination of enforcement authority supported by the police or public prosecution service is responsible for enforcement. Italy is unusual in having set up a specific branch of the Carabiniere, known as the Operational Ecology Unit, to enforce environmental legislation. In England and Wales, the Environment Agency takes legal action directly through the courts.

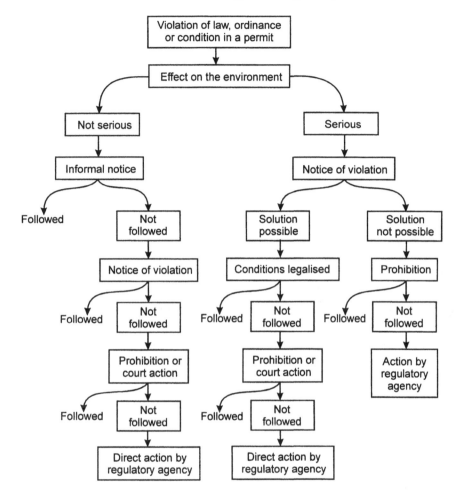

Figure 5.2 A typical decision tree as used by the Danish Environmental Protection Agency (Based on Danish EPA, 1995)

5.4.9 Compliance assessment

A necessary precursor to the enforcement process is the availability of data from monitoring and inspection visits. All countries use routine sampling of effluents combined with laboratory analysis and reporting. Permits must contain provisions for the collection of samples and specifications for sampling points. A significant number of countries rely upon a measure of self-monitoring in order to provide data beyond those that are required by the regulator and, more importantly, to ensure that the operator takes sufficient interest in his own effluent system by the requirement to take, examine and report upon his own samples. Where self-monitoring is used, suitable safe-guards are required to prevent fraudulent data being reported, such as quality control systems for management (for example BS 7750; see section 5.4.7) and laboratory work (such as accreditation of methods and laboratory procedures under International Organization for Standardization (ISO) or Comité Européen de Normalisation (CEN) standards or equivalent).

Sampling regimes must be established and recognised by the discharger as being an important part of the quality control and regulation of the discharge. Many countries specify the numbers of samples to be taken over fixed time periods (daily, weekly, monthly, yearly) based on the size of the discharge, its nature and the sensitivity of the receiving watercourse. Sampling frequencies vary from once a day for larger discharges to once a month or less for smaller ones.

Site inspections should also take place on a regular, unannounced, basis to examine the works and its discharge. In Europe such visits may take place annually or more frequently. In some developing countries frequencies of once a year to once a month can be attained for large polluting industries.

Although the assessment of compliance is seen primarily as a means of measuring the conformity with regulations, it can also provide information about the achievement of the objectives of an investment programme. Used in this way, data accumulated for enforcement can be used positively to judge whether the investment was sufficient and the designs of the plants, for example, were correct. Assessment of compliance can also be used for forward planning by identifying shortfalls that need correction by further investment or through improved operating procedures.

5.5   Non-point source pollution

5.5.1   Identification of sources
It is more difficult to control non-point source pollution than defined discharges. Even though stringent controls may be placed on industrial and municipal sewage discharges, environmental water quality may not improve to the extent expected. This may be due to diffuse pollution caused by agriculture or by urban run-off. The first problem lies in the identification of sources. The catchment inventory approach is recommended and is already used in a number of countries.

In New South Wales, Australia, for example, the principle of environmental auditing has been applied to the identification of diffuse pollution. In order to identify non-point source pollution and its relationship with land-use activities, a geographical information system is used to hold and to relate data associated with land use (e.g. cropping intensity, vegetation clearance and soil erosion information). Water quality data are entered in order to estimate the effects of agricultural activities on water quality so that pollution control policies can be devised.

Work in the catchment of the River Danube on nutrient balances indicates that the input of nitrogen and phosphorus from diffuse sources, mainly agriculture, is as significant as that from sewage works. Those areas which use sewage on land, either as a disposal route or for soil conditioning, may also be contributing to diffuse pollution.

5.5.2 Agricultural sources

The major causes of concern associated with agricultural pollution are: organic matter (which often leads to nutrient enrichment of water bodies) including the disposal of solid organic wastes and slurries from livestock, effluents from silage clamps and, in some situations, domestic effluents from farmstead septic tanks; pesticides and fertilisers; and soil erosion.

In the UK, a regulation to control the storage of silage, slurry and agricultural fuel oil was passed under the Water Resources Act 1991 and has been effective in improving the design of these facilities. This sets out minimum design features together with guidance for spreading the waste products on land. The NRA has also produced detailed guidance on the general problems of farm waste management (NRA, 1992) and pressed for the use of waste management plans for individual farms (which must be submitted when applications are made for any farm grants available from the government). Farm waste plans contain the following details:

- An outline of the proposals to deal with farm wastes, including full details of equipment to be used.
- A description of the present effluent arrangements.
- Production figures for the effluent.
- Details of land application proposals.
- How the system is to be managed.
- Contingency planning.
- A field plan which includes an assessment of pollution risk from slurry run-off.

A further approach, which has precedents in many situations, is the issue of Codes of Practice or Best Practice Guidelines. A Code of Good Agricultural Practice for the Protection of Water has been in operation since 1991 in the UK and has been granted statutory status. This means that it can be taken into account in any legal proceedings. The code sets out detailed guidance on: the principles which should be adopted for storing and applying livestock wastes (and other organic wastes) to land, the alternatives available for the design of slurry stores and advice on their maintenance, the importance of separating clean and dirty water and the choice of disposal systems for dirty water, manures and silage production storage and management, storing and using fertilisers and fuel oil, advice on the use of sheep dips and pesticides, and information on the effects of farming practices on nitrate production.

A recognition that diffuse pollution can result from forestry operations has resulted in the issue of a similar code of practice for forestry operations known as "Forests and Water Guidelines" (Foresty Commission, 1991). The code covers such issues as the precautions to be taken in cultivation and drainage works (including detailed advice on ploughing procedures and the use of precautionary buffer strips), planting near streams, forest road construction, harvesting, and the use of pesticides and fertilisers.

A lesser known activity, but one which can be classified as agricultural, is fish farming. In some countries this activity is now an important source of food and its development can give rise to diffuse pollution problems. For example, in Norway the problem has become sufficiently important for the State Pollution Control Authorities to issue instructions and guidance to minimise pollution from such activities. Guidance relates to, for example, the siting of farms, the control of feeding rates, precautions for the use of anti-fouling agents, antibiotics and insecticides, and the correct manner with which to deal with dead fish and offal. Permits are issued to control fish farming. Similar problems occur in Denmark, and to a lesser extent in Scotland. The pollution control authorities in these countries also offer advice.

Nutrient control

The control of nutrients is an important issue throughout much of the world, both from a public health perspective and to keep natural waters free from eutrophication. The most widely used water quality standard for nitrate (NO_3^-) is the 50 mg l^{-1} limit adopted by WHO as a precautionary level to safeguard babies from the risks of contracting methaemoglobinaemia (WHO, 1993). Most national authorities regard the 50 mg l^{-1} concentration as a realistic target in relation to eutrophication and, therefore, programmes aimed at controlling eutrophication often use this value as an EQS. Whereas there are process techniques available to remove nitrate from drinking water after it has been abstracted (allowing higher levels to be tolerated in raw water used for potable supply) the eutrophication problem is universally dependent on the control of nitrate sources. There is an added and unexpected health implication related to eutrophication, particularly in lakes. In a number of countries in the world, the phenomenon of excessive growth of blue-green algae (notably *Microcystis* spp.) has caused concern where recreational pursuits take place on the lakes and also where the water is withdrawn for public supply. This is due to the recognition that such algae produce a number of toxins which, if ingested, can cause liver damage. There are a number of well documented cases where animals are known to have died as the result of drinking water heavily laden with these algae (e.g. Australia, South Africa, UK) and, for this reason alone, nutrient control is justified.

The most common source of the nutrients nitrogen and phosphorus is agriculture, and this is closely followed in the industrialised world by sewage effluents. The reduction of nitrogen and phosphorus from agriculture relies upon changes to farming practices because they give rise to diffuse sources.

Nitrogen. Ploughing of grassland and other crops, particularly during autumn, leads to the release of large quantities of soil nitrogen and, therefore, a general move towards permanent pasture regimes assists in lowering nitrate leaching. When this is not possible, the use of short-term rotational crops to take up nitrogen, followed by their harvesting and subsequent removal from the catchment, is helpful. Animal wastes should be used carefully, avoiding over-use and direct run-off into water courses; but wherever possible they should be used in place of synthetic fertilisers. Use of all types of fertilisers should be carefully controlled and matched to crop requirements.

In Europe, legal control exists through the Directive Concerning the Protection of Waters against Pollution caused by Nitrates from Agricultural Sources (91/676/EEC). The purpose of this Directive is to reduce and to prevent pollution of fresh surface water, estuarine and coastal waters whch arises from diffuse sources of nitrates. Within two years, member states have to identify and designate vulnerable zones, i.e. all areas of land draining into waters affected by pollution and that contribute to the pollution. Annex 1 to the Directive gives criteria which can be used to identify vulnerable zones. Action plans must be presented to improve the situation in these zones by specifying periods when land application of fertilisers is prohibited, by quantifying criteria for land application rates and by limiting use according to codes of good agricultural practice. Annex 2 of the Directive establishes voluntary codes.

Under the Water Resources Act, the UK government has introduced a further measure to reduce nitrate pollution through the identification of nitrate-sensitive areas. These reflect a specific clause in the legislation, Section 92, which permits the identification of such areas, and allows compensation payments to farmers in exchange for a reduction in the amount of nitrogenous fertiliser used. Such areas have been associated mainly with individual groundwater zones where nitrate concentrations have exceeded the standard of 50 mg l^{-1}. Farmers enter into such agreements on a voluntary basis, for a five year period, and are set limits on the amount and timing of fertiliser application. In addition, green crops have to be planted in winter to minimise nitrate leaching.

The Agri-Environment Initiative of the EU is also a voluntary scheme introduced in 1994 to encourage farmers to reduce the use of nitrates. Those doing so receive compensatory payment.

Phosphorus. A key issue controlling phosphorus input from agriculture is the need to prevent erosion from field surfaces. Phosphate tends to bind to soil particles which, when washed from fields into watercourses, become a source of phosphate in suspended form and in deposited sediments. Sediments act as a long-term source of phosphate by releasing it (i.e. by redissolution) under certain environmental conditions. Physical removal of the sediment layer, in order to remove the bound phosphate from the catchment, has been tried in a number of locations around the world. Some success has been achieved in lowering phosphate levels in the Norfolk Broads in England by a combination of the diversion of effluents containing phosphorus out of the area, phosphorus stripping at sewage treatment works, and by the dredging of 1 m of sediment. Concentrations below the target of 100 μg l^{-1} of phosphorus were reached (RCEP, 1992).

Pesticides

Pesticides represent a particularly difficult area of pollution control activity, not only because the environmental effects in relation to aquatic flora and fauna are important but because human health issues have a very important bearing upon the nature of the controls applied. There are several thousands of formulations of insecticides, herbicides and fungicides in common use and, therefore, the potential for water pollution is very high. There are also very stringent limits for water used for public supply and, consequently, the control of pollution by pesticides is crucial in water supply catchments.

Most pesticides in waters are derived from agricultural use and it is therefore difficult to regulate their input to water bodies. Regulation can only take place by prevention, i.e. by indirect controls on their manufacture, storage and use. Approval for the use of pesticides is granted in the UK by government, following expert assessment of safety and the environmental risks. Authorisation is harmonised throughout Europe by the Directive concerning the Placing of Plant Protection Products on the Market (91/414/EEC) in which uniform principles for the authorisation process are adopted by member states. The active ingredients in pesticides are approved by the European Union and placed on an approved list. This Directive allows authorisation provided the pesticide is not expected to occur in groundwater at concentrations above 0.1 μg l^{-1}. Further controls are also placed on products by Directives such as that on Classification, Packaging and Labelling of Dangerous Substances.

Once a pesticide is in use, it is controlled by safety legislation, such as the UK Food and Environmental Protection Act of 1985. A large number of guidance manuals have been issued on the safety precautions to be taken. These manuals usually indicate pollution prevention precautions and include advice on storage, on the disposal of unused material and application.

In addition to specific legislation, a number of pesticides appear in other EU directives and in reduction programmes agreed in international protocols such as the Declaration on the North Sea. Several have also been totally banned because of their environmental impacts.

5.5.3 Urban sources

The major sources of urban pollution are urban stormwater run-off discharged through road drains or combined sewer outfalls, industrial area drainage discharged through surface water drainage systems (including spills of chemicals and oil) and refuse or solid waste drainage.

Run-off from roads

Urban pollution occurs largely as a result of run-off from roads. Road surfaces are generally impermeable and thus any polluting material falling on them is, eventually, washed into a receiving watercourse or finds its way into groundwater. Such pollutants arise from many sources, the most important of which are traffic and maintenance operations. Traffic generates pollutants from vehicle emissions, including volatile solids, polynuclear aromatic hydrocarbons derived from unburned fuel, lead compounds and hydrocarbons. On main roads, leaks from lubrication systems provide a continuous source of fluid hydrocarbons. Abrasion of tyres during normal wear releases zinc, lead and hydrocarbons. Research work in Germany has indicated that tyre abrasion on motorways can release typically 572 g ha^{-1} a^{-1} lead, 120 g ha^{-1} a^{-1} chromium and 115 g ha^{-1} a^{-1} nickel (Muschack, 1990). Corrosion of vehicles also contributes quantities of metals, including chromium and lead.

Road maintenance, particularly de-icing, is an important source of pollution, e.g. salt and urea. The impurities in road-grade salt can contribute to water quality deterioration. Roadside weed control also leads to diffuse sources of pesticide pollution. In addition to the pollution arising directly from road use and maintenance, road drains accept pollutants from atmospheric deposition, agricultural activities (after heavy rainfall) and general littering. Animal wastes, rich in bacteria, can also accumulate and contribute to the high levels of micro-organisms found in some waters. Oil pollution associated with vehicle maintenance is a specific problem in many areas.

It is impractical and virtually impossible to control the quantity or quality of road run-off by normal regulatory means. As a result it is necessary to rely on good design of drainage systems with adequate built-in protection and on maintenance procedures which minimise the risks of pollution. Various studies have been carried out to determine the most appropriate measures, and these have resulted in the following guidance (CIRIA, 1994):

- Gully pots, filter drains and soak-aways, which are all commonly used, can assist the removal of sediment but, unless maintained properly, can also pose a threat to surface and groundwaters. Infiltration basins and trenches can remove suspended material and possibly some dissolved pollutants, but can also be a threat to water quality. Detention tanks, storage ponds and sedimentation tanks operated at the end of the drainage system are successful in removing sediments. Lagoons and purpose built wetlands are capable of treating many potential pollutants, largely through the action of the associated vegetation.
- Liaison between the regulator and the highway authority is essential to ensure that suitable systems are installed when road schemes and urban development is planned.

In the UK, the Department of Transport has issued design guidance and codes of good practice for routine and winter maintenance which include information for pollution prevention.

Urban pollution from separate drainage systems
In towns and cities, drainage systems can be of two types, combined sewers or separate sewers (see Chapter 3). In the first case drainage from roads, rooftops and similar impermeable surfaces is accepted into the foul sewerage network for treatment at sewage works along with domestic sewage and industrial wastes. In the second case, domestic sewage and industrial waste is separated for treatment, and the wastewater arising from rainfall run-off is discharged directly to watercourses without treatment. Drains in heavily urbanised areas may accept a variety of pollutants from rooftops, lorry loading bays, industrial sites and even from illegal connections to the surface water system. Publicity and inspections are needed to ensure that unexpected pollution does not arise from these sources.

Pollutant loads discharged from urban drainage systems vary depending on local rainfall patterns, the variety of materials entering the sewer network, and the processes of mixing and degradation that occur in the sewers. The impact of the pollutant load is also complicated by variations in flow and quality in the receiving watercourse. Careful planning of the sewerage network is required to address these problems. Various modelling tools, such as rainfall modelling, sewer and sewage treatment modelling and river quality modelling, are now available to assist in this planning process. A simplified model SIMPOL (Spreadsheet Simplified Urban Pollution Model) combining many of the key processes has also been developed to test rapidly the performance of potential solutions and to identify rainfall events which could lead to important impacts on river water quality (FWR, 1994).

Nutrient control
A number of specific measures have been adopted to deal with nutrient problems from sewage effluents. Legal instruments relate primarily to the control of point sources of nitrate and phosphate from sewage works — the most recent international measure being the adoption of the EC Urban Wastewater Treatment Directive in 1991 (91/271/EEC). The nutrient control measures therein comprise a requirement to add tertiary treatment (nitrate and phosphate removal) to plants which discharge into "sensitive areas". These are defined by reference to the eutrophic state of the water receiving the effluent. In the UK, it has been decided to concentrate on phosphate removal for treatment plants discharging to inland waters and nitrate removal for treatment plants where the effluents enter marine waters (representing the critical elements for eutrophication in the respective situations). Other countries, such as Germany, intend to add nutrient removal to all of its larger sewage works.

In addition to agricultural sources, phosphate occurs in sewage effluents as the result of its use as a detergent "builder" (creating optimum conditions in the wash water for the surfactants to operate). Sodium tripolyphosphate is commonly used; this compound breaks down to orthophosphate which can be used a nutrient source by aquatic plants. Control of the phosphate at source is not yet practised and removal relies on "stripping" of the orthophosphate from the sewage works effluent by chemical treatment. Sewage works can remove about 40 per cent of incoming phosphate, but removal of at least 90 per cent is often required to bring about a change in the trophic state of a receiving water body. Reductions in the polyphosphate content of some detergents is also assisting in this process. The importance of phosphates is recognised by the EC Urban Wastewater Treatment Directive by the inclusion of phosphate limits for discharges to eutrophic waters (2 mg l^{-1} for populations between 10,000 and 100,000 and 1 mg l^{-1} for populations greater than 100,000). The criteria for recognising eutrophication have been defined in a publicly available document which includes proposed EQSs for phosphate in different waters. This guidance has been used to identify eutrophic waters throughout the UK and enables the regulators to indicate where phosphate stripping from sewage works effluent is justified.

5.5.4   Catchment management planning

In several countries, including the UK, Canada and the USA, the process of catchment or watershed planning has been introduced. This has resulted in a recognition of the importance of non-point sources of pollution and of the need to manage it as carefully as point sources of pollution. Catchment management plans are designed to cover a catchment of a river and its tributaries and any associated groundwater flows. A catchment is a discrete area of land which has a common drainage system. The surface water catchment is defined by the topography of the land although this may not coincide with any associated groundwater catchment (which is influenced by the underlying strata). The principles of catchment management planning, however, apply to both. In terms of the impact of activities within them, catchments are largely self-contained, manageable units, although such activities can affect downstream areas.

Catchment management plans are designed to be strategic in approach and to take into account regional and national policies and all activities likely to have an impact on the watercourse. The key attributes of a properly prepared plan are set out in a consultation report for all interested parties (local industries, public bodies and the local population). Following the consultation stage a final plan is prepared for adoption. The consultation report should contain:

- Records of physical attributes and catchment uses.
- Proposed environmental targets.
- Comparisons of targets with current status of the water environment.
- Identification of issues and options for addressing them.

Its preparation will involve undertaking consultations on the uses, targets, issues and options, the preparation of action plans to address the issues, implementing the actions, and monitoring and reviewing the plan.

The final plan should contain a future vision for the catchment (having taken note of the results of the consultation exercises), an overview of the catchment and action plans. As a result of the plan the main polluting sources should be identified and solutions agreed which can be achieved with available resources and to a timescale agreed by all concerned.

5.5.5 Laws and regulations

Non-point sources arise mainly from agriculture, but urban road run-off, effluents from contaminated land and effluents from storm sewage overflows are also known to contribute to the problem. In some countries the latter would be classified as point sources although they are difficult to control.

Laws aimed at controlling non-point source pollution are extremely difficult to frame. General pollution offences (e.g. as expressed in the Water Resources Act in the UK) deal adequately with accidental or deliberate pollution incidents but not with the insidious, unquantifiable, land run-off problem, nor with highway drains (where the pollutant enters the water course through a fixed pipe) where discharge authorisations are not practical. Most countries issue "codes of practice" or "pollution control guidelines" and stress the importance of collaboration between the regulators and the likely polluters.

Pollution prevention is ultimately a question of land use and, therefore, statutes related to planning need to take account of this. In Australia, the New South Wales Environmental and Planning Assessment Act 1979 contains a range of provisions relating to pollution prevention and control. These provisions include a wide range of environmental protection measures to be taken in conjunction with proposed development and the comprehensive use of environmental planning instruments.

The Canadian Government takes the view that regulatory control of non-point source pollution is extremely difficult and has, therefore, undertaken some novel initiatives to deal with such problems. For example, the Government has attempted to regulate the entire pesticide industry in the belief that strict control of use will reduce the incidence of non-point pesticide pollution. The Pest Control Products Act regulates the distribution of pesticides nationwide and further legislation, in the form of the Pest Management Regulatory System, controls the pesticide industry itself. The Canada Clean Water Act facilitates the creation of federal and provincial agreements to address water quality and resource management through, in some cases, a system of taxes and subsidies designed to encourage the agricultural industry to implement best management practices.

In the UK the situation is similar. The basic water pollution Acts, although they contain provisions to deal with incidents once they have occurred, are limited in their proactive provisions. They are reliant on co-operation being established with the agricultural and other industries in order to promote best practice.

The designation of areas of land as "water protection zones" is a possible legislative option enabling restrictions to be applied to practices which are considered to pose a risk to water quality in order to reduce such risks. In England and Wales the application of water protection zones is accomplished through the use of Section 93 of the Water Resources Act, 1991. Zones may be designated by the Secretary of State. The first such zone is currently the subject of a Public Inquiry and may be established for the River Dee in 1997. In the zone, industrial and agricultural activities may be curtailed, and requirements placed on operators to take precautions. Such precautions would be legally enforceable due to the statutory nature of the zone designation. The effectiveness of such an approach has yet to be assessed.

Under the Water Pollution Control Ordinance of 1980, the Environmental Protection Department divided the waters of Hong Kong into 10 water control zones which must meet strict water quality standards. Nine of these are fully operational at present and only the urban areas fringing Victoria Harbour remain to be brought under control between 1995 and 1997. It is an offence to discharge wastes into such an area but where effluents are inevitable a Technical Memorandum sets out the required standards.

According to Article 19 of the German Water Management Law (Wasserhaushaltgesetz), water protection zones can be established in which certain activities are prohibited and in which the property owners can be obligated to tolerate certain intrusions by the pollution control authorities (e.g. for the taking of samples of water). Compensation provisions are also available in some situations.

5.6   Groundwater protection

Groundwater usually requires special efforts to protect it from pollution. Although general pollution control laws for discharges and measures taken to prevent non-point source pollution on land can apply equally to groundwater protection, practically any activity on the surface can have an effect on the quality of underground water. Being out of sight, it is not always apparent that damage has been, or is being, done to the groundwater resource. The need to prevent groundwater pollution is important because of the very high proportion of groundwater resources that are used for potable supply. This has been recognised in the EU by the proposal to set up a groundwater action and water resources management programme based on the precautionary principle and on the principles of prevention, rectification at source and "polluter pays". The action programme is expected to emphasise the need for national administrative systems to manage groundwater, preventative measures, general provisions for handling harmful substances safely and provisions to promote agricultural practices consistent with groundwater protection. A key part of preventative measures for groundwater is the identification of groundwater reserves and potentially polluting activities.

A groundwater protection policy has been written for England and Wales. A key objective has been to devise a framework which covers all types of threat to groundwater, whether large or small, from point or diffuse sources, and by both conservative and degradable pollutants. The policy, which is published as a guidance note and issued to all authorities whose work has a relevance to the issue (such as planning authorities, waste regulatory authorities and others) contains a classification of groundwater in terms of vulnerability, a definition of source protection zones, and statements on how activities may be controlled to reduce or to eliminate the risks of pollution occurring by those activities.

Factors which together define the vulnerability of groundwater are the presence and nature of the overlying soil, the presence and nature of drift, the nature of the strata and the depth of the unsaturated zone. Since these measures relate to the whole of the groundwater resource they are referred to as groundwater resource protection. A distinction needs to be made between the general protection of the resource and specific protection which may be needed for individual groundwater abstractions. It is possible to define the catchment area for a particular abstraction with information on the aquifer and on the rates of abstraction. A protection policy defines groundwater source protection zones: an inner zone, defined as a 50 day travel time from a pollutant input to the abstraction; an outer source protection zone, defined as a 400 day travel time; and a total source catchment zone. This approach enables different levels of protection to be applied at varying points in the catchment. Vulnerability maps are prepared for the overall resource, but not for individual groundwater sources. The policy sets out guidance for taking pollution prevention measures covering a number of key situations where it is necessary for the regulatory authorities to consider their potential impact on aquifers. These include:

- The control of groundwater abstractions.
- The physical disturbance of aquifers and groundwater flow.
- The impact of waste disposal to land.
- Problems associated with contaminated land.
- The disposal of slurries and liquid effluents to land.
- The control of discharges to underground strata.
- Diffuse pollution of groundwater.
- Developments which may pose a threat to groundwater quality.

The basic approach of the policy is that of developing a co-operative approach to solving potential problems and of preventing future ones by collaboration.

A similar approach has been taken in Brazil where a vulnerability map, based on 31 aquifer units with six levels of vulnerability index, was developed for the state of São Paulo. Critical areas for groundwater pollution were determined by comparing the vulnerability map with a potential contaminant load map drawn up on the basis of records of industrial activity, cities, mining activities and waste disposal sites. The concept of groundwater pollution risk was based on the interaction between the potential pollution load and the vulnerability derived from the natural characteristics of the strata.

Section 13(1) of the Canadian Environmental Protection Act applies specifically to groundwater. It contains a general prohibition that "*no person shall discharge a contaminant or cause or permit the discharge of a contaminant into the natural environment that causes or is likely to cause an adverse effect*". The term discharge includes leaks, escapes and spills likely to affect groundwater. Contamination must be reported to the Ministry of the Environment which has powers to take action, including cleaning-up. Various other sections of this Act allow orders to be issued to clean-up discharges from waste disposal sites (Part V) and leakage or spills from other facilities such as storage tanks (Part IX). The penalties are very high where non-compliance is detected.

5.7   Transboundary pollution

The problem of transboundary pollution occurs where water bodies, such as the Rivers Rhine and Danube, flow through or border more than one country. Water quality in one country may depend upon the effectiveness of controls in another country. In a similar way seas such as the Baltic and North Sea, which are practically enclosed, require pollution control action to be taken by all surrounding countries in order to guarantee improvements in water quality.

More than 100 conventions, treaties and other arrangements have been concluded amongst European countries to strengthen co-operation on transboundary waters at bilateral, multilateral and pan-European levels. These agreements bear witness to the concern and interest of European countries to prevent the deterioration of water quality in transboundary waters. Following provision of the Convention on the Protection and Use of Transboundary Watercourses and International Lakes (the ECE Water Convention) (UNECE, 1994), some long-established bilateral and multilateral agreements have recently been revised, supplemented and updated to meet the urgent need for integrated water management, including the control of trans-boundary water pollution.

Examples of multilateral agreements include the Convention of Cooperation for the Protection and Sustainable Use of the River Danube, the Agreement on the Protection of the Scheldt, and the Agreement on the Protection of the Meuse, all of which were signed in 1994. These agreements fall within the framework of the ECE Water Convention. Examples of new bilateral agreements, which are also based on provisions of the ECE Water Convention, are the 1992 Agreement on the Joint Use and Protection of Transboundary Waters (Kazakhastan and Russian Federation), the 1992 Agreement on the Joint Use and Protection of Transboundary Waters (Russian Federation and Ukraine), the 1994 Agreement on Water Management Relations (Croatia and Hungary), and the 1994 Agreement on the Joint Use and Protection of Trans-boundary Waters (Republic of Moldova and Ukraine).

An important element of co-operation under several transboundary water agreements is the development of concerted action programmes to reduce pollution loads. Examples include the action programmes drawn up under the auspices of the International Commission for the Protection of the Rhine against Pollution (1987), the International Commissions for the Protection of the Moselle and Saar (1990), and the International Commission for the Protection of the Elbe (1991). These programmes provide detailed measures for reduction of discharges of pollutants from industries and the municipal sector, reduction of inputs of pollutants from diffuse sources, reduction of the risk of accidents through reinforced security and improvement of hydro-logical and morphological conditions in the respective rivers.

Countries bordering other water bodies may be guided by common elements of these conventions, agreements and action programmes when developing their international legal instruments. Common elements of such agreements include: taking action to improve the riverine ecosystem in such a way that higher organisms which were once present would return, to guarantee the production of drinking water, to reduce the pollution of the water by hazardous substances to such a level that sediment can be used on land without causing harm, and to protect the marine environment against the negative effects of the river waters.

To achieve the objectives of the ECE Water Convention, future parties will require strengthened capabilities to comply with its provisions. These capabilities concern, for example, the use of the best available technology for the treatment of industrial wastewaters containing hazardous substances, water-saving technology, reliable measurement systems on industrial outlets and waters, as well as advanced laboratory equipment and analytical techniques. Most of these measures require substantial resources. Programmes for assistance, particularly for countries with economies in transition, are to be further developed. They will aim at exchanging relevant information, the results of research and development, water management practices and instruments, and at providing training.

Controls on the movement of waste may indirectly assist with transfrontier water pollution control. The EC has set out its position in Directives and Regulations, such as in Directive 84/631/EEC on the supervision and control within the EU of the transfrontier shipment of hazardous waste together with subsequent amending Directives, in Regulation 259/93/EEC on the supervision and control of shipments of waste within, into and out of, the EU and in the establishment of a recognised list of hazardous waste by Council Decision on 15 December 1994.

5.8   Conclusions

There are a large number of potential legal and regulatory instruments which are available for pollution prevention and control, and examples of which can be found in operation in many industrialised countries. Developing countries need to examine these in the context of their capability to deliver the end result without over-stretching their resources. A balanced view must be taken as to the standards which should be set as targets. Finance must be made available to enable industries, municipalities, farmers and others to meet the targets. This is best achieved by allowing full participation in the decision-making process by those likely to be affected by the standards imposed, by adopting a multi-faceted approach to the use of the various instruments and by adopting an appropriately phased programme, matching the availability of finance and resources to standards introduced over a number of years.

5.9   References

CIRIA 1994 *Control of Pollution from Highway Dainage Discharges*. Report 142, Construction Industry Research and Information Association, London.

Danish EPA 1995 *Inspection and Enforcement of Environmental Legislation in some EU Countries and Regions*. EU Network for the Implementation and Enforcement of Environmental Law, Danish Environmental Protection Agency, Copenhagen.

DOE 1994 *Reducing Emissions of Hazardous Chemicals to the Environment*. Discussion paper of the Department of the Environment, Her Majesty's Stationery Office, London.

DOE 1995 *Risk Reduction for Existing Substances*. Department of the Environment, London.

EEC 1992 *European Community Environmental Legislation*. Volumes 1–7, L2985, Office for Official Publications of the European Communities, Luxembourg.

Forestry Commission 1991 *Forests and Water Guidelines*. UK Forestry Commission, Her Majesty's Stationery Office, London.

FWR 1994 *Urban Pollution Management Manual*. Report FR/CL0002. Foundation for Water Research, Marlow, Bucks, 129–40.

HMIP 1991 *Chief Inspectors Guidance Notes to Inspectors* Environmental Protection Act 1990 Process Guidance Notes (IPR Series). Her Majesty's Stationery Office, London.

Kletz, T.A. 1988 *Learning from Accidents in Industry*. Butterworths.

Muschack, W. 1990 Pollution of street runoff by traffic and local conditions. *Sci. Tot. Envir.*, **93**, 419–31.

NRA 1992 *The Influence of Agriculture on the Quality of Natural Waters in England and Wales*. Water Quality Series No. 6. National Rivers Authority, Bristol.

NRA 1994a *Water Quality Objectives: Procedures used by the National Rivers Authority for the Purposes of the Surace Waters (River Ecosystem) (Classification) Regulations 1994*. National Rivers Authority, Bristol.

NRA 1994b *Discharge Consents Manual* (Volumes 024A and 024B). National Rivers Authority, Bristol.

NRA 1994c *Discharge Consents and Compliance — the NRA's Approach to Control of Discharges to Water*. Water Quality Series No. 17. National Rivers Authority, Bristol, 27 pp.

OECD 1994 *OECD Pollution Prevention and Control Group, Summary of Member Country Information on Policies for Applying BAT/EQO in Environmental Regulation of Point Sources*. Organisation for Economic Co-operation and Development, Paris.

RCEP 1992 *Royal Commission on Environmental Pollution, Freshwater Quality, Sixteenth Report, Comnd 1966*. Her Majesty's Stationery Office, London, 65–67.

UNECE 1993 *Protection of Water Resources and Aquatic Ecosystems*. Water Series No. 1. United Nations, New York.

UNECE 1994 *Convention on the Protection and Use of Transboundary Watercourses and International Lakes*. United Nations, New York.

WHO   1993   *Guidelines   for   Drinking-Water   Quality.   Volume   1   Recommendations.* Second edition. World Health Organization, Geneva.

Chapter 6[*]

ECONOMIC INSTRUMENTS

6.1   Introduction

In 1972 the Organisation for Economic Co-operation and Development (OECD) adopted the polluter-pays-principle. This principle, which was later adopted as official policy by the European Union (EU), expresses the central notion of environmental economics, i.e. that the cost of pollution should be internalised. Since the principle was introduced it has been extended to include resource use and, thus, the polluter and the user should pay (OECD, 1994b). The introduction of the polluter-pays-principle has also stimulated growing interest world-wide in applying economic instruments. When properly applied they have, in theory, the potential for encouraging cost-effective measures and innovation in pollution control technology. Moreover, water quality is one of the few environmental policy areas where economic instruments already play a significant role in OECD countries and in transitional economies. The purpose of this chapter is to review the most commonly used economic instruments for controlling water pollution, to highlight practical considerations in applying them to water pollution, to suggest criteria for selecting the most appropriate instruments, and to discuss implications for applying them in developing countries and in transitional economies that do not already use them.

6.2   Why use economic instruments?

Economic or market-based instruments rely on market forces and changes in relative prices to modify the behaviour of public and private polluters in a way that supports environmental protection or improvement. They represent one of the two principle strategic approaches to pollution control. The other main approach is regulatory, often referred to as "command and control" (CAC). Regulatory tools influence environmental outcomes by regulating processes or products, limiting the discharge of specified pollutants, and by

---

[*]   *This chapter was prepared by J.D. Bernstein*

restricting certain polluting activities to specific times or areas. Another means of influencing polluter behaviour is through persuasion. In the case of polluting industries, this approach may involve voluntary agreements to undertake pollution control measures. In the case of consumers, it may involve public education and information campaigns to influence patterns of consumption and waste disposal. This approach is applied in countries such as The Netherlands, Japan and Indonesia.

Since the inception of environmental policy in most industrial countries, governments have tended to use these instruments as their main strategy for controlling pollution. Many countries, however, are becoming aware that regulatory instruments are inefficient for achieving most pollution control objectives, and that the level of expenditure required to comply with increasingly stringent environmental laws and regulation is becoming a major cost of production. In the USA, for example, the US Environmental Protection Agency (EPA) estimates that the proportion of Gross National Product (GNP) devoted to environmental protection can be expected to grow from 1.7 per cent in 1990 to nearly 3 per cent by the year 2000, and that most of these costs will be borne by the private sector (US EPA, 1991). An increasing number of governments are, therefore, investigating alternative mechanisms to achieve the most cost-effective means for controlling pollution that will not place excessive financial burdens on businesses and individuals, and that will not undermine economic development.

In contrast to regulatory instruments, economic instruments have the potential to make pollution control economically advantageous to commercial organisations and to lower pollution abatement costs. They can be applied to a wide range of environmental problems and can involve varying degrees of incentives, information, and administrative capacity for effective implementation and enforcement. The principal types of economic instruments used for controlling pollution are:

- *Pricing.* Marginal cost pricing can reduce excessive water use and consequent pollution as well as ensure the sustainability of water treatment programmes. Water tariffs or charges set at a level that covers the costs for collection and treatment can induce commercial organisations to adopt water-saving technologies, including water recycling and reuse systems, and to minimise or eliminate waste products that would otherwise be discharged into the effluent stream. In Thailand, for example, many hotels along the country's eastern coast are treating and recycling their water for landscape irrigation because the cost of freshwater now exceeds the cost of treatment (Foster, 1992). Before considering the use of other instruments in environmental policy, it is advisable for countries to evaluate their water pricing policies because such policies can encourage over-use and water degradation.

- *Pollution charges.* A pollution charge or tax can be defined as a "price" to be paid on the use of the environment. The four main types of charges used for controlling pollution are: (i) effluent charges, i.e. charges which are based on the quantity and/or quality of the discharged pollutants, (ii) user charges, i.e. fees paid for the use of collective treatment facilities, (iii) product charges, i.e. charges levied on products that are harmful to the environment when used as an input to the production process, consumed, or disposed of, and (iv) administrative charges, i.e. fees paid to authorities for such purposes as chemical registration or financing licensing and pollution control activities.

- *Marketable permits.* Under this approach, a responsible authority sets maximum limits on the total allowable emissions of a pollutant. It then allocates this total amount among the sources of the pollutant by issuing permits that authorise industrial plants or other sources to emit a stipulated amount of pollutant over a specified period of time. After their initial distribution, permits can be bought and sold. The trades can be external (between different enterprises) or internal (between different plants within the same organisations).

- *Subsidies.* These include tax incentives (accelerated depreciation, partial expensing, investment tax credits, tax exemptions/deferrals), grants and low interest loans designed to induce polluters to reduce the quantity of their discharges by investing in various types of pollution control measures. The removal of a subsidy is another effective tool for controlling pollution. In many countries, for example, irrigation water is provided free of charge, which encourages farmers to over-irrigate, resulting in salinisation and/or water logging.

- *Deposit-refund systems.* Under this approach, consumers pay a surcharge when purchasing a potentially polluting product. When the consumers or users of the product return it to an approved centre for recycling or proper disposal, their deposit is refunded. This instrument is applied to products that are either durable and reusable or not consumed or dissipated during consumption, such as drink containers, automobile batteries and pesticide containers.

- *Enforcement incentives.* These instruments are penalties designed to induce polluters to comply with environmental standards and regulations. They include non-compliance fees (i.e. fines) charged to polluters when their discharges exceed accepted levels, performance bonds (payments made to regulatory authorities before a potentially polluting activity is undertaken, and then returned when the environmental performance is proven to be acceptable), and liability assignment, which provides incentives to actual or potential polluters to protect the environment by making them liable for any damage they cause. This chapter only addresses fines because they are the most commonly used enforcement incentives, particularly in the area of water pollution control.

Although economic instruments have several advantages over direct regulation, applying them to pollution control does not, and should not, preclude the use of regulatory instruments. In most cases, economic instruments supplement the existing regulatory framework, with ambient standards remaining the objectives for both. By selecting the right mix of regulatory and economic instruments, and in some cases other types of instruments such as property rights or educational approaches, policy makers can combine the positive elements of both approaches.

The main advantage of the regulatory approach is that, when properly implemented and enforced, regulation affords a reasonable degree of predictability about how much pollution will be reduced. In theory, the advantages of economic instruments are:

- They allow commercial organisations and individuals to respond flexibly and independently in line with market prices in order to meet environmental management objectives at the least cost.
- They provide a continuing incentive for commercial organisations to reduce pollution and therefore to develop and adopt new pollution control technologies and processes to minimise waste.
- They have the ability to raise revenue (in the case of charges) in order to finance pollution control activities.
- They accommodate the growth of existing industries and the entry of new ones more than would otherwise be possible under a regulatory approach.
- They reduce compliance and administrative costs for both government and industry. For example, the use of environmental taxes or tradable permits eliminates the need for government certification of production processes and technologies. They also eliminate the government's need for large amounts of information to determine the most feasible and appropriate level of control for each regulated plant or product.

The advantages of economic instruments offset the main drawback of the regulatory approach, i.e. regulatory tools can be economically inefficient and excessively costly to implement. For example, under the regulatory approach, all commercial organisations would be subject to the same emission standards regardless of their pollution abatement costs. Ideally, only the larger polluters would install pollution control equipment; the large scale of their operations makes the cost of pollution control per unit of output lower than that for small-scale polluters. The regulatory approach also tends to discourage innovation in pollution control technology. It gives little or no financial incentive to organisations to exceed their control targets. This is a particular disadvantage where the development of a new control technique could be subsequently held as the future standard but without allowing any opportunity to benefit from the innovation. Moreover, compliance in most cases depends on the enforcement capacity of the regulatory agency and the number of organisations or individuals being regulated. The greater the number of organisations or enterprises to be regulated, the more difficult it is to enforce the regulations properly. Economic instruments, by contrast, are better suited to a larger number of point and non-point sources of pollution.

While economic instruments can be more cost-effective than regulatory instruments and more appropriate for dealing with numerous point and non-point sources, the economic or market-based approach to pollution control also has its own drawbacks. The major weaknesses of economic instruments are:

- Their effects on environmental quality are not as predictable as those under a traditional regulatory approach because polluters may choose their own solutions.
- In the case of pollution charges, some polluters opt to pollute and to pay a charge if the charge is not set at the appropriate level.
- They usually require sophisticated institutions to implement and enforce them properly, particularly in the case of charges and tradable permits.

In addition to these drawbacks, both government agencies and individual polluters have resisted the introduction of economic instruments. Regulatory agencies, for example, have objected to them largely because they afford them less control over polluters. Industry and other polluters have resisted them because they feel that they have greater negotiating power over the design and implementation of regulations than they do over charges. Industries also view economic instruments as additional constraints (where they supplement existing regulations). For example, charges impose a financial burden beyond the cost of complying with regulations. A further deterrent to using economic instruments is their, often complicated, implementation requirements. The main difficulties relate to setting prices for environmental resources and estimating the full extent of environmental damage.

6.3 Applying economic instruments

Despite the general resistance of countries to using economic instruments in environmental management, water pollution control is one of the few environmental policy areas where they have played a relatively significant role. Charges for the collection and treatment of water are well established in most industrial countries. In many countries, charges also are applied to polluters who discharge their effluent directly into open water. In addition, combinations of direct regulation and economic instruments, particularly charges, have produced positive results in terms of revenue raising and pollution control.

The remainder of this section discusses how specific instruments are used in controlling water pollution. Among these instruments, water pricing, effluent charges, user charges, and subsidies are the principal economic instruments used in this respect by both industrialised and developing countries.

6.3.1 Pricing

Water pricing policies can be an effective tool for reducing pollution; not only by promoting water conservation, but by raising funds to support pollution control programmes. Mexico City, for example, has increased the price for industrial water consumption. This has discouraged the establishment of water intensive industries in the Mexico City Metropolitan Area (MAMA) and encouraged water conservation by making recycling an attractive proposition. It has also promoted the use of water saving technologies (World Bank, 1994). As has been demonstrated in Mexico City, where wastewater standards are defined in terms of pollutant concentrations, pollution charges and standards should be co-ordinated carefully with water prices to ensure effective pollution control. If water prices are low, polluters can meet the standard by dilution — leading to higher water use without reducing the overall pollution load.

6.3.2 Effluent charges

Several countries apply effluent charges in order to finance necessary measures for wastewater collection and purification, and to provide financial incentives for reducing discharges of effluent. The charge can be based either on the actual quality and quantity of wastewater (determined through yearly or more frequent monitoring by the responsible administrative body or through self-monitoring by the polluter), or on a substitute based on information on the output, treatment levels and number of employees within an organisation. In some cases, a flat rate is charged. Successful implementation of a charge system depends on four key factors (OECD, 1991):

- Recognising the fundamental characteristics of the environmental problem.
- Choosing a competent authority to legislate, implement, and monitor the tax.
- Establishing a suitable tax base.
- Setting an appropriate tax rate.

The experience of most of countries applying water effluent charges, e.g. France, Germany, Italy, and Central and Eastern European countries, indicates that charges are set far below the level required to induce polluters to reduce their discharges, although they do raise revenue for pollution control purposes. By contrast, in The Netherlands, the water effluent charge, which was designed as a tool for revenue raising only, has also served as an incentive because of the high charge rates. The Netherlands also adopted the following approach to reduce the need for large amounts of information to assess the fees to be charged:

- Households and small industrial polluters producing less than 10 pollution equivalents (pe) are not charged for the actual pollution they cause. Having relatively few opportunities to limit discharges, this category of polluters is of minor importance to the instrument's regulating power. The great benefit is that this allows the executive bodies to reduce drastically the amount of information required. Fixed rates are used instead.
- Charges for medium-sized polluters (10–100 pe) are not based on samples of their effluent but according to a coefficient table prepared by experts. This permits the probable amount of pollution to be estimated accurately for each branch of industry or sector on the basis of easily obtainable data, such as the amount of water used by the production plant and the amount of raw materials it processes. Nonetheless, the incentive to reduce pollution remains intact. Companies that believe they are overrated on the coefficient table can request their effluent to be sampled and then charged on the basis of the results (Braceros and Schuddeboom, 1994).

As demonstrated by effluent charge systems in numerous countries (Box 6.1), these systems are most successful when combined with regulation, when applied to stationary pollution sources and when marginal abatement costs vary amongst polluters (the wider the variation, the greater the cost-saving potential). Other determinants of success are the feasibility of monitoring effluents (either by direct monitoring or proxy variables), the ability of polluters to react to the charge, the ability of pollution control authorities to assess appropriate fees, and the potential for polluters to reduce emissions and to change their behaviour. Russia's pollution charge system demonstrates how administrative weaknesses can undermine environmental effectiveness (Box 6.2).

In Mexico, an effluent charge is directly tied to regulation, but its design and implementation could also be improved. The Federal Water Charges Law in Mexico establishes water pollution charges applicable to all discharges to national waters that exceed the applicable standard. The charges are based on volume of flow, discharges of conventional pollutants (suspended solids and chemical oxygen demand (COD)), the costs of pollution abatement and regional water scarcity. The charge, however, does not take into account the

Box 6.1   Examples of effluent charge systems

Brazil
In Brazil, four States are experimenting with effluent charges in the form of an industrial sewage tariff based on pollutant content. Although the formulae adopted to define the tariff levels vary among States, cost recovery is the objective in all cases. In the State of Rio de Janeiro, the local environmental protection agency Fundaçao de Tecnologia de Saneamento Ambiental (FEEMA) is responsible for tax collection. It is creating an effluent charge to be approved by the State government. The charge will be levied on all polluters and will be based on the volume and concentration of the effluent, including BOD and heavy metals. Tariff rates will be calculated to recover the budgetary needs of the State agency. In the case of Rio de Janeiro, the budget of the state agencies is so low, at present, that the administration relies on revenue raising approaches to fulfil its funding requirements. Revenues are usually distributed for such functions as pollution abatement, financing of administrative costs, monitoring enforcement and educational campaigns.

France
To manage its water resources and to halt or reduce growing river pollution, the French government decided in 1964 to apply economic instruments to supplement its regulations. At the same time, the planning and financing water management responsibilities of the country were devolved to new operational agencies, i.e. river basin committees and water agencies. These institutions, created in the six river basins, play an essential role in water planning and controlling domestic and industrial pollution. The creation of these agencies made it necessary to take a consistent approach to pollution so that charges could be established on the basis of a small number of clearly defined variables. Initially the basis for the fee consisted of two variables: the weight of suspended matter and the weight of organic matter. Both were considered priorities, representing the most visible type of pollution, and the means to tackle them were also known. Much later, when new pollution variables began to cause concern or when techniques for evaluating and eliminating them became available, the basis for assessment was gradually extended (e.g. to include salinity, nitrogen, phosphorous, halogenated hydrocarbons, toxic and other metals). In each case, the aim was to use charges as an incentive to reduce pollution caused by the variable in question and to avoid charges being transferred to users who are not responsible for increased levels of pollution. The rates are set by each agency board and approved by the corresponding river basin committee. Their values are determined in such a way that the income from charges balances the financial assistance provided, while avoiding excessive discrepancies between charges to the various charge

Continued

Box 6.1   Continued

payers. The charge is also a source of information about users' activities, offering more precise knowledge of how water is used and a better understanding of the natural environment. The quantities of pollution discharged by a user, which is impractical to measure for each one, are assessed at a flat rate according to a national scale based on the type of activity (in the case of industry) or number of inhabitants (in the case of urban centres). The amount of pollution produced by a particular industrial establishment is measured only at the operator's or agency's request. When this occurs, measurements are taken by a laboratory approved by the agency and the costs are borne by the party making the request. The agencies also are authorised to promote measures to conserve water supplies. In addition to the pollution charge, therefore, a charge is levied on the basis of the volume of water taken by each user. Charge payers may choose between a flat-rate assessment of the volume of water they use and metering (the income from this type of charge is generally much less than the income from pollution charges). The law gives agencies a dual role in promoting water protection in their particular river basin, providing financial assistance for works of common interest and conducting studies and research in water-related matters. In the same way, polluters are taxed when their activity is harmful to the environment and polluters receive an award, in the form of subsidies, when their actions are beneficial to the environment.

Germany
The German Effluent Charge Law authorises States to levy charges on direct discharges of specified effluents into public waters. Commercial organisations and households discharging into municipal sewerage facilities are not charged directly. The pollutants considered for the purposes of effluent charges are settleable solids, COD, cadmium, mercury and toxicity to fish. In setting the charge base, the law established the right to discharge and includes all of the physical, chemical and biological data and monitoring procedures pertaining to wastewater quality. For each organisation, the State also specifies a total discharge based on historical volumes of wastewater allowable per year. Since the effluent charge is combined with a permit procedure, the maximum effluent level is also specified. The actual effluent discharged by the organisation must be of a quality equal to, or higher than, the minimum requirements laid out in the regulation. The taxable base is specified in terms of concentration per cubic meter of discharge volume or per tonne of product produced. An organisation's discharge is then converted into damage units using coefficients provided in the law. The tax liability is determined by multiplying the number of damage units by the tax rate per damage unit. This tax rate is

Continued

Box 6.1   Continued

revised annually based on an established increment. To provide an incentive to limit pollution loads, higher charges are imposed per damage unit if organisations exceed the permit limit. These excesses are allowed only twice a year. Lower charges per damage unit are used to compute the total tax liability for those who discharge below permit limits.

Korea
The emission charge system combines elements of regulation and market-based incentives and applies to both air and water discharges. The charge is applied to organisations who are operating facilities that do not meet emission/effluent standards. The charge rate, however, is not directly linked to the level of excess discharges, nor is there an upper limit on the amount of the levy. In practice, however, charge rates have sometimes been set lower than the operating costs of a pollution treatment facility and so organisations have been known to under-use their treatment plants at the risk of being detected and fined. Another limitation of the system is that it does not encourage over attainment.

The Netherlands
The charge on water pollution can be imposed on everyone who emits waste, polluting or noxious substances directly or indirectly into surface water, or into a collectively-used water purification plant. The charge can be levied by public authorities or by Water Boards, i.e. non-governmental bodies governed by councils in which affected interests are represented. The charge can be based on the quantity and/or quality of the pollutants. In practice, the charge is applied to discharges of oxygen consuming substances and heavy metals (only for emissions into non-State waters). Both kinds of pollution are expressed in so called "population equivalents" (pe). The number of pes for households and small enterprises is fixed by the authorities. The emissions of larger organisations are assessed by means of a table of emission coefficients, or can be measured individually. Only in the latter case is an incentive effect to be expected. The water pollution charge has primarily a financial purpose; it is intended to finance the costs of water purification. The charge rate for authorities is relatively low because the State does not exploit its own water treatment plants. Apart from being an important source of finance for water purification plants, the water pollution charge also has had a strong incentive effect. In the 20 years since its existence, both the quality of water and the number of treatment plants have risen considerably.

Sources: Hahn, 1989; Cadiou and Duc, 1994; Freitas, 1994; O'Connor, 1994

Box 6.2   Administrative problems in Russia's pollution charge
                 programme

In 1991–92, Russia adopted pollution charges for air emissions, water effluents and waste disposal. The rates were determined on the basis of maximum permitted concentrations and reflected the desire to mitigate environmental health and other pollution risks. Although, initially, the charges were intended to induce optimum pollution levels, charge rates were calculated to generate enough revenues to finance critical projects, such as the construction of water treatment facilities and the clean-up of hazardous waste sites. Within this context, the charge system worked to the satisfaction of national and local authorities. However, several administrative weaknesses in the programme undermined its capacity to encourage effectiveness in changing polluting behaviour. These weaknesses can be summarised as follows:

- The lack of an appropriate system (equipment, methods, personnel) for monitoring discharges.
- Inadequate equipment and expertise of inspection personnel responsible for identifying and punishing violators.
- Inability to enforce the collection of charges due to uncertainty and contra-dictions in the legislation.
- Absence of a clear assignment of responsibility between the federal and territorial levels.
- Absence of clear regulations spelling out how to distribute environmental costs among polluters, the federal and regional budgets, and the federal and regional environmental funds.
- Unresolved questions regarding economic liability for environmental damage resulting from an enterprise's previous and current technologies.
- Insufficient institutional support, including a lack of special staff training and a special implementation programme.
- Excessively complicated charge systems, partly because of the inclusion of hundreds of types of pollutants and the need to calculate precise charges.
- Erosion of the pollution charges by inflation. The 500 per cent increase in charge rates in 1992 was insufficient to offset inflation.

Nevertheless, the pollution charge system has become the cornerstone of environmental protection programmes in Russia. Since 1992, agreements between polluters and the environmental protection authorities have created the legal basis for the collection of charges. Such agreements specify the permitted level of discharge, base rates and penalty rates for each pollutant discharged, as well as the schedule of charge payments.

Source: National Academy of Public Administration, 1994

effluent's toxicity or the quality of the receiving body of water. The objective of the pollution charge is to encourage organisations to comply with effluent standards, and only those organisations that do not comply are subject to a charge. Those that do not comply but have a plan to control emissions can obtain an exemption for up to two years. The tax base has three components: the excess of COD emissions above the standard, the excess of suspended solids emissions above the standard and a volume component. The volume component is applied whenever the organisation is in violation of any of the pollutants for which it is subject to a standard, even when that organisation is in compliance with COD and suspended solids. For each of these three components, there are charges that depend on the zone in which the firm is located.

In practice, the implementation and impact of Mexico's effluent charge have been very limited. The total revenue collected from the charge in 1993 was only US$ 5.6 million, a very small proportion of the potential revenue. Just for one region, the potential tax yield is estimated to be US$ 35 million and would induce a pollution abatement of more than 70 per cent (World Bank, 1994). Although Mexico's water pollution charge is a positive initiative, its design and implementation can be improved in two ways. Firstly, separate charges for suspended solids are not necessary because the abatement of other substances (e.g. COD) normally leads to a relatively high abatement of suspended solids. Secondly, the volume component could be removed because it provides an incentive to increase pollutant concentrations because it is the largest component when estimating the pollution charge. Additional ways to improve the charge would be to include charges for heavy metals and to exclude suspended solids, as well as to vary the charge according to the quality of the receiving water body.

Although effluent charges are among the most commonly applied economic instruments, experience in many countries indicates that they are often set at too low a level to act as an effective deterrent to pollution. Most polluters prefer to pay the charge rather than to change their polluting behaviour. Consequently, the principal function of most effluent charge systems is to raise revenue. In several countries where charges are widely applied (e.g. China, Japan, Indonesia, Korea, Poland, Russia, Thailand), governments deposit revenues from pollution charges and taxes into environmental funds that provide loans and grants to municipalities or to local enterprises for the purchase of abatement equipment and the introduction of clean technologies (Box 6.3).

Box 6.3   Examples of environmental funds

China

To help bring industrial pollution under control, a revolving loan fund was established that provides below-market financing for pollution control efforts by local, mostly small and medium size enterprises. The loans are financed by proceeds from waste discharge fees. The basic fee is charged for releases up to a specified concentration, above which a penalty fee is imposed. The funds are administered by the provincial or municipal environmental protection bureau and directed by a board of representatives from the local economic planning, finance and environmental bureaus. To qualify, the industrial enterprise and target pollutants must be listed as part of the area's pollution control strategy. Loans are extended for 50–80 per cent of project costs; grants are for 10–30 per cent of costs.

Korea

The Environmental Pollution Prevention Fund is financed, in part, from Government contributions and, in part, from fines (or pollution charges) levied on organisations found to be exceeding emission standards. The fund, which was established in 1983, is administered by the semi-governmental Environmental Management Corporation. The resources for the fund are used to provide long-term, low-interest loans for pollution control investments, as well as to compensate pollution victims.

Thailand

In October 1991, Thailand launched an Environmental Fund with an initial capital contribution by the Government of roughly US$ 200 million. Partial grants and low interest loans from the fund are made available to municipalities, sanitary districts and private businesses which are required to set up treatment facilities. The city of Pattaya is the first to use this fund for its central wastewater treatment plant.

Indonesia

A Pollution Abatement Fund was established to provide US$ 300 million to banks to finance loans to companies investing in pollution control equipment or hiring environmental consultants.

Poland

The national environmental fund finances most environmental investments. Sources of revenue for the fund include air and water pollution charges, water-use charges and waste charges. The funds are allocated through grants and interest-free (and other soft) loans to support air and water pollution control as well as for other environmental management purposes (soil protection, monitoring, education).

Russia

According to a regulation issued in June 1992, environmental funds should apply their revenues from pollution charges to a wide variety of environmental activities. Among other uses, they can be applied to implement regional and inter-regional projects for: improving environmental and human health, conducting research and designing projects in the areas of pollution control, clean-up and treatment; to support enterprises, research and development organisations and individuals that introduce environmental-friendly equipment; to the design of computer systems for environmental monitoring; and to construct or share in the construction of treatment and other protective facilities. A World Bank loan to the Russian Federation is supporting the establishment of a National Pollution Abatement Facility (NPAF) which will fund economically and financially viable pollution abatement projects.

Source: Lovei, 1994; O'Connor, 1994; Kaosa-ard and Kositrat, 1994

6.3.3   User charges

User charges may be variable (i.e. linked to water consumption or property values), fixed or some combination of the two and they are assessed on both municipal and industrial discharges into public sewerage (Box 6.4). Experience in numerous countries suggests that the effectiveness of these charges in controlling pollution requires the setting of appropriate charges and ensuring the existence of necessary institutional capacity for monitoring discharges and enforcing regulations.

In Izmir and Istanbul, Turkey, for example, sewerage charges (wastewater charges) are assessed on industrial discharges into the sewer systems. These charges are significant because they motivate factories to treat industrial effluents. Enterprises face two costs: treatment costs and disposal costs (sewer charges). Generally, high sewerage charges encourage full treatment of industrial wastewaters such that they are suitable for discharge to surface waters, thereby eliminating sewerage charges. Low sewerage charges, by contrast, encourage only sufficient pre-treatment of wastewaters to make them suitable for discharge to the municipal sewer system. In this way, the enterprises minimise their treatment costs. When seeking to minimise their costs, therefore, the decision of an organisation to apply pre-treatment or full treatment will be a direct response to the level of the sewer charge. Nonetheless, the problem of illegal discharges complicates the application of an optimal tariff in Izmir and Istanbul. If the sewer charge is too high, firms may seek to avoid it by illegally discharging wastewater. Thus, the ability to monitor industrial polluters and to enforce pollution standards is critical (Kosmo, 1989).

Experience in the eastern part (Suzano) of São Paulo, Brazil, also demonstrates the importance of establishing sewerage charges at the appropriate level before public investment in sewage treatment. It also demonstrates the need for contracts that commit industrial users to the scheme, as well as demonstrating that the building of a treatment plan for, basically, one industry by the public sector is inadvisable. In this case, a sewage treatment plant was being constructed largely to treat the wastes of a local paper mill. About 90 per cent of the capacity of the plant was expected to be used by this company. Due to an unacceptably high tariff level set by the State sanitation company SABESP (Basic Sanitation Company of the State of São Paulo), the paper company chose not to connect to the new sewage treatment plant and constructed its own treatment facility at a lower cost. Consequently, the Suzano treatment plant operated at only 10 per cent of its full capacity for several years because it was necessary to phase investments in residential sewer networks.

Box 6.4   Examples of user charges

Canada
The sewage charge levied on domestic users may be based on residential property values or calculated according to a formula that includes consumption (in m^3). A flat rate residential sewage tax is also used.

Colombia
In Cali, sewerage tariffs are set at 60 per cent of the water tariff, in Cartegena 50 per cent and in Bogota 30 per cent.

Sweden
Municipalities levy a charge for treatment of sewage water. The charge consists of two elements: a fixed charge and a variable charge related to consumption. The charge appears to be effective because the numbers of households and smaller industries attached to the sewer system and extended water treatment facilities are growing. The charge has some incentive effect, in that industries try to reduce water use when extending or renewing their plants, although this could give rise to higher pollution concentrations. In some municipalities, a redistribution occurs because enterprises pay a relatively high charge, implying a subsidy to households.

Thailand
To control pollution, industrial enterprises discharging effluent are required to pay service fees to a central wastewater treatment facility or to set up their own treatment facilities. The revenues from the fees are used to cover the operating costs of the treatment facility.

USA
Towns receiving federal grants for the construction of sewer systems are required by the Water Pollution Control Act to recover their operating costs and part of the capital costs from their users, through municipal sewage treatment user charges. A number of States charge flat permit fees that entitle the permit recipient to discharge wastewater. For example, California levies a wastewater discharge permit fee, based on type and volume of discharged pollutants.

Source: OECD, 1989, 1994

A groundwater charge (or abstraction fee) can be used to discourage excessive pumping of aquifers which can result in salinisation and other types of groundwater contamination (as well as land subsidence). In the Netherlands, the provinces can levy a groundwater charge from those who extract groundwater, based on the amount of the resource extracted. The revenues can be used for research, necessary groundwater management and for compensation payments when damage caused by a drop in the groundwater level cannot be attributed to a specific individual abstractor (OECD, 1994a). In common with many effluent charge systems, this charge is too low to have any significant incentive or economic effect.

6.3.4   Product charges

Product charges can be applied to products that will pollute surface water or groundwaters before, during, or after consumption. They are best applied to products that are consumed or used in large quantities and in diffuse patterns (e.g. fertilisers, pesticides, lubricant oils). A special type of product charge is tax differentiation. Product price differentials can be applied in order to discourage the use of polluting products and to encourage consumption of cleaner alternatives. When a product is highly toxic, and when its use should be drastically or completely reduced, a partial or total ban is preferable to product charges.

Product charges can act as a substitute for emission charges whenever it is not feasible to apply direct charges to pollution. The rates of product charges should reflect the environmental costs associated with each step of the product life-cycle. The rates are fixed but can be re-calculated if the charge lacks incentive power. The effectiveness of a charge on polluting products or product inputs will generally depend on the elasticity of the demand for that product. For example, where input costs are a small fraction of total costs, doubling or tripling the price through an input tax is unlikely to have a significant effect on consumption, unless there are suitably priced substitutes. If less polluting substitutes are available, small increases in input prices may induce substitution and innovation over the longer term (Moore *et al.*, 1989). Revenues from product charges can be used to treat pollution from the product directly, to provide for recycling of the used product or for other budgetary purposes.

6.3.5   Marketable permits

Setting up effective marketable permit programmes involves establishing rules and procedures for defining the trading area or zone, for distributing the initial set of permits (e.g. direct allocation by a regulatory agency, grandfathering, various types of auctions), for defining, managing and facilitating permissible trading after the initial allocation, and for carrying out monitoring and enforcement activities. Tradable permit systems work best where (OECD, 1991):

- The number of pollution sources is large enough to establish a well functioning market.
- The sources of pollution are well defined.
- The amount of pollution generated by each source is easily computed.
- There are differences in the marginal costs of pollution control among the various sources.
- There is potential for technical innovation.
- The environmental impact is not dependent on the location of the source and time of year.

Marketable permits are not as effective for controlling water pollution as other instruments because water pollution is directly tied to location and time of year. Where they have been applied to this purpose, they have not produced impressive results.

In the USA, for example, the state of Wisconsin implemented a programme to control biochemical oxygen demand (BOD) in the Fox River. The flexibility of the programme allowed limited trading of marketable discharge permits. Organisations were issued five-year permits that defined their waste load allocation, which in turn defined the initial distribution of permits for each organisation. Although early studies indicated several potentially profitable trades involving large cost savings (in the order of US$ 7 million), there has been only one trade and actual cost savings have been minimal since the programme began in 1981 (Hahn, 1989). Stringent restrictions have significantly inhibited trading under this programme (Oates, 1988). Numerous administrative requirements also add to the cost of trading and lower the incentive for facilities to participate. Some costs can be attributed to the small number of organisations involved and others to the absence of brokering or banking functions (Anderson *et al.*, 1989). In many developing countries, the absence of well-functioning markets would place further constraints on effective trading.

6.3.6 Subsidies
Numerous countries make available tax reductions, grants or low interest loans to mitigate those water pollution abatement or prevention costs that must be borne by polluters (Box 6.5). Policy makers tend to favour these instruments because they ease the transition to a more stringent regulatory environment (especially for established polluting enterprises) and because there may be an economic justification for applying them where there are clear positive externalities associated with private investment in pollution control. Nonetheless, there are some disadvantages to using them. First,

Box 6.5   Examples of subsidies for water pollution control

France
River basin agencies may provide financial assistance in the form of grants or loans in addition to any other assistance that may be obtained from, for example, the government, region or department. The total amount of assistance must not exceed 80 per cent. Grants are the most common form of financial assistance. Where loans are involved, they are generally for a period of 10–125 years and the interest rate is lower than the market rate. In the Seine-Normandie river basin, for example, the interest rate is equal to half the rate of the Credit Local de France.

Indonesia
The Environmental Impact Management Agency (BAPEDAL), with support from Japan, has established a five-year US$ 103 million soft loan programme for industrial organisations investing in waste treatment. Loans are made available on a first-come, first-served basis and are for a period of between 2 and 30 years with a grace period of 1–5 years and an average interest rate of 14 per cent per year (well below market lending rates). The loan programme should facilitate the implementation of the Government's PROKASH, or clean rivers programme.

Korea
Two provisions under the Tax Exemption and Reduction Control law provide direct and indirect incentives for pollution control. First, there is a direct investment tax credit of 3 per cent (or 10 per cent for equipment made in Korea) of the value of the investment which is restricted to facilities for increasing productivity, energy-saving facilities, anti-pollution facilities, facilities for preventing industrial hazards and other specified facilities. More indirectly, for persons starting a business using technology, there is a choice between accelerated depreciation of 30 per cent (50 per cent in the case of machinery manufactured in Korea) of the asset's acquisition price in the fiscal year of acquisition or an investment credit at the rate of 3 per cent (or 10 per cent in the case of machinery made in Korea) of the value of the investment for new assets.

Philippines
The Environmental Code enacted in 1977 allowed half of the tariff and compensating tax on imported pollution control equipment to be waived for a period of years from the date of enactment. The code also made available

Continued

subsidies can result in inefficiencies by encouraging over-investment in pollution control or over-expansion of the polluting activity. For example, large subsidy shares in the investment costs of pollution control, as implemented in the United States Construction Grants Program, can induce plant

Box 6.5   Continued

rebates for domestically produced equipment and a deduction for certain pollution control research.

Taiwan
The government offers a range of subsidies. Among activities eligible for subsidy are acquisition of land for waste treatment facilities and the installation of pollution control equipment. A real estate tax concession is also offered for the relocation of a polluting facility and a number of other tax concessions are offered for pollution control investments, including duty free importation of pollution control equipment, corporate income tax reduction for purchasing such equipment, two-year accelerated depreciation for pollution control facilities, and a 20 per cent profit tax reduction for research and development on pollution control.

Thailand
Partial grants and low interest loans are made available from the Environment Fund to local administrations and private businesses required to set up treatment facilities. Other subsidies include the reduction of import duties to no greater than 10 per cent for equipment used for any treatment facilities. During 1984–89, however, only 130.9 million baht (US$ 5.14 million) worth of wastewater treatment equipment had been imported under this incentive.

Turkey
The Government has provided subsidised credit for relocating polluting industries to alternative industrial zones. For example, leather tanneries relocating to the Maltepe Industrial Zone north of Izmir would be entitled to subsidised interest rates of 35 per cent for general loans and 22 per cent for construction and infrastructure investment, implying negative real interest rates at an 80 per cent annual rate of inflation. This is a clear incentive because interest costs in 1988 and 1989 accounted for 20 per cent of total investment expenditures. The Government also has offered a 40 per cent tax deduction on investment for tanneries relocating to another industrial zone during the first two years of estate construction and a 7 per cent reimbursement on investment for small and medium-scale tanneries.

Sources: Kosmo, 1990; Cadiou and Duc, 1994; Kaosa-ard and Kositrat, 1994; O'Connor, 1994

operators to design capital intensive facilities with excessive capacity. They also are not consistent with the polluter-pays-principle because the general taxpayer subsidises the control costs of specific polluters. Moreover, subsidies pose a drain on government resources (O'Connor, 1994).

Subsidies, in general, should be selective and should be provided on a temporary basis. In many cases governments subsidise small and medium size enterprises because they suffer a competitive disadvantage when they adopt environmental control technologies where there are economies of scale. The problems of small enterprises may be especially acute in the case of process changes aimed at reducing waste rather than end-of-pipe treatment technologies. While the latter can be added on without disrupting the production process, the former may require the temporary shutdown of the production process during conversion or retrofitting. When introducing process changes, an organisation also may encounter costly start-up problems. While a large enterprise, with several processes running in parallel, may be able to make changes incrementally, small enterprises must face all-or-nothing decisions and face considerably higher financial risks than the larger enterprises. Therefore, even where such subsidies are not justified on the basis of efficiency, they may address equity concerns (O'Connor, 1994).

The removal of water or other types of subsidies can also have a positive effect on water quality. For example, the removal of a water subsidy can lead enterprises and residential users to conserve water and thereby reduce the amount of pollutants they discharge into the effluent stream. Ensuring marginal cost pricing for water can even help to ensure the sustainability of a water treatment programme. Similarly, the removal of subsidies on pesticides and chemical fertilisers can reduce water pollution, particularly groundwater contamination, and the poisoning of aquatic life through run-off into water systems. For residential polluters, however, water subsidies may have to be maintained in order to support the economically weaker segments of the population, particularly the urban poor. Nonetheless, a free-ride situation of a totally free resource is not sustainable. The poor should be required to pay a small charge for water (which should be increased incrementally) not only to cover the costs of water treatment, but also to promote water conservation.

6.3.7 Deposit–refund system

Although not a principal instrument for controlling water pollution, deposit–refund systems can be applied to this purpose if potentially polluting products which are not consumed or dissipated during consumption, such as pesticide containers, can be returned to an approved centre for proper disposal or recycling. Establishing successful deposit-refund systems requires products that are easy to identify and handle and users and consumers that are able and willing to take part in the scheme. It often also requires new organisational arrangements for handling the collection and recycling of products and substances as well as for managing the financial arrangements, and a national or state authority to establish the system. The advantages of deposit–refund systems are that most of the management responsibility remains with the private sector and incentives are in place for third parties to establish return services when users do not participate. A major disadvantage of this approach is that the costs of managing deposit–refund programmes, i.e. administrative, collection, recycling, and disposal expenditures, fall to the private sector.

6.3.8 Enforcement incentives

Penalties for failing to meet environmental standards are commonly-used instruments to encourage dischargers to comply with environmental standards and regulations. In Mexico, fines are set according to the severity of pollution and adjusted for inflation; repeated offences lead to plant closure. Combined with public pressure, these measures have been effective in controlling surface water pollution. In Argentina, by contrast, fines for discharging into water bodies without treatment are set too low to achieve the environmental objectives (Box 6.6).

6.4 Choosing between instruments

As illustrated in several of the examples above, economic instruments are rarely used alone to manage water pollution. The focus of any policy debate should not be weighing the relative advantages and disadvantages of economic and regulatory instruments, but instead the most important issue is to find the appropriate mix of instruments that would best respond to the special characteristics of each problem and locality, together with specific operators whose behaviour needs changing, and the desired behavioural response.

For effective water pollution control, pollution charges and standards have to be combined carefully with water prices which should be high enough to cover all costs and provide an incentive for water conservation and recycling. In this way, the incentive to achieve standards by dilution is reduced, resulting in less liquid effluent being discharged into rivers and streams.

In selecting instruments, policy makers need to take into account the nature of the environmental problem and its causes, as well as practical, economic, and political realities. In determining the most appropriate instruments, each country needs to establish clear and transparent criteria upon which to base its selection. In developing countries, where there are extremely limited financial resources and weak institutional capacity, the two most important criteria are cost-effectiveness and administrative feasibility. Other criteria include equity, consistency with other objectives, flexibility and transparency.

Box 6.6   Enforcement incentives in Buenos Aires Provincial

The Law Protecting Water Bodies that Supply and Receive Effluents in the Buenos Aires Provincial prohibits any discharges into water bodies (or to the air) without treatment. In practice, this means that industries must obtain a license to operate. In 1986 the law was modified to enable the application of fines to industries that do not comply with the legislation, according to the extent of the violation. The municipality would be responsible for imposing fines that would then be set aside for its own operations. The municipality also had the right to close production plants temporarily or permanently. The process of imposing these fines, however, is very slow. The fines are extremely low and can be applied "as many times as necessary" and, as a result, industries find it cheaper to pay the monthly fine rather than to adopt pollution control measures. Although this has financial benefits for the municipality, it undermines the main objective of the fine, which is environmental protection.

Source: Margulis, 1994

6.4.1   Cost-effectiveness

In selecting instruments, it is important to select those that achieve the desired outcome at the least possible cost and with a total cost that does not exceed the expected benefits. In theory, market-based policies offer the "least-cost" solution to environmental problems, but there is relatively little experience in using them, particularly for pollution charges on industry. Overall, the optimal instrument is one that leads to the so called "win–win" solutions, i.e. improvements in the environment and other sectors of the economy occur simultaneously and therefore do not involve difficult development–environment trade-offs. Although there will be winners and losers in almost all environmental decisions, some actions can bring about substantial social benefits with a minimum of cost, such as accelerating provision of clean water and sanitation.

6.4.2   Administrative and financial feasibility

An instrument should be selected only if the responsible agencies are prepared to deal with the often complex procedures required for implementing them properly, such as billing and collecting taxes and charges, measuring emissions, determining environmental effects, and taking the necessary enforcement action for non-compliance. All of these require good co-ordination between government agencies. Instruments that require strong enforcement capacity or a high rate of voluntary compliance are difficult to implement.

6.4.3   Consistency with other objectives

The chosen instrument should be consistent with other policies and instruments within or external to the sector. For example, the application of the instrument should not lead to cross-media pollution or conflict with relevant national laws, international agreements, treaties or principles. Moreover, no system of pollution charges or other economic instruments can change the underlying political climate. If a government gives priority to maintaining production and employment, then environmental policies that threaten these goals will be ignored. In addition, adopting policies that are not enforced will merely undermine the credibility of the environmental authorities and the government in general.

6.4.4   Equity

Equity considerations should be carefully balanced with environmental factors when selecting instruments. A major policy question when considering any tax system is who, ultimately, will bear the burden of the tax? Or, does the tax fall proportionately more on the rich or the poor? Most proposals for environmental taxes involve either taxes on environmentally harmful consumption or taxes paid by industrial polluters that may be passed on to consumers through higher prices. Poor people spend a larger percentage of their income on consumption of goods than do the wealthy and, therefore, consumption-based taxes affect the poor disproportionately. To avoid this situation, policy makers should ensure appropriate sharing of the costs and benefits of environmental protection, paying particular attention to the poor. For example, requiring private organisations to absorb the full costs of pollution abatement shifts the burden from those who normally suffer from environmental degradation (usually the poor) to those responsible for causing it (i.e. industry) and, eventually, the consumer of polluting goods.

6.4.5   Transparency

The process of adopting and implementing standards must be transparent so that enterprises can adapt to changing regulatory conditions. Enterprises and other stakeholders are more likely to comply with instruments when they understand how they were derived. In the case of an environmental charge, the polluter knows both the costs of investing in pollution abatement and the tax that would need to be paid if current levels of pollution continue. By contrast, in a tradable permit system, the polluter does not have advance knowledge of the price that the market might assign to permits in the future.

6.4.6 Flexibility
The flexibility of the instrument in adapting to a changing environment can be an important consideration where there are changing local conditions. For example, depending on local political conditions, changing a charge rate may be more easily accomplished than changing legislation, except of course if the rates are set within the legislation. Environmental taxes also confer, on producers and consumers, the flexibility needed to minimise the costs of achieving a given goal. Faced with an emission tax, for example, each enterprise can compare various ways of reducing emissions and choose the solutions that match its own circumstances. The various measures include changing the product mix, modifying production technologies and installing equipment that can filter or clean end-of-pipe discharges. To the extent that different organisations can have different costs for pollution abatement, a charge can encourage those facing lower abatement costs to go further in cleaning up their operations.

6.5   Application in developing countries
Despite growing evidence that environmental degradation is an important socio-economic problem, governments in developing countries have been unsuccessful in stopping it. A common argument is that environmental control is too costly and that countries should concentrate on other development priorities. Underlying such thinking may be a lack of information and insufficient awareness of the true costs involved, together with inertia, lobbying by powerful interest groups, and limited public support and participation. Even where there is strong political will, governments may not be able to act effectively because of institutional deficiencies. Under these unfavourable circumstances, therefore, opportunities for the effective application of economic instruments in developing countries can be very limited. Where they are contemplated, however, policy makers should take into account the following factors:

- *Weak institutional capacity*. Economic instruments cannot be implemented successfully without pre-existing appropriate standards and effective administrative, monitoring, and enforcement capacities. Moreover, there is little difference, if any, in the monitoring and enforcement capability required of government for regulatory and economic instruments. If there is uncertain monitoring and weak enforcement, there is little or no reason for an organisation to report its discharges and pay a fee. Similarly, if discharges are normally made without a permit, organisations will not be motivated to purchase permits or to engage in emission trading. Without existing regulations that establish baseline treatment standards for different kinds of discharges, it will be difficult to determine initial allocations of marketable permits. Moreover, subsidies for less than the total cost of pollution abatement activities will not influence organisations that have no other reason to change their practices. In addition the use of charges for industrial wastewater discharges into municipal sewer systems will be limited.

- *Inadequate co-ordination.* Institutional co-ordination is an important prerequisite for the effective application of most economic instruments. In the case of water management, however, there is often a traditional rivalry between the environmental and water and sanitation agencies. This may be due to a number of reasons such as political power and differing goals and perspectives. Nonetheless, the structure of an effluent charge system involves parameters and information that are more in the domain of the environmental agencies, while the implementation of the system is largely the responsibility of the water and sanitation companies. Unless the relevant agencies are well co-ordinated, the application of effluent charges will be undermined (Margulis, 1994).
- *Economic instability.* Economic stability is critical for the effectiveness of economic instruments. Although regulatory instruments probably depend less on the level of economic stability in a country, charges and taxes are highly dependent on it. For example, Brazil has not been using economic instruments as often as the institutional and legal frameworks would allow, largely because of its unstable economic situation. The fiscal system in the country is very complex and the collection of duties very deficient, and therefore the creation of an environmental tax would only complicate and weaken the system further (Margulis 1994).
- *Government resistance or inertia.* In some countries, there is a general perception by environmental agencies that the use of economic instruments will not only weaken their control over polluters, but that they will have to share their control with economic ministries, who are usually responsible for creating new taxes or charges. The application of economic instruments, therefore, is likely to make environmental agencies even weaker than they already are in most countries. Moreover, the results in terms of pollution levels would be less certain. In other countries, where regulators have relied on standards, inspections and penalties for managing pollution, there is a reluctance to try a new approach unless it is clearly demonstrated to be better than the existing regulatory system.
- *Resistance by polluters.* In developed countries, as in industrial ones, industrial polluters often have resisted economic instruments because they

believe that they have greater negotiating power over the design and implementation of regulations than they do over economic instruments. Moreover, local industries rightly assume that it is easier to avoid compliance with a standard where there is poor monitoring and enforcement capacity, than to avoid fiscal and incentive mechanisms where there is less flexibility.

6.6   Conclusions

Finding the right mix of policy instruments can help to ensure effective water pollution control. In developing countries, cost-effectiveness and administrative capacity are the two most important criteria for selecting them. In every country, however, water pricing policies that may be encouraging over-use and water degradation should be considered first. Although the experience in applying other economic instruments remains limited, particularly in developing countries, there is evidence that effluent and user charges have the most potential for effective application by helping to pay for environmental improvement. Nonetheless, they are not sufficient for achieving water quality objectives. They should be accompanied by investment in wastewater treatment facilities and, locally, by appropriate regulatory instruments as well as programmes to persuade water users to change their polluting behaviour.

6.7   References

Andersen, R. C., Hofmann, L.A. and Rusin. M. 1989 *The Use of Economic Incentive Mechanisms in Environmental Management*. American Petroleum Institute, Washington, D.C.

Braceros H. and Schuddeboom, J. 1994 *A Survey of Effluent Charges and Other Economic Instruments in Dutch Environmental Policy.*

Cadiou, A. and Duc, N.T. 1994 The use of pollution charges in water management in France. In: *Applying Economic Instruments to Environmental Policies in OECD and Dynamic Non-Member Economies.* Organisation for Economic Cooperation and Development, Paris.

Foster, J.D. 1992 The role of the city in environmental management. Paper prepared for USAID Office of Housing and Urban Programs Workshop, Bangkok, Thailand.

Freitas, M.D. 1994 Policy instruments for water management in Brazil. In: *Applying Economic Instruments to Environmental Policies in OECD and Dynamic Non-Member Economies.* Organisation for Economic Co-operation and Development, Paris.

Hahn, R.W. 1989 Economic prescriptions for environmental problems: how the patient followed the doctor's orders. *J. Econ.. Perspec.,* **3**(2).

Kaosa-ard, Mi. and Kositrat, N. 1994 Economic instruments for water resource management in Thailand. In: *Applying Economic Instruments to Environmental Policies in OECD and Dynamic Non-Member Economies.* Organisation for Economic Co-operation and Development, Paris.

Kosmo, M. 1989 Economic incentives and industrial pollution in developing countries. World Bank Environment Department, Division Working Paper No. 1989–2, World Bank, Washington, D.C.

Lovei, M. 1994 Pollution abatement financing: theory and practice. Draft written by Pollution and Economics Division of the Environment Department (ENVPE), The World Bank, Washington, D.C.

Margulis, S. 1994 The use of economic instruments in environmental policies: the experiences of Brazil, Mexico, Chile and Argentina. In: *Applying Economic Instruments to Environmental Policies in OECD and Dynamic Non-Member Economies.* Organisation for Economic Co-operation and Development, Paris.

Moore, J.L. *et al.*, 1989 *Using Incentives for Environmental Protection: An Overview.* Library of Congress, Washington, D.C.

National Academy of Public Administration 1994 *The Environment Goes to the Market.* National Academy of Public Administration, Washington, D.C.

Oates, W.E. 1988 The role of economic incentives in environmental policy. Paper presented at the AEA Session on "Economics and the Environment", December.

O'Connor, D. 1994 The use of economic instruments in environmental management: the East Asian experience. In: *Applying Economic Instruments to Environmental Policies in OECD and Dynamic Non-Member Economies.* Organisation for Economic Co-operation and Development, Paris.

OECD 1989 *Economic Instruments for Environmental Protection.* Organisation for Economic Co-operation and Development, Paris.

OECD 1991 *Environmental Policy: How to Apply Economic Instruments.* Organisation for Economic Co-operation and Development, Paris.

OECD 1994a *Environment and Taxation: The Cases of The Netherlands, Sweden and the United States.* Organisation for Economic Co-operation and Development, Paris.

OECD 1994b *Managing the Environment: The Role of Economic Instruments.* Organisation for Economic Co-operation and Development, Paris.

US EPA 1991 *Economic Incentives: Options for Environmental Protection.* US Environmental Protection Agency, Washington, D.C.

World Bank 1994 Mexico: Integrated Pollution Management. Draft report prepared by Country Department II, Latin America and the Caribbean Regional Office, World Bank, Washington, D.C.

Chapter 7[*]

FINANCING WASTEWATER MANAGEMENT

7.1   Introduction

Urban sanitation is a priority issue for cities everywhere. Major deficiencies in the provision of this basic service contribute to environmental health problems and the degradation of scarce water resources. The rapid growth of cities and the accompanying concentration of population leads to increasing amounts of human wastes that need to be managed safely. The relative success in providing cities with usable water has led to greater volumes of wastewater requiring management, both domestic and industrial. As population densities in cities increase, the volumes of wastewater generated per household exceed the infiltration capacity of local soils and require greater drainage capacity and the introduction of sewer systems. Wastewaters flowing out of cities can, in turn, affect downstream water resources and threaten their sustainable use.

The mix of problems and the capacity to deal with these sanitation problems varies amongst cities and countries. Table 7.1 provides a simple typology of the problems according to national economic development levels. Confronting these problems requires an ability to face a number of challenges, including different environmental health challenges as well as financial, institutional and technical challenges.

7.2   The challenges of urban sanitation

The environmental health challenges facing the urban sanitation subsector in developing countries are of two types (Serageldin, 1994). First, there is the "old agenda" of providing all urban households with adequate sanitation services. Second, there is the "new agenda" of managing urban wastewater safely and protecting the quality of vital water resources for present and future populations. The relative importance of each agenda normally depends

*   *This chapter was prepared by C.R. Bartone and based on Bartone (1995).
The views expressed are solely those of the author and do not necessarily
represent the views of the World Bank or its affiliates*

Table 7.1   Economic–environmental typology of urban sanitation problems

Urban sanitation problems	Lower-income countries (< US$ 650 per capita)	Lower middle-income countries (US$ 650–2,500 per capita)	Upper middle-income countries (US$ 2,500–6,500 per capita)	Upper-income countries (> US$ 6,500 per capita)
Access to basic sanitation services	Low coverage, especially for urban poor; mainly non-sewered options	Low access for urban poor; increasing use of sewerage	Generally accep-table coverage; higher sewerage levels	Good coverage; mainly sewered
Wastewater treatment	Virtually no treatment	Few treatment facilities; poorly operated	Increasing treat-ment capacity; operational deficiencies	Generally high treatment levels; major investments over past 30 years
Water pollution issues	Health problems from inadequate sanitation and raw domestic sewage "in the streets"	Severe health problems from untreated municipal discharge	Severe pollution problems from poorly treated municipal and mixed industrial discharges	Primarily concerned with amenity value and toxic substances

Source: Adapted from Bartone, et al., 1994

upon the level of development as illustrated in Table 7.1, although these two "agendas" coexist in most cities of the developing world, even in some of the most modern cities.

7.2.1   Basic sanitation services for urban households

The provision of sanitation services, including sewerage, has not kept pace with population growth in urban areas. Despite this, the significant progress that was achieved by countries during the 1980s has resulted in a 50 per cent increase in the number of urban people with adequate sanitation facilities (see Figure 7.1). These achievements, although impressive, were not sufficient because the number of people without adequate sanitation actually increased by 70 million in the same period, and as many remained unserved as were provided with service. The results of a recent survey by the World Health Organization (WHO) and the United Nations Children's Fund (UNICEF) in 63 countries are shown in Figure 7.2 (WHO/UNICEF, 1993). These results distinguish between the type of sanitation services reaching the upper and lower income urban populations.

The health consequences of the service shortfalls are enormous and fall most heavily on the urban poor. In most low-income communities, the pollut-ant of primary concern is human excreta. It has been reported by WHO that 3.2 million children under the age of five die each year in the developing world from diarrhoeal diseases, largely as a result of poor sanitation,

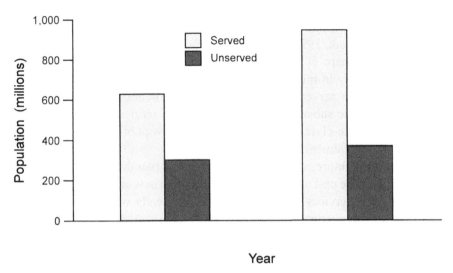

Figure 7.1 Access to urban sanitation in developing countries, 1980–90 (After World Bank, 1992)

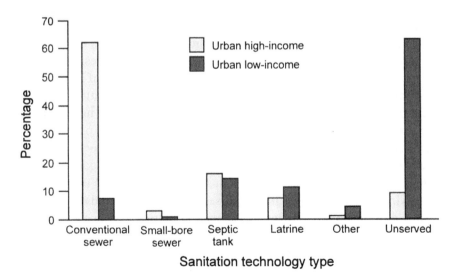

Figure 7.2 Urban sanitation by technology type and income (After WHO/UNICEF, 1993)

contaminated drinking water and associated problems of food hygiene (WHO, 1992). Infectious and parasitic diseases linked to contaminated water are the third leading cause of productive years lost to morbidity and mortality in the developing world (World Bank, 1993a). Diarrhoeal death rates are typically about 60 per cent lower among children living in households with adequate water and sanitation facilities than those in households without such facilities (World Bank, 1992).

An increasing share of urban sanitation services are being provided by sewerage, especially in middle-income countries. About 40 per cent of the urban population is served by sewers. User contributions, however, have been low and public subsidies for these household services have benefited primarily the middle-class and rich. This has left few public resources to be spent on sewage treatment and safe disposal.

Looking to the future, the challenge of the next two decades dwarfs the progress made in the past decade; some 1,300 million new urban residents will require sanitation services in addition to those presently without service. In total, this is roughly six times the increase in service provided during the 1980s. Clearly, the aim of providing all urban households with adequate sanitation services still poses large financial, institutional and technical challenges.

7.2.2 Urban wastewater management and pollution control

A "new agenda" of environmentally sustainable development has emerged forcefully, and appropriately, in recent years. One aspect of sustainable development is the quality of the water environment which is seen as a global concern about sustainable water resources. The situation in cities in developing countries is especially acute. Even in middle-income countries, sewage is rarely treated. Buenos Aires, for example, treats only 2 per cent of its sewage, a percentage that is typical for the middle-income countries of Latin America. There is also the problem of uncontrolled industrial discharges into municipal sewers, increasing organic loads and introducing a range of chemical contaminants that can damage sewers, interrupt treatment processes, and create toxic and other hazards. As shown in Figure 7.3, water quality is far worse in developing countries than in industrialised countries. Furthermore, while environmental quality in industrialised countries improved through the 1980s, it did not improve in middle-income countries, and even declined sharply in lower-income countries.

The costs of this degradation can be seen in many ways. The vast majority of rivers in and around cities in developing countries are little more than open sewers. Not only do these degrade the aesthetic quality of life in the city, but they constitute a reservoir for cholera and other water-related diseases. The cause of the major outbreak of cholera in Peru in 1991 could be traced to inadequate urban sanitation and water contamination. It cost the Peruvian economy over US$ 150 million in 1991–92 in direct and indirect health impacts (WASH, 1993). Similarly, the otherwise inexplicable persistence of typhoid in Santiago over four decades has been attributed to the pollution of

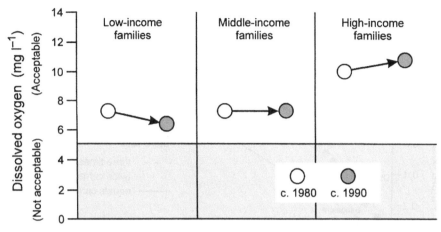

Figure 7.3 Dissolved oxygen concentrations in rivers in developing and developed countries (After World Bank, 1992)

irrigation waters by untreated metropolitan discharges (Ferreccio, 1995). Energetic emergency measures, taken as a result of the Latin American cholera outbreak in 1991, prevented the spread of cholera in Santiago and brought typhoid under control with estimated savings in direct and indirect health costs in the order of US$ 77 million (World Bank, 1994c). The costs of urban water pollution also create an additional burden for cities in the form of higher water supply costs (Figure 7.4). In metropolitan Lima, for example, the cost of upstream pollution has increased water treatment costs by about 30 per cent. In Shanghai, China, water intakes had to be moved upstream more than 40 km at a cost of about US$ 300 million (World Bank, 1992).

7.2.3   Connection between sanitation services and environmental issues

To understand the connection between sanitation services and environmental issues, it is necessary to consider the sequence in which people demand water supply and sanitation services. For a family which migrates into a shanty-town, the first environmental priority is to secure an adequate water supply at reasonable cost. This is followed shortly by the need to secure a private, convenient and sanitary place for defecation. Families show a high willingness to pay for these household or private services, in part because the alternatives are so costly. Accordingly, they pressure local and national governments to provide such services, and in the early stages of economic development much external assistance goes to meeting the strong demand for these services. The very success in meeting these primary needs, however, gives rise to a second generation of demands, namely for the removal of

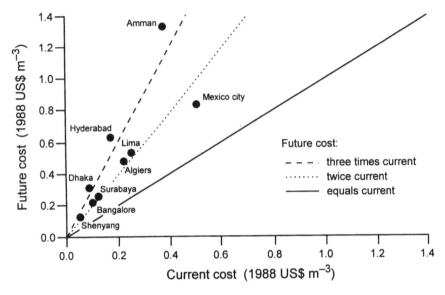

Figure 7.4 How the cost of water supply is increasing (After World Bank, 1992)

wastewater from the household, then from the neighbourhood and then from the city. As cities succeed in meeting this demand another problem arises, namely the protection of the environment from the degrading effects of such large and concentrated pollution loads.

This succession of demands has been observed in the historic experience of the industrialised countries and in the contemporary experience of developing countries. Thus it is no surprise that the portfolio of external assistance agencies has focused heavily on the provision of water supply. For example, World Bank lending for water and sanitation over the past 30 years has only included about 15 per cent for sanitation and sewerage, with most of this spent on sewage collection and only a small fraction spent on treatment. In a description of the Orangi Pilot Project in Karachi, Pakistan, Hasan (1995) describes how forcefully poor people demand environmental services, once the primary demand for water supply is met, and how it is possible to respond to the challenge of these new demands.

7.3 The financial challenges

Completing the supply of basic sanitation services and making progress on wastewater management and pollution control creates major financial challenges for developing countries. Mobilising the necessary financial resources requires both recognising the need for an urban sanitation subsector and reliance on new ways of financing urban sanitation, sewerage and wastewater management.

7.3.1   Responding to the demands of households and communities

In recent years there has been a remarkable consensus on market-friendly and environment-friendly policies for managing water resources and for delivering water and sanitation services on an efficient, equitable and sustainable basis. At the heart of this consensus are three closely related guiding principles expressed at the 1992 Dublin International Conference on Water and the Environment, namely:

- *The ecosystem principle.* Planners and policy makers at all levels should take a holistic approach linking social and economic management with protection of natural systems.
- *The institutional principle.* Water development and management should be based on a participatory approach, involving user, planners and policy makers at all levels, with decisions taken at the lowest appropriate level.
- *The instrument principle.* Water has an economic value in all its competing uses and should be recognised as an economic good.

The challenge facing the urban sanitation subsector is to put these general principles into operation and to translate them into practice on the ground. The new consensus gives prime importance to a central principle of public finance, i.e. that efficiency and equity both require that private resources should be used for financing private goods and that public resources should be used only for financing public goods. Implicit in this principle is a belief that social units themselves, whether households, commercial organisations, urban communities or river basin associations, are in the best position to weigh the costs and benefits of different levels of investment. The vital issue in the application of this principle to the urban sanitation subsector is the definition of the decision unit and the definition of what is internal (private) and external (public) to that unit.

It is useful to think of the different levels at which such units may be defined, as illustrated in Figure 7.5. For each level, the demand for sanitation services must be understood, and each social unit should pay for the direct service benefits it receives. To illustrate the application of this emerging ideal, it is necessary to consider how urban sanitation should be financed.

7.3.2   Sanitation, sewerage and wastewater management

The benefits from improved sanitation, and therefore the appropriate financing arrangements, are complex. At the lowest level (see Figure 7.5), households

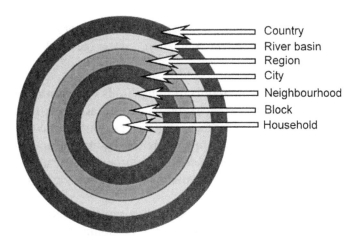

Country
River basin
Region
City
Neighbourhood
Block
Household

Figure 7.5 Levels of decision-making on water and sanitation (After Serageldin, 1994)

place high value on sanitation services that provide them with a private, convenient and odour-free facility which removes excreta and wastewater from the property or confines it appropriately on-site. However, there are clearly benefits which accrue at a more aggregate level and are, therefore, "externalities" from the point of view of the household. Willingness-to-pay studies (see, for example, Ducci (1991)) have shown consistently that households are willing to pay for the first category of service benefits, but have little or no interest in paying for external (environmental) benefits that they consider beyond their concern.

At the next level (i.e. the block) households in a particular block value services which remove excreta from the block as a whole. Moving up a level, to that of the neighbourhood, residents value services which remove excreta and wastewater from the neighbourhood, or which render these wastes innocuous through treatment. Similarly, at the level of the city, the removal and/or treatment of wastes from the city and its surroundings are valued. Cities, however, do not exist in isolation — wastes discharged from one city pollute the water supply of downstream cities and of other users. Accordingly, groups of cities (as well as farms and industries and others) in a river basin can perceive the collective benefit of environmental improvement. Finally, because the health and well-being of a nation as a whole may be affected by environmental degradation in one particular river basin, there are sometimes additional national economic, health and environmental benefits from wastewater management in that basin. The example of typhoid in Santiago (World Bank, 1994c; Ferreccio, 1995) illustrates the latter point.

The fundamental principle of public finance is that costs should be assigned to different levels in this hierarchy according to the benefits accruing at the different levels. This suggests that the financing of sanitation, sewerage and wastewater treatment should be allocated approximately as follows:

- Households pay the cost incurred in providing on-site facilities (bathrooms, toilets, sewerage connections).
- The residents of a block collectively pay the additional cost incurred in collecting the wastes from individual homes and transporting these to the boundary of the block.
- The residents of a neighbourhood collectively pay the additional cost incurred in collecting the wastes from blocks and transporting these to the boundary of the neighbourhood (or of treating the neighbourhood wastes).
- The residents of a city collectively pay the additional cost incurred in collecting the wastes from blocks and transporting these to the boundary of the city (or of treating the city wastes).
- The stakeholders in a river basin (cities, farmers, industries and environmentalists) collectively assess the value of different levels of water quality within a basin and decide on the level of quality they wish to pay for, and on the distribution of responsibility for paying for the necessary treatment and water quality management activities.
- The nation, for the achievement of broader public health or environmental benefits, may decide to pay collectively for meeting more stringent treatment standards.

Sanitation and sewerage
Although there are complicating factors to be taken into account (including transaction costs of collection of revenues at different levels and the interconnectedness of several of the benefits), the principles discussed above are reflected both in the way some industrialised countries finance sewerage investments and in the most innovative and appropriate forms of subsector financing observed in developing countries. In many communities in the USA, for example, households and commercial organisations pay for sewer connections, primary sewer networks are financed by a sewer levy charged to all property owners along the streets served, and secondary sewers and major collectors and interceptors are often financed by improvement levies on all property owners in the serviced areas.

Innovative sewerage financing schemes are now being observed in developing country cities. In Orangi, an informal urban settlement in Karachi, a hierarchical scheme for financing sewerage services has developed in

Box 7.1　　The condominial sewerage system in Brazil

The "condominial" system is the brainchild of Jose Carlos de Melo, a socially committed engineer from Recife. The name condominial was given for two reasons. First, a block of houses was treated like a horizontal apartment building (or condominial in Portuguese) (see figure). Second, "Condominial" was a popular Brazilian soap opera and associated with the best in urban life. As is evident in the figure, the result is a radically different layout, with a shorter grid of smaller and shallower "feeder" sewers running through the backyards and with the effects of shallower connections to the mains rippling through the system. These innovations cut construction costs to between 20 and 30 per cent of those of a conventional system.

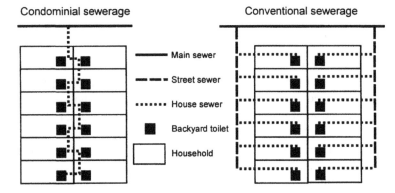

The more fundamental and radical innovation, however, is the active involvement of the population in choosing their level of service, and in operating and maintaining the "feeder" infrastructure. The key elements are that families can choose to continue with their current sanitation system, to

Continued

which households pay the costs of their "on-lot" (i.e. on-site) services (e.g. latrines and septic tanks), the primary sewers are paid for by the households along the "lane" (public passageway between rows of houses), contiguous "lanes" pool their resources to pay for neighbourhood sewers, and the city (via the Municipal Development Authority) pays for trunk sewers (Hasan, 1995). The arrangements for financing condominial sewers by the urban poor in Brazil (see Box 7.1) follow a remarkably similar pattern; households pay for the on-site costs, blocks pay for the block sewers (and decide what level of

Box 7.1 Continued

connect to a conventional waterborne system or to connect to a condominial system. If a family chooses to connect to a condominial system, it has to pay a connection charge, which can be financed by the water company, and a monthly tariff. If on the other hand, the family wants a conventional connection, it has to pay an initial cost and a monthly tariff (both of which are about three times higher) reflecting the different capital and operating costs. Families are free to continue with their current system, which usually means a holding tank discharging into an open street drain. In most cases, however, those families who, initially, chose not to connect eventually end up connecting. Either they succumb to heavy pressure from their neighbours or they find the build-up of wastewater in and around their houses intolerable once the (connected) neighbours fill in the rest of the open drain.

Individual households are responsible for maintaining the feeder sewers, with the formal agency maintaining only the trunk mains. This increases the communities' sense of responsibility for the system. Also, the misuse of any portion of the feeder system, for example by putting solid waste down the toilet, soon shows up in a blockage in the neighbour's portion of the sewer. The rapid, direct and informed feedback to the misuser virtually eliminates the need to educate the users of the system in the "acceptable and unacceptable" and results in fewer blockages than in conventional systems. Finally, because of the greatly reduced responsibility of the wastewater utility, its operating costs are sharply reduced.

The condominial system is now providing service to hundreds of thousands of urban people in northeast Brazil and is being replicated on a large scale throughout the country. The danger, however, is that the clever engineering is seen as "the system". Where the community and organisational aspects have been missing, the technology has worked poorly (as in Joinville, Santa Catarina) or not at all (as in the Baixada Fluminense in Rio de Janeiro).

Source: Briscoe, 1993; de Melo, 1985

service they want from these), with the water company or municipality paying for the trunk sewers.

Lack of access to credit may impede investment in sanitation, drainage and other essential urban environmental services, especially in small cities and towns. This problem has been overcome in some cases by creating special municipal development funds or rotating funds to finance environmental investments. For example, the World Bank has supported the creation of municipal development funds in the State of Minas Gerais, Brazil, for

---

Box 7.2　　Co-operative Housing Foundation Sanitation Loan Programme
in Honduras

Noting the need and demand for sanitary improvements, the Co-operative
Housing Foundation (CHF), an international NGO, helped to establish a lending programme for various types of latrines and toilets, showers and laundry
and wash areas. A sanitation loan fund was created to make small, short-term
loans that are affordable to informal settlement residents around Tegucigalpa.
Loans range in size from US$ 100–400 and are made through local non-
governmental organisations (NGOs) (i.e. non-traditional finance organisations).
The loans are based on several important principles, which include matching
the loan amount with the expected result and securing the loan through
community-based mechanisms (for example by co-signing) rather than the
traditional mortgage approach. The key elements of the Honduras model are:
- It is responsive to individual and community demand.
- It includes a sustainable revolving loan programme.
- It emphasises local NGO capacity enhancement.
- It seeks to stimulate the local economy.
- A range of technologies are offered.
- Health education is a condition (integral part) of the loan.

Source: Hermanson, 1994

---

environmental improvements in small cities and towns, and in Mexico
for municipal water supply, sewerage and solid waste investments in intermediate cities.

Similarly, poor urban households need mechanisms to finance sewer
connections and in-home sanitary facilities. Some cities provide credit to
poor households for these investments that can be paid off in instalment payments (not subsidised) over periods of three to five years. Where there are
well-managed water and sewerage utilities, the instalment payments can be
collected as part of the monthly water bill. In some cases, households can provide "sweat equity" (labour inputs provided by the community for self-help
construction schemes) or even make partial payment in the form of construction materials. A special sanitation credit fund has been established in
Honduras (Box 7.2) for poor urban households, fashioned along the lines of
the well-known Grameen rural credit bank in Bangladesh. Such experiences
show that the urban poor will invest in a healthier environment if they can
spread the initial costs over time. Similarly, innovative schemes for providing

urban households access to credit for sanitation investments have been demonstrated in Lesotho (Blackett, 1994) and in Burkina Faso (Ouayoro, 1995).

Wastewater treatment
Even when the appropriate financing and institutional principles are followed, very difficult issues can still arise with respect to the financing of wastewater treatment facilities. In industrial countries, two very different models are used.

In many industrialised countries, the approach followed has been to set universal environmental standards and then to raise the funds necessary to finance the required investments. It is becoming increasingly evident that such an approach is proving to be very expensive and not financially feasible, even in the richest countries of the world. In the UK, the target date for compliance with the water quality standards of the European Union (EU) is being reviewed as customers' bills rise astronomically to pay the huge costs involved (over US$ 60,000 million this decade). In the USA, US$ 56,000 million in federal construction grants were provided to local governments from 1972–89 to build mandated secondary treatment facilities, but these grants have now been eliminated (and replaced by State revolving funds for loans to municipalities) at the same time that increasingly stringent environmental standards are being proposed. Many local governments are now refusing to comply with the unfunded mandates of the Federal Government (Austin, 1994). The city of San Diego, for example, has refused to spend US$ 5,000 million on federally-mandated secondary treatment, arguing that it is more cost-effective to use long, coastal outfalls for sewage disposal. San Diego brought suit against the Federal Government and recently won its case in the federal courts (Mearns, 1994). The US National Research Council has advocated a change in which costs and benefits are both taken into account in the management of sewage, with a shift to a water quality-based approach at the coastal zone, watershed or basin level (National Research Council, 1993).

In a few countries, a different model has been developed. In these countries, river basin institutions have been put into place which:
- Ensure broad participation in the setting of standards, and in making the trade-offs between cost and water quality.
- Ensure that available resources are spent on those investments which yield the highest environmental return.
- Use economic instruments to encourage users and polluters to reduce the adverse environmental impacts of their activities.

These institutional arrangements are described more fully below. In river basins in Germany and France, and more recently in Brazil, river basin financing and management models are applied in order to raise resources for wastewater treatment and water quality management from users and polluters in the basin. The stakeholders, including users and polluters as well as citizens' groups, are involved in deciding the level of resources to be raised and the consequent level of environmental quality they wish to "purchase". This system has proved to be efficient, robust and flexible in meeting the financing needs of the densely industrialised Ruhr Valley for 80 years, and for the whole of France since the early 1960s (see Box 7.3).

There is growing evidence that if such participatory agencies were developed, people would be willing to pay substantial amounts for environmental improvement, even in developing countries (Serageldin, 1994). In the state of Espirito Santo in Brazil, a household survey showed that families were willing to pay 1.4 times the cost of sewage collection systems, but 2.3 times the higher cost of a sewage collection and treatment system. In the Rio Doce Valley, an industrial basin of nearly three million people in south-east Brazil, a river basin authority (like those in France) is in the process of being developed. Stakeholders have indicated that they are willing to pay about US$ 1,000 million over a five-year period for environmental improvement. In the Philippines, recent surveys show that households are often prepared to make substantial payments for investments which will improve the quality of nearby lakes and rivers.

For developing countries, the implications of the experience of industrialised countries are clear. Even rich countries manage to treat only a part of their sewage, e.g. only 52 per cent of sewage is treated in France and only 66 per cent in Canada. As in the USA, Japan and France, most countries have provided some form of environmental grants to municipalities in order to achieve their present levels of treatment. Given the very low initial levels in developing countries (e.g. only about 2 per cent of wastewater was treated in Latin America at the beginning of the decade) and the vital importance of improving the quality of the aquatic environment, an approach is needed that simultaneously makes the best use of available resources and provides incentives to polluters to reduce the loads they impose on surface and groundwaters.

An effluent tax is one form of incentive that is used in many countries, ranging from France, Germany and The Netherlands to China and Mexico. It can be applied to any dischargers, cities or industries, with two benefits; it induces waste reduction and treatment and can provide a source of revenue for financing wastewater treatment investments (see Chapter 6). The dramatic impact of the Dutch effluent tax on industrial discharges is

Box 7.3   Water resource financing through river basin agencies in Germany and France

The Ruhrverband
The Ruhr Basin, which has a population of about five million, contains the densest agglomeration of industrial and housing estates in Germany. The Ruhrverband is a self-governing public body which has managed water in the Ruhr Basin for 80 years. There are 985 users and polluters of water (including communities, districts, and trade and industrial enterprises) who are "Associates" of the Ruhrverband. The highest decision-making body of the Ruhrverband is the assembly of associates, which has the fundamental task of setting the budget (of about US$ 400 million annually), fixing standards and deciding on the charges to be levied on users and polluters. The Ruhrverband itself is responsible for the "trunk infrastructure" (the design, construction and operation of reservoirs and waste treatment facilities), while communities are responsible for the "feeder infrastructure" (the collection of wastewater).

The French River Basin Financing Agencies
In the 1950s it became evident that France needed a new water resources management structure capable of managing the emerging problems of water quality and quantity successfully. The French modelled their system closely on the principles of the Ruhrverband, but applied these principles on a national basis. Each of the six river basins in France is governed by a Basin Committee, also known as a "Water Parliament", which comprises between 60 and 110 persons who represent all stakeholders, i.e. national, regional and local government, industrial and agricultural interests and citizens. The Basin Committee is supported by a technical and financial Basin Agency. The fundamental technical tasks of the Basin Agency are to determine how any particular level of financial resources should be spent (e.g. where treatment plants should be located and what level of treatment should be undertaken) so that environmental benefits are maximised, and what degree of environmental quality any particular level of financial resources can "buy". On the basis of this information, the Water Parliament decides on the desirable combination of costs and environmental quality for their (basin) society, and how this will be financed, relying heavily on charges levied on users and polluters. The fundamental financial task of the Basin Agency is to administer the collection and distribution of these revenues.

In the French system, in contrast to the Ruhrverband, most of the resources that are collected are passed back to municipalities and industries for investments in the agreed-upon water and wastewater management facilities.

Source: Briscoe and Garn, 1994

Table 7.2   Impact of the effluent tax system introduced in the Netherlands on pollution loads (10^6 population equivalents)

	1969	1975	1980	1985
Domestic discharges	12.5	13.3	14.3	14.5
Industrial discharges	33.0	19.7	13.7	11.3
Total discharges	45.5	33.0	28.0	25.8
Removed by wastewater treatment plants	5.5	8.7	12.6	14.5
Remaining pollution	40.0	24.3	15.4	11.3

Source: Jansen (1991)

described by Jansen (1991). The results given in Table 7.2 show that overall industrial effluent loads decreased by two-thirds between 1969, when an effluent tax was first applied, and 1985 (falling from 33 million to 11 million population equivalents). The experience of China in the application of an industrial effluent tax for financing industrial wastewater management improvements has been described by Suzhen (1995). In France and Mexico, the effluent tax is applied equally to municipal and industrial effluents, thus encouraging local investment in municipal wastewater treatment plants. An effluent tax, however, should be used in combination with municipal sewer use charges in order to ensure that industries do not escape paying for their discharges by passing the cost on to the municipality, as well as to ensure that the municipal sewerage authority has sufficient revenues to build and to operate sewerage and treatment works.

7.3.3   Community participation
The aspiration of most urban households, including the urban poor, is to have access to cost-effective and affordable sanitation services via public or private utilities. Consequently, they would be willing to participate, as responsible users, by paying the appropriate service charges. In the cities of many developing countries, however, such services are not yet universally accessible and poor communities must, themselves, get involved in the planning and delivery of sanitation and sewerage options.

The examples of the condominial sewer system in Brazil and the Orangi Pilot Project indicate an important institutional approach to community participation in which a productive partnership is formed between community groups and the municipal government or the utility. Often, such a system involves public provision of the external or trunk infrastructure, which may be operated by either the public or private sector, and the community providing and managing the internal or feeder infrastructure. The link between feeder and trunk infrastructure is essential for the evacuation and disposal of human waste collected by the community, but it is too easily overlooked. Many forms of community participation are possible for the provision of sanitation and sewerage services, such as:

- Information gathering on community conditions, needs and impact assessments.
- Articulation of, and advocacy for, local preferences and priorities.
- Consultations concerning programmes, projects and policies.
- Involvement in the selection and design of interventions.
- Contribution of "sweat equity" or management of project implementation.
- Information dissemination.
- Monitoring and evaluation of interventions.

Promoting and enabling community participation can take many forms. Where political will exists, governments may promote participation and create the conditions under which communities and households, as well as NGOs and the private sector, can play their appropriate roles. The World Bank-financed PROSANEAR project in Brazil (Box 7.4), for example, provides a framework and the resources for municipalities and utilities to experiment with innovative technical and institutional arrangements for providing sanitation services to the urban poor. When such government support is absent, alternative approaches have commonly been used to stimulate community involvement and to build the necessary political will. First, NGOs or community-based organisations (CBOs) often play a catalytic role in mobilising communities and forming partnerships. In one of the largest scale examples involving an NGO, Sulabh Shauchalaya International began, in 1970, promoting the construction of pour-flush latrines in Delhi and other Indian cities, and over a period of 20 years assisted in building over 660,000 private latrines and 2,500 public toilet complexes with community participation and government support (NIUA, 1990). Second, consultations and town meetings are increasingly used as a forum to discuss and agree on environmental priorities, and to propose participatory solutions (Bartone *et al.*, 1994). Finally, communities may engage in public protests or legal actions as a means of building a constituency of the urban poor, and applying pressure on local governments and utilities for dialogue and action. The Orangi Pilot Project (see section 7.2.3) had its origins in the discontent of local residents with excreta and wastewater overflowing in the streets as a result of the failure of the Karachi Development Authority to provide adequate sewerage (Hasan, 1995).

Box 7.4   The PROSANEAR Project in Brazil

The World Bank, in collaboration with the Brazilian Government, has financed the PROSANEAR project as a means of addressing the complex issues of water and sanitation service provision in low income neighbourhoods. The project tests technical and institutional solutions in these *favelas*, without any pre-established "plan" in terms of service levels, delivery systems and targets. About US$ 100 million of investments are providing water and sanitation infrastructure to about 800,000 *favela* residents in 11 cities, using a radically different approach compared with other projects. State water and sewerage companies are encouraged to try out flexible, adaptive and participatory project designs, so that projects are based on what the poor residents want and are willing to pay for.

The PROSANEAR project, which reached its peak implementation period during 1992–95, provided convincing evidence of the advantages of following a participatory and flexible approach. At the very least the per capita investment costs have averaged about one half the investment cost "ceilings" of US$ 140 for sewerage that the state water and sewerage companies were allowed by the project loan agreements. These dramatic reductions in costs can be attributed to several factors:

- Sub-projects were encouraged to build upon localised, but significant, Brazilian experiences of the past two decades with intermediate technical solutions.
- State companies were required by project rules to consult with CBOs (such as church groups, resident associations and women's' groups) at every stage, from design to construction.
- Participation was further re-enforced by requiring the state companies to award project design consultancies to consortia of engineering companies and companies or NGOs specialising in community participation, rather than just to the former.
- Project design consultants and state water company engineers were actively supervised by the national project management team (in Caixa Economica Federal), so that proposals on service levels, technology, construction schedules, cost recovery arrangements, billing and other details were finalised only after active negotiations with communities.
- Close supervision of bidding documents ensured that construction contracts were competitive and that construction companies were fully accountable to local communities.

An interesting feature of the PROSANEAR project has been that diverse institutional routes were taken to finalise sub-project designs. At the risk of oversimplification, three models can be identified. One class of "community organisation" models worked out project designs in consultation with leaders of existing community organisations, and then the details with actual beneficiaries. A second class of "direct consultations" models, reached agreement directly between design engineers and affected beneficiaries, with community leaders and organisations retaining a consultative role. In both models, conflicts of interests between the state company and CBOs were resolved through negotiations. The project design consultants functioned as facilitators, with community meetings serving as a type of market surrogate institution. In the third class of "pedagogic" models, training in participatory methods and hygiene education were advocated as the means of raising awareness and building up the ability of the poor communities to confront the established powers and special interest groups.

Source: World Bank, 1994a; Project Supervision Reports

7.3.4   A role for the private sector

Financial resources can also be mobilised through the private sector; poor service provision by the public sector often suggests a need for increasing partnerships with the private sector. Private sector participation, however, is only one possible opportunity; it is not a panacea. In situations in which existing sanitation service delivery is either too costly or inadequate, private sector participation should be examined as a means of enhancing efficiency and lowering costs, and of expanding the resources available for service delivery.

In deciding whether to involve the private sector, it is important to assess several key factors which have been summarised by the *Infrastructure for Development: World Development Report, 1994* (World Bank, 1994a). Introducing competition is the most important step in creating conditions for greater efficiency by both private and public operators; some services can be split into separate operations to help create contestable markets. The principle of accountability to the public should be maintained through transparent contractual agreements that are open to public scrutiny and should help to minimise risks to public welfare, create real competition, ensure efficiency, and promote self-financing. Paradoxically, public sector capacity may have to be strengthened in order to achieve effective private sector participation which requires public sector agencies with sufficient capacity to prepare bidding documents and performance indicators, assess proposed outputs and costs, administer the contracting process, and regulate contract performance.

In Mexico, municipalities are granting concessions to the private sector to build and operate wastewater treatment plants, both as a means of financing investments in plants through the private sector and to overcome problems with weak local operating capacity. The Puerto Vallarta wastewater treatment plant was the first of many new plants to come on line in the past few years (Martin, 1995). An important point to remember in cases such as Puerto Vallarta is that the private sector performs the necessary function of mobilising financing for needed investments, but the investments made together with operations, maintenance and depreciation costs will all have to be recovered through tariffs charged to domestic and industrial customers. Another innovative example is a concession to 26 industries in the Vallejo area of Mexico City to form a new enterprise, Aguas Industriales del Vallejo, to rehabilitate and expand with its own funds an old municipal wastewater treatment plant, treat up to 200 l s^{-1} of sewage, and sell the treated water to shareholders at 75 per cent of the public utility water tariff (IFC, 1992).

Box 7.5   The Strategic Sanitation Plan for Kumasi, Ghana

Kumasi has had 3 master plans in the last 40 years but still has no comprehensive sewerage system. Meanwhile, sanitary conditions continue to deteriorate as the population grows. The residents of Kumasi already pay about US$ 1 million a year to have only 10 per cent of their waste removed from their immediate environment. The current system of human waste management in Kumasi is inadequate; most of the waste removed from public and bucket latrines ends up in nearby streams and in vacant lots within the city limits, creating an environment prone to the spread of disease. With increasing rapid urbanisation and competition for limited resources, there is the fear that the already poor sanitary conditions will worsen if no urgent and rational actions are taken.

In response to the inadequate sanitation conditions prevailing in the city, the Kumasi Metropolitan Area Waste Management Unit, with the assistance of the United Nations Development Programme (UNDP)/World Bank Regional Water and Sanitation Group for West Africa, prepared a Strategic Sanitation Plan (SSP). The SSP reflects the willingness of the Kumasi Metropolitan Assembly to take the institutional and financial actions needed to ensure delivery of affordable sanitation services to all segments of the population by the year 2000. The plan differs from a traditional master plan in that it:

- Tailors recommended technical options to each type of housing in the city.
- Considers user preferences and willingness-to-pay.
- Uses a relatively short planning horizon (10–15 years), emphasising actions that can be taken now.
- Breaks the overall plan into projects that can be implemented independently but which together provide full coverage.

The SSP moves away from reliance on conventional sewerage alone and considers a range of proven technologies that address the needs of all segments of the urban population, recognising resource constraints, and pays due attention to the willingness and capacity of users to pay for improved services.

The strategic planning process being used in Kumasi is dynamic and the SSP itself will evolve as experience is gained. This iterative process began with a pilot project funded by UNDP in which the various technical, institutional and financial issues that are proposed in the SSP are being evaluated and refined. The pilot project is, in fact, the first phase of city-wide implementation to be supported by a World Bank-financed project.

Source: Whittington *et al.*, 1992; KMA, 1993

7.4   Strategic planning and policies for sustainable sanitation services

Applying a strategic planning approach to urban sanitation problems should result in choosing the right policy instruments, agreeing priorities, selecting appropriate standards for service provision, and developing strategic investment and cost recovery programmes. The question of appropriate service standards is a particularly vexing one that, in the end, should be answered by considering user preferences and willingness-to-pay. In a large city with many pockets of poverty, service standards are likely to be spatially differentiated because many households cannot afford conventional sewerage without massive government subsidies. The Kumasi Strategic Sanitation Plan (Box 7.5) provides an example of a differentiated plan matching housing types, income levels and user preference; the plan recommends that sewers be used in tenement areas, latrines in the indigenous areas, and flush toilet/septic tank systems in high income and new government areas. Willingness-to-pay surveys were carried out (Whittington *et al.*, 1992), and the results were used to help define differentiated financing options. Explicit subsidies were targeted to the city's low-income population.

Municipal wastewater treatment is a particularly costly and long-term undertaking so that sound strategic planning and policies for treatment are of special importance. The recently endorsed Environmental Action Programme for Central and Eastern Europe (CEE), formulated with the assistance of the World Bank (1994b), recognises that the CEE countries will require a plan to move towards Western European standards over a period of 15–25 years as financial resources become available. Although urban sewerage levels in the CEE are generally adequate, 40 per cent of the population are not, at present, served by wastewater treatment plants. The domestic pollution load represents 60–80 per cent of the combined municipal and industrial organic waste load in many CEE cities. Furthermore, many of the existing plants are currently overloaded, poorly operated and maintained, or bypassed. The following is a checklist of policy questions posed in the CEE Action Programme to be answered before proceeding with municipal wastewater investments:

- Have measures been taken to reduce domestic and industrial water consumption?
- Has industrial wastewater been pre-treated?
- Is it possible to reuse or recycle wastewater?
- Can the proposed investment be analysed in a river basin context? If so, have the merits of the investment been compared with the benefits from different kinds of investments in other parts of the river basin? (Note that a least-cost solution to achieve improved water quality may involve different, or no, treatment at different locations.)
- Has the most cost-effective treatment option been used to achieve the desired ambient water quality?

- Has there been an economic analysis to assess the benefits (in terms of ambient water quality) that could be achieved by phasing investments over 10 years or more?

7.4.1   Cost-effective technologies

Developing country cities are beginning to recognise that poor urban residents cannot afford, nor do they necessarily want or need, costly conventional sewerage. Beyond the dense urban centres, the average household cost of conventional sewerage may range from US$ 300–1,000. This is clearly too expensive for many households with annual incomes well below US$ 300. Fortunately, a broad range of cost-effective technological options are available to respond to the demands of urban consumers beyond the urban centre, with the potential to reduce costs to the order of US$ 100 per household. The UNDP/World Bank, Water and Sanitation Program has worked with many countries over the past decade to develop, demonstrate, document and replicate many of these low-cost sanitation options. The examples drawn upon throughout this chapter illustrate many of the options available to households (e.g. ventilated improved pit (VIP) latrines in Lesotho, Sulabh pour-flush latrines in India, condominial sewers in Brazil and simplified sewerage in Pakistan), as well as the supporting institutional and financial systems that make possible the wide-scale application of these options.

Wastewater treatment technologies also have a wide range of costs. Conventional treatment processes may cost US$ 0.25–0.50 per cubic metre (Figure 7.6). If non-conventional options can be used, it may be possible to cut these costs by at least one-half. Promising low-cost treatment approaches, especially for small and intermediate cities, range from natural treatment systems (such as waste stabilisation ponds, engineered wetlands systems and even ocean outfalls), to decentralised treatment systems (such as are used in Curitiba, Brazil), to new treatment processes (for example anaerobic treatment processes such as the upflow anaerobic sludge blanket (UASB) reactors presently operating in cities in India, Colombia and Brazil). In large cities, land or other constraints may result in conventional treatment being the most cost-effective approach for achieving the desired water quality objectives, although this should always be a decision resulting from an economic analysis. Lifetime costing should always be used to compare and to choose among treatment options, because operations and maintenance constitute a major share of the costs.

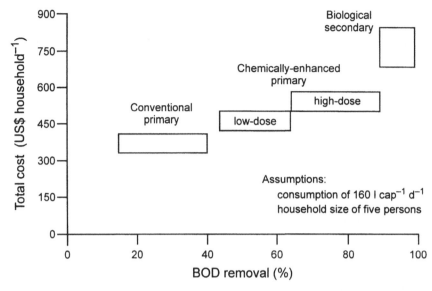

Figure 7.6 The costs of conventional sewage treatment (After National Research Council, 1993)

7.4.2   Conservation and reuse of scarce resources

Cornerstone ecological principles for sustainable cities include the conservation of resources and the minimisation and recycling of wastes. Translating these principles into urban policies for wastewater management should emphasise the strategic importance of water conservation and wastewater reclamation and reuse in cities. Successful conservation and reuse policies, moreover, need to achieve a balance between ecological, public health and economic and financial concerns.

Pricing and demand management are important instruments for encouraging efficient domestic and industrial water-use practices and for reducing wastewater volumes and loads. Water and sewerage fees can induce urban organisations to adopt water-saving technologies, including water recycling and reuse systems, and to minimise or eliminate waste products that would otherwise end up in the effluent stream. In addition to price-based incentives, demand management programmes should include educational and technical components, such as water conservation campaigns, advice to consumers, and promotion, distribution or sale of water-saving devices like "six-litre" toilets which use less than half the volume of water per flush than a standard toilet (World Bank, 1993b).

Wastewater reclamation and reuse is increasingly recognised as a water resources management and environmental protection strategy, especially in arid and semi-arid regions (see Chapter 4). The use of reclaimed urban waste-water for non-potable purposes, such as in-city landscape irrigation and industry or for peri-urban agriculture and aquaculture, offers a new and reli-able resource that can be substituted for existing freshwater sources. Water pollution control efforts can make available treated effluents that can be an economical source of water supply when compared with the increasing expense of developing new sources of water (e.g. Asano, 1994). Conversely, in developing countries only recently embarking on major wastewater treat-ment investments, reuse has the potential to reduce the cost to municipalities of wastewater disposal. A framework for the economic and financial analysis of reuse projects has been provided by Khouri *et al.* (1994) in a planning guide that integrates economic, environmental and health concerns with agronomic concerns for the sound management of crops, soil and water.

7.5   Conclusions

This chapter has identified a number of financial and related challenges facing cities, and countries, as they seek to meet the growing demand of the urban population for sanitation and sewerage services, and for improved wastewater management.

First, cities need to complete the "old agenda" of extending sanitation services to the entire urban population. It is clear that the bulk of the finance for this can, and should, come from users. Achieving this requires provision of the services that people want and are willing to pay for. To assist poor urban households in meeting their sanitation needs, innovative credit mechanisms will also be required. Institutional arrangements should be founded on the principle of shared responsibility, with the devolution of decision-making to the lowest appropriate level; service delivery institutions should be responsive and accountable to users. In many cases, this will involve local partnerships to ensure effective community participation in serv-ice delivery and financing, and a greater role for the private sector in mobilising investment resources. On the technical side, cities should consider strategic sanitation planning in order to match service options to user incomes and preferences, and they should adopt cost-effective technologies to deliver the desired services.

Second, developing country cities are being called upon to embark on the "new agenda" of wastewater treatment and water quality management while still dealing with the "old agenda". This represents an enormous finan-cial challenge, as has been illustrated by the recent experience of the industrialised countries. Difficult choices are being forced on national and local authorities about the level of investment to make in preserving the aquatic environment, about who should pay, and about how to spend available resources. Resource limitations and difficult trade-offs in developing countries reinforce the need to make strategic choices that simultaneously make the best use of available resources and provide incentives to dischargers to reduce their pollution loads, such as using economic instruments like water pricing and pollution taxes. New institutional arrangements are needed, such as river basin associations, that enable stakeholder participation in making the difficult decisions about environmental quality, financing and the allocation of responsibilities for action. Ideally, such arrangements should respect the principle of non-interference in the functioning of municipalities while creating the enabling conditions for them to act as good environmental citizens, for example through financially self-sufficient water and sewerage utilities. New planning approaches are also needed, such as the adoption of strategic planning and policies that establish long-term environmental goals, that identify critical immediate actions and that determine sustainable means of implementation. Finally, greater reliance on conservation and reuse in wastewater management also depends on pricing and demand management.

The challenges are great, but the evidence shows that they are not insurmountable. Meeting them requires the political will and support of urban residents to adopt appropriate investment and cost-recovery policies as well as sustaining the implementation of strategic actions.

7.6   References

Asano, T. 1994 Reusing urban wastewater — an alternative and a reliable water resource. *Wat. Int.,* **19**(1), 36–42.

Austin, T. Roiled waters: water politics in the 1990s. *Civ. Eng.,* **64**(7), 49–51.

Bartone, C. 1995 An overview of urban wastewater and sanitation: responding to growing household and community demand. In: Serageldin, I., Cohen, M. and Sivaramakrishnan, K. [Eds] *The Human Face of the Urban Environment: Proceedings of the Second Annual World Bank Conference on Environmentally Sustainable Development.* Environmentally Sustainable Development Proceedings Series No. 6. World Bank, Washington, D.C., 139–49.

Bartone, C., Bernstein, J., Leitmann, J. and Eigen, J. 1994 *Toward Environmental Strategies for Cities: Policy Considerations for Urban Environmental Management in Developing Countries.* Urban Management Programme Policy Paper No. 18. The World Bank, Washington, D.C.

Blackett, I.C. 1994 *Low-Cost Urban Sanitation in Lesotho.* Water and Sanitation Discussion Paper Series No. 10, UNDP/World Bank Water and Sanitation Program, Washington, D.C.

Briscoe, J. 1993 When the cup is half full. *Environment*, **35**(4), 7–37.

Briscoe, J. and Garn, M. 1994 *Financing Agenda 21: Freshwater.* Transport, Water and Urban Development Department Paper No. TWU OR5. World Bank, Washington, D.C.

de Melo, J.C.R. 1985 Sistemas condominial. *Engen. Sanit.* **24**(2), 237–38.

Ducci, J. 1991 Valuacion contingente y proyectos de Alcantrillado Sanitario: Resumen de algunos estudios de caso. Conference paper for Seminario Evaluacion Economica de Proyectos: La Utilizacion del Metodo de Evaluacion Contigente, Bogota, 6–31 May.

Ferreccio, C. 1995 Santiago, Chile: avoiding an epidemic. In: Serageldin, I., Cohen, M. and Sivaramakrishnan, K. [Eds] *The Human Face of the Urban Environment: Proceedings of the Second Annual World Bank Conference on Environmentally Sustainable Development.* Environmentally Sustainable Development Proceedings Series No. 6. World Bank, Washington, D.C., 160–62.

Hasan, A. 1995 Replicating the low-cost sanitation programme administered by the Orangi Pilot Project in Karachi, Pakistan. In: Serageldin, I., Cohen, M. and Sivaramakrishnan, K. [Eds] *The Human Face of the Urban Environment: Proceedings of the Second Annual World Bank Conference on Environmentally Sustainable Development.* Environmentally Sustainable Development Proceedings Series No. 6. World Bank, Washington, D.C., 150–53.

Hermanson, J.A. 1994 New partnerships for a healthier environment. Paper presented to the International Medical Services for Health (INMED) 5th Millennium Conference: Urban Health Challenges for the 21st Century, Washington, D.C., 22 June.

IFC 1992 *Investing in the Environment: Business Opportunities in Developing Countries.* International Finance Corporation, Washington, D.C.

Jansen, H.M.A. 1991 West European experiences with environmental funds. Institute for Environmental Studies, The Hague, The Netherlands, mimeo, January 1991.

Khouri, N., Kalbermatten, J.M. and Bartone, C.R. 1994 *Reuse of Wastewater in Agriculture: A Guide for Planners.* UNDP/World Bank Water and Sanitation Program, Washington, D.C.

KMA 1993 *Strategic Sanitation Plan for Kumasi.* Draft report, January 1993. Kumasi Metropolitan Assembly, Kumasi, Ghana.

Martin, J. 1995 Sistemas BOOT para plantas de tratamiento de aguas servidas: El caso de Puerto Vallarta, Mexico. Conference paper for International

Seminar on Treatment and Reuse of Urban Wastewater, Santiago, Chile, May 8-12.

Mearns, A.J. 1994 How clean is clean? The battle for Point Loma. *Wat. Env. Res.*, **66**(5), 667–668.

NIUA 1990 *A Revolution in Low Cost Sanitation: Sulabh International.* Case study prepared for the Mega-Cities Project and the Urban Management Program. National Institute of Urban Affairs, New Delhi.

National Research Council 1993 *Managing Wastewater in Coastal Urban Areas.* National Academy Press, Washington, D.C.

Ouayoro, E. 1995 Ouagadougou low-cost sanitation and public information program. In: Serageldin, I., Cohen, M. and Sivaramakrishnan, K. [Eds] *The Human Face of the Urban Environment: Proceedings of the Second Annual World Bank Conference on Environmentally Sustainable Development.* Environmentally Sustainable Development Proceedings Series No. 6. World Bank, Washington, D.C., 154–59.

Serageldin, I. 1994 Water supply, sanitation and environmental sustainability: the financing challenge. Keynote address to The Ministerial Conference on Drinking Water and Environmental Sanitation: Implementing Agenda 21, Noordwijk, the Netherlands, 22–23 March.

Suzhen, Y. 1995 Strategies for controlling industrial wastewater pollution in Beijing. In: Serageldin, I., Cohen, M. and Sivaramakrishnan, K. [Eds] *The Human Face of the Urban Environment: Proceedings of the Second Annual World Bank Conference on Environmentally Sustainable Development.* Environmentally Sustainable Development Proceedings Series No. 6. World Bank, Washington, D.C., 163–68.

WASH 1993 *The Economic Impact of the Cholera Epidemic in Peru: An Application of the Cost of Illness Methodology.* WASH Field Report No. 415. Water and Sanitation for Health, Washington, D.C.

Whittington, D., Lauria, D.T., Wright, A.M., Choe, K., Hughes, J.A. and Swarna, V. 1992 *Household Demand for Improved Sanitation Services: A Case Study of Kumasi, Ghana.* UNDP/World Bank Water and Sanitation Program Report No. 3. Washington, D.C.

World Bank 1992 *Development and the Environment: World Development Report, 1992.* Oxford University Press, New York.

World Bank 1993a *Investing in Health: World Development Report, 1993.* Oxford University Press, New York.

World Bank 1993b Water resources management. Policy Paper, World Bank, Washington, D.C.

World Bank 1994a *Infrastructure for Development: World Development Report, 1994*. Oxford University Press, New York.

World Bank 1994b *Environmental Action Programme for Central and Eastern Europe*. Technical Department, Europe and Central Asia, Middle East and North Africa Regions, Report No. 10603-ECA. World Bank, Washington, D.C.

World Bank 1994c *Managing Environmental Problems in Chile: Economic Analysis of Selected Issues*. Environment and Urban Development Division, Country Department I, Latin American and the Caribbean Region, Report No. 13061-CH. World Bank, Washington, D.C.

WHO 1989 *Health Guidelines for the Use of Wastewater in Agriculture and Aquaculture*. Technical Report Series 778, World Health Organization, Geneva.

WHO 1992 *Our Planet, Our Health*. World Health Organization, Geneva.

WHO/UNICEF 1993 *Water Supply and Sanitation Monitoring Report 1993: Sector Status as of 31 December 1991*. World Health Organization/United Nations Children's Fund Joint Monitoring Programme, Geneva.

Chapter 8[*]

INSTITUTIONAL ARRANGEMENTS

8.1   Introduction

Water pollution control is typically one of the responsibilities of a government as it aims to protect the environment for the good of the general public. Governments undertake to do this by establishing an appropriate set of organisations and launching specific programmes. These interventions aim at achieving national, or even regional, objectives that include, for example, enhanced economic productivity, public health and well-being (all of which should, ideally, form part of a sustainable development strategy). To meet these objectives resources are mobilised, notably financial resources (capital from local people, government and the market), physical resources (raw materials and agricultural products), environmental resources (such as water) and human resources (the active time and capabilities of people). These resources are scarce and have an associated cost, therefore their use must be efficient, that is maximum output (e.g. highest water quality) must be achieved at minimum resource input. Alternatively, it may be more important to organise the pollution control sector in such a way that governmental policy is implemented effectively; for example that wastewater treatment plants are actually built and operated or that sanitation facilities, once constructed, are actually used and remain maintained. Effective implementation can be extremely difficult, especially for pollution control. In reality, wastewater control always receives the lowest priority, although its infrastructure is at least as expensive as that for water supply.

Water is an environmental resource with a profound impact on public health, economic activity and environmental (and ecosystem) quality. Therefore, the prerequisite for any sustainable development scenario is that the organisations that are assigned with water management actually possess the capability to carry out this task. A well-balanced arrangement of flexible, dynamic organisations and other related institutions is the best assurance that

---

[*]   *This chapter was prepared by G.J. Alaerts*

unpolluted water resources remain available in the future, that the right quantity and quality of water are delivered to the water users (including the ecosystems), and that people can live in a healthy environment. These organisations, however, can only execute these functions if they have access to an appropriate financial base to expand and maintain the infrastructure, to attract qualified professionals, and to prepare well for the future.

8.2   The water pollution control sub-sector

The organisational structure and the administrative procedures to implement water pollution control are very much determined by the characteristics of the sub-sector and the functions to be performed. These differ between countries, as well as over time. Over the past decades, industrialised countries have learnt that water resources, although finite, must keep satisfying a variety of user demands (such as water supply, irrigation, amenity) and that they need protection (ICWE, 1992; World Bank, 1993). They have also learnt that different types of pollution (e.g. domestic or industrial) demand specific approaches and that pollution prevention is more cost-effective than the removal of the pollutants by end-of-pipe treatment (see Chapter 3). In addition, water pollution control is intricately linked to the work of other sub-sectors, particularly environmental management, water resources management, industrial development, and land use and urban management.

The water pollution control sub-sector typically concerns itself with four functions that are relatively distinct and that require specific expertise (see Chapter 1):

- Water quality management of water resources such as rivers, lakes and wetlands. This involves setting of operational quality standards for the receiving water as well as for the waste discharged, and integrated planning in order to achieve water quality levels that allow appropriate water use (e.g. for the production of drinking water, fish cultivation, navigation) (see Chapters 2 and 5).
- Regulation of general quality standards for health, water and the environment. Regulation and setting of standards for industrial sewage treatment and stimulation of waste minimisation and pollution prevention instead of conventional "end-of-pipe" approaches.
- Organisation, construction and management of on-site sanitation in rural and peri-urban areas.
- Collection and off-site centralised treatment of domestic sewage, including its planning, construction and management.

The physical and socio-economic conditions of a country dictate which functions must take priority and hence determine the preferred institutional arrangement. Sometimes these functions are best served by two or more separate entities, because each function requires a specific mandate, organisational structure and procedures, as well as specific technical expertise.

The first two functions listed above are of a regulatory nature and the last two are executive. In most countries, setting discharge and water quality regulations has proved to be the easiest (and cheaper) aspect. The execution of the, relatively more, capital-intensive investment programmes in cities and towns has been much more difficult to achieve or even to initiate. In addition, in many countries, much of the new wastewater infrastructure ends up poorly operated and maintained, thereby lowering its effectiveness dramatically. Large and comparatively wealthy industries are often the first to build and operate treatment plants, while the majority of smaller industries find it exceedingly difficult to comply with standards.

On-site sanitation comprises a set of distinct activities. Much of the work is carried out by house-owners who have to invest in the construction of septic tanks or pit latrines. The maintenance, mainly desludging and disposal and treatment of the sludge, is usually carried out by private contractors. The sector organisations are responsible for ensuring that government targets are met by devising adequate building regulations and city ordinances, and through a strong, facilitating role. In most countries this is also an arduous task.

8.3   Institutions and organisations
Before discussing the role of institutions and organisations in water pollution control activities, it is first necessary to distinguish between them and to recognise that the function of all institutional factors goes well beyond the boundaries of the common, typical "sector organisations". Institutions are defined as the "rules" in any kind of social structure, i.e. the laws, regulations and their enforcement, agreements and procedures (see for example Uphoff, 1986; Israel, 1987; de Capitani and North, 1994). Organisations are a particular type of institution and are composed of groups of people with a common objective. Organisations can be formalised, such as "official" sector organisations with operational objectives, their own budget and professional staff (such as water departments in Government Ministries, Water Boards, Environmental Protection Agencies, laboratories, consultant companies) or they can be informal and less well described (such as "the public", the "customers" who pay for a water service, the socio-economic distinct groups in a village or town community).

The success achieved when implementing a government's policy for water pollution control primarily depends on the suitability of the chosen institutional arrangement. Other factors are also important prerequisites, such as availability of capital, of technology and of human resources (expertise). Generally, however, the maximum benefit can only be generated from available resources by an "optimum" institutional arrangement that makes the resources work effectively for the sub-sector. This "optimum" depends on the characteristics of the sub-sector, which differ from those of other water-using sub-sectors, such as water supply or hydropower, and the requirements of the country. Good institutional arrangements are essential to liberate and to develop resources further; for example to make more finance available by increasing the willingness of customers and citizens to pay for sewerage services or to educate and train the professional staff.

A sector can only prepare and manage its programmes properly if all institutions are appropriately involved in the three main phases; planning, implementation (construction), and operation and maintenance linked with cost recovery. Although this is normal for formal organisations such as government departments, it is also true for all other institutions that are indirectly implicated and will affect, in one way or another, the water pollution programme. Examples of such institutions are:

- Policies and regulations that determine tariff-setting and taxation. These commonly fall outside the jurisdiction of pollution control organisations, although their success depends on their financial strength. Responsibility for decision-making commonly lies with the Ministry of Finance, in municipalities or amongst the politicians.
- Enforcement of regulations and laws. Any pollution control law is only as strong as the will and the capability of the law enforcement institutions.
- Human resources and development of expertise. Pollution control is technically complicated and, therefore, education and research institutions must be able to support a national pollution control policy.
- Mechanisms to render organisations more responsive to customer demands, flexible and accountable. This generally requires devolution of decision-making and financial autonomy to the most appropriate, lower levels of administrative government. It can also lead to the inclusion of private partners. Rules that stifle initiative and good performance should be removed (deregulation) and replaced by other regulations that, typically, are based more on performance. Again, the required institutional framework is determined outside the environmental or water sector.
- Mechanisms that enable the definition of the economic value to the nation of good water quality. This requires a full appreciation and understanding of water uses and their significance for the nation's long-term sustainable development.

A crucial institution to the success of water pollution control is the group of people that will "benefit" from it. World-wide, numerous water supply and sanitation schemes have failed completely, or partially, because the designated users (and financial supporters) of the new infrastructure were not consulted about whether they valued the initiative and would be willing to contribute for its proper operation. Thus, inadequate involvement of the users during the planning phase created a situation with a lack of demand. Provision of a service, such as a clean environment, is not merely a question of meeting a presumed demand from customers. Without a clearly expressed demand, customers are not committed to the infrastructure and they will fail to use it properly or to pay a reasonable compensation for it. An existing demand may be insufficiently developed, for example, because prospective customers have not recognised the long-term benefits of the service (good public health or education) or because they may prefer "purchasing status" (increasing their consumer goods) rather than investing in the long-term benefits. Consequently, demand may need to be developed.

8.4 Criteria and determinants

No fixed, optimum model for institutional arrangements exists that would suit all countries, at all times. The organisations that would fulfil the requirements best in a given country and in a particular period of its development, depend on the local characteristics, i.e. the hydrogeology and topography, industrialisation, culture, economy and the natural environment. The institutional arrangement of a sub-sector will have to adjust continuously because the institutional environment around the sub-sector changes so much. Preferably this arrangement should prepare for and facilitate continuing change. Inevitably, institutional arrangements are very case specific; what works for one country in a given period may be detrimental to another. Nevertheless, experience suggests that good arrangements consist of a number of standard institutional components (e.g. organisation types, financial measures) that perform well in different arrangements. The determinants for these arrangements are usually external boundary conditions with which the sub-sector has to be able to work. Criteria are often derived from business and public administration and specify how a successful sector, and performing organisations, should be managed.

8.4.1 Prioritising functions and setting mandates of organisations

First of all, the priority issues for water pollution control in the medium term (with a planning horizon of 10–20 years) need to be determined. Countries with a high population density and high industrial output require a different approach from others which are predominantly rural and less industrialised. In the same way, arid regions may put a high priority on water conservation and re-use. Other regions may have to cope with the diverse effects of multifarious wastewater constituents that have long-term deleterious effects, sometimes at locations very distant from the discharge point. For example, the nutrients discharged by households along the Rhine River in Switzerland cause algal blooms along the Danish North Sea coast triggering oxygen deficiency and fish kills, and polychlorinated biphenyls (PCBs) discharged in Europe may, over a period of years, accumulate in the fatty tissue of seals near the North Pole. Institutional arrangements must reflect environmental priorities.

It is commonly assumed that water pollution control requires the same institutional arrangements as for water supply. However, often this is not the case. In many countries, domestic wastewater collection and treatment are administered within the same organisation as water supply, for example in India, Uganda, China, Brazil (some regions), Mozambique, Yemen, the Philippines, and England and Wales. In other countries, separate organisations have been created, such as in Indonesia (for the urban areas), Colombia, Argentina, and most West African and Western European countries. The executive functions for large infrastructure development, and for its management, commonly fall with an engineering-based government department, board, authority or enterprise. These can take many forms (see section 8.5). By contrast, the executive function of on-site sanitation is often best associated with urban management authorities that hold the mandate for land-use planning and housing regulations. Most urban authorities, unfortunately, show little interest in, or understanding of, water pollution control. In addition, they feel less accountable to the national goals of environmental management and, typically, limit their interventions to removing the local pollution to the border of the city. Similarly, urban planning authorities can force industries and workshops to move out from the inhabited areas into designated industrial zones, where they are (in theory) best equipped to separate and contain domestic and industrial wastewater flows (a condition for adequate water pollution control). The function of water quality management is often carried out by a government department but in many instances the management function has been taken up by the infrastructure organisation, especially when it covers a territory large enough to encompass a whole natural water system (e.g. a river basin). Finally, regulatory functions are typically the responsibility of a national government ministry (health or environment) although in some cases they are delegated to a full government

---

Box 8.1   Operation and maintenance and cost recovery are two sides of the same coin

The World Bank, when monitoring projects, insists on good accounting and financial procedures. However, financial indicators such as cost recovery ratio and balance of payment can, when monitored over four or five years, hide structural weaknesses. An organisation can spend most of the recovered charges on hiring unqualified staff, while at the same time postponing essential maintenance. Thus it may as well remain totally unprepared for imminent major problems (such as eutrophication in a lake that should provide millions with good drinking water). The monitoring of key financial indicators is only appropriate if complemented with data on institutional performance, particularly capacity to improve in the future.

---

agency (such as the Environmental Protection Agencies in the USA and China, and the Pollution Control Board in India).

A second major consideration concerns the prioritisation of investment (construction) or operation and management (O&M). Sustainability is served by institutions that ensure the infrastructure serves a long, active life. Well-operated and maintained devices minimise resource losses due to spillage, breakage and leakage. Poor O&M also leads to a poor service to the consumer. Clogged drains and pumps, and treatment works that are out of order, provide an unreliable and low-level service that severely reduces the consumer's and citizen's willingness to pay.

In many countries, the O&M of the water infrastructure is very weak. This is worrying because it renders many water organisations unable to recover the costs (including asset depreciation) of their water supply operations, let alone their sewerage operations. The consensus of opinion suggests that, in a healthy sub-sector, the water organisations should be able, in the long run, to recover full costs from their consumers. In many developing countries, the organisations need to be re-orientated and retrained to execute this task more efficiently (see section 8.5.8). Wastewater infrastructure, in particular, is an unpopular item on the budgets of authorities and citizens alike. As of today, wastewater treatment costs in several European countries have still not been fully recovered from consumers. Operation and maintenance is an expensive, yet unforgiving, item on the budget of any enterprise and is often neglected at the expense of the cost-recovery performance shown in an enterprise's accounts (Box 8.1). In many instances, a well-defined construction mandate (typical for many organisations in developing countries) is not particularly compatible with a cost recovery and O&M mandate. Often, a concentrated investment effort necessitates setting up a devoted organisation for a specific time period (see for example Case Study I, India, and section 8.5.5 for Aquafin in Belgium).

8.4.2   Scale and scope of organisations and decentralisation

The required sector organisations can be of different scale and scope. The scale reflects the typical size of the area for which the organisation has a mandate. This can range from small, such as a city quarter or village, to very large, the size of a country or state of over 100 million inhabitants within the country, e.g. India). The scope of the organisation defines whether it concentrates on (an aspect of) water pollution control or whether it also covers other utilities. Other utilities can be more or less related to wastewater, such as water supply, drainage, water quality management, river basin management, power generation and/or distribution, public transportation, environment protection. Importantly, because much O&M and cost recovery is physically associated with fine-detailed reticulated networks and individualised households, decentralisation or devolution of responsibilities to the lowest appropriate administrative level is an important guideline (ICWE, 1992). Part of the local network or infrastructure can then be entrusted to a local water users association.

Determining the preferred scale and scope depends on the local characteristics of the water sector, the possible interactions with developments in other sectors such as power, and the identified priorities; it also depends on the national policy on state organisation (see section 8.5). In many European countries there is, at present, a process of concentration (scale increase, sometimes with a broadening of scope). The rationale behind this development is that wastewater management, together with water supply, is increasingly complex in respect of technical expertise and water resources management. To cope with this, the organisations need strong and expensive central engineering and laboratory facilities, they need to be able to raise large sums of money, and they must be in a position to co-ordinate the works in a whole region efficiently. Interestingly, within a period of barely 15 years, England and Wales have changed the scale and scope of their water-related organisations twice (see section 8.5.1). Figure 8.1 provides an overview of possible situations.

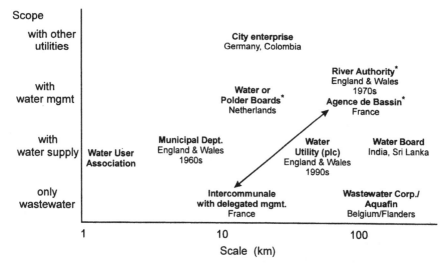

Figure 8.1 Examples of scale and scope of the organisation responsible for waste-water management. Organisations with a purely regulatory function are excluded. The water quality management function is covered by the organisations marked with an asterisk. The double arrow connects, for France, the two complementary organisations that together cover the sector

8.4.3   Deregulation and regulation and enterprise autonomy

Institutional architecture should from one perspective ensure consistency of policy over the whole territory, and from the other it should allow for sufficient flexibility, particularly in order to respond to local issues and demands and to adapt to changing conditions in the country. The first requirement calls for a centralised, top-down approach, with adequate control from the top. The second, however, tends to put more responsibility at the local levels and calls for more local and sub-sectoral autonomy. While accepting that much of the work needs to be carried out by a variety of organisations at different levels, governments tend to keep control by means of regulations. For example, governments define national health and environmental quality standards and personnel structures in the public service, decide on the targets for pollution control achievements, set price structures and may attribute the market mechanisms a major or minor role and, importantly, decide on who will take the important decisions. Experience over the past decades has shown that too much regulation is inefficient, it creates its own distortions and stifles initiatives for improvement.

Mechanisms to reduce the level of top-down regulation include:

- Decentralisation and devolution of decision making to lower administrative levels, including the right to raise finance (e.g. through tariffs).
- Wastewater utilities, and in some cases water quality management organisations, allowed to operate as autonomous entities, i.e. they can decide on tariff structures and personnel management without explicit interference by the local or central government.
- Involve private partners to carry out (part of the) management, bring in finance, or buy the assets (infrastructure, land, the organisation) and operate them as a private company. These alternatives, with increasing private sector involvement, are called leasing, concession and privatisation.
- Identify (waste)water rights and allow their owners to trade them on the basis of their market values.
- Avoid introduction of measures such as subsidies or taxes that may distort the price-value ratio of the water as it is perceived by the water user.
- Apply financial (dis)incentives rather than inflexible command-and-control regulations to control, for example, waste discharges (see Chapter 6).

Although the purpose of deregulation is to allow decision-making outside direct government control, national government does retain an important policy making and monitoring function and, in particular, is responsible for the functioning of the sectoral organisations. Deregulation, therefore, must be compensated by other types of regulation. Typical regulations include:

- Installing mutual control amongst the organisations by creating open competition, such as by tendering out all government contracts to private, as well as to semi-governmental, enterprises.
- Installing mutual control amongst the organisations by creating watchdog organisations and balancing the power of one organisation with that of another; for example by putting a powerful, objective regulatory agency in place (as in England and Wales following privatisation, see section 8.5.1). Whatever the situation, an executive organisation should be prevented from empowering and regulating itself (as was the situation with the Water Authorities in England and Wales in the 1970s, see section 8.5.1) because this creates internal conflicts of interest.
- Ensuring that utilities which benefit from a higher degree of autonomy are also more accountable to their clients, to their shareholders (commonly local government) and to the national government with respect to their support for achieving national goals.
- Preventing monopoly and cartel formation. Recent European Union (EU) legislation forbids cartel formation and attempts to break up monopolies, including those of the water services.

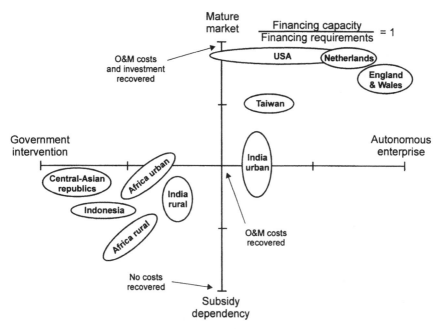

Figure 8.2 The relationship between national water sector organisations as a function of their autonomy and the development of the water services "market". A "mature" market implies that the willingness-to-pay of the consumers balances the financing requirements.

The degree of desired autonomy for an organisation is related to the "maturity" of the market, i.e. the willingness of the consumers to pay for the service. Figure 8.2 charts the relationship of a number of national institutional arrangements with respect to the degree of autonomy in their waste(water) sector organisation and the maturity of the market. A proportionality becomes apparent where local organisations are more autonomous where the market is mature and the demand is more developed. Arguably, England and Wales have the highest degree of autonomy, because their organisations are privatised and operate as independent companies. Most probably, maturity and autonomy must be developed in a co-ordinated fashion and must mutually reinforce each other. An organisation which is suddenly cut off from regular subsidies has no option other than to educate its consumers. Autonomy is measured by the absence of political interference in an organisation and not simply by its "name"; for example, city departments in Western Europe are allowed more true managerial autonomy than governmental enterprises in developing countries.

8.4.4 Capable organisations

Sector organisations can only perform well if they are properly managed, guided and staffed. This implies that:

- Management must offer leadership, to ensure that the organisation and its staff have a clear and shared view of their purpose and how this will be achieved.
- Staff must be adequate and with the right combination of levels of expertise.
- Personnel management must be dynamic, stimulating loyalty and minimising operational cost.

Instruments to further this include career development and salary measures to motivate staff to improve their performance, education and training (see section 8.5.8), and management consultancy. In France, it is argued that the system of delegated management (see section 8.5.2) allows municipal governments to concentrate on policy making and essential tasks, while technical management is left to private organisations that are more expert and better equipped for this purpose.

Sustainable institutions, in addition, possess built-in capacity to monitor critically the overall contribution of the sub-sector to the achievement of the nation's goals, and to influence these goals for the better, for example by introducing the economic replacement value of water and environmental quality in national economic planning, and by demonstrating the economic value of water for sustainable economic development. Such institutions possess the internal mechanisms that enable them to review the management performance and the effectiveness of the separate organisations and institutional measures. Ideally, an organisation should be allowed to operate in an institutional environment such that, without government interference, it gives maximum performance under its present mandate, it learns from errors and improves on its weaknesses, and it is able to identify the future requirements of the sector and to propose the new concomitant institutional arrangements (even if that means abolishing the organisation and replacing it with another).

8.5   Examples of institutional arrangements

8.5.1   England and Wales

In recent years England and Wales have gone through four phases of institutional arrangements. Before 1972, water pollution control infrastructure was under the responsibility of, and was owned by, local government departments, and was often combined with the water supply sub-sector. This led to serious inefficiencies because each municipality had its own small treatment plant and there was no critical mass of technical expertise and financial support. Regulation and water quality management rested with Inspectorates and the River Authorities (one for each of the nine major river basins).

Between 1972 and 1982 nine Water Authorities were created and all infra-structure, with the exception of local sewerage, was transferred to the new authorities in order to increase the scale of the organisations and to bring all water management functions into single entities. This led to the merger of many sub-sectors, including drainage and river management, and brought the regulatory and executive functions together, thus broadening their scope (for more detail see Okun, 1977). The newly created organisations proved too large and unfocused, struggling with internal conflicts of interest, and unable to generate sufficient investment to meet increasing environmental quality standards.

Between 1982 and 1989, the Water Authorities were made more business orientated in order to increase their efficiency as well as their effectiveness. In addition, they were placed primarily under the supervision of the national environment ministry. Preparations were made for privatisation. After 1989, the Government sold the water supply and wastewater infrastructure of the Water Authorities to public and private investors. These private enterprises remain operating in the same river basins. One of their main tasks is to gener-ate finance for the overdue expansion and modernisation of the water and wastewater infrastructure in order to meet the strict EU environmental direc-tives. As a result, tariffs have been raised. The regulatory and water quality management functions were taken over by the National Rivers Authority (NRA), which is also responsible for river management, and by the Inspecto-rates of the environment and of health. The enterprises are allowed to operate as monopolies within their region and, therefore, the new Office of Water (Ofwat) was created as a financial regulator (under the Ministry of Environ-ment) to ensure that water companies meet government policy, and that they do not exploit their monopolistic position at the expense of the citizens or the nations. It is a matter of continuing debate whether this arrangement is considered successful.

In 1996 the water quality regulatory function of the NRA was merged with air and soil quality regulatory functions from the Inspectorates to create an American-style environmental protection agency (known as the Environment Agency).

8.5.2   France

In 1982, the French state structure was fundamentally altered by a decentralisation law that devolved a substantial part of the central government to local government. Traditionally, France had been strongly centralised, but the municipalities were now attributed more responsibility for infrastructure planning and financing. In addition, economic development and water management required a new regional approach with more integration between sectors. Thus, the new law allowed municipalities and Départements (counties) to develop appropriate institutions.

Wastewater collection and treatment is the responsibility of municipalities, which commonly make joint-ventures (intercommunales) to execute this task. However, in most cases the actual management (operation, maintenance and cost recovery) is delegated to private enterprises. Five such companies operate in France and compete with each other during the frequent public tendering of contracts, for example for operation and maintenance, all over the country. Such contracts are very specific, stipulating what the municipality wants the contractor to achieve in a given period of time (5–20 years) and the associated performance parameters. A water price is agreed, from which the contractor has to recover costs and pay a lease fee to the municipality. The contractor can carry out management tasks on the infrastructure owned by the municipality (lease), or it can also provide financing for investment which reverts after a suitable period to municipal ownership (concession) (Lorrain, 1995). Water quality management and regulation is carried out by the Agences de Bassin (river basin boards) which carry out planning, collect fees for abstraction and pollution of the water resources, and also provide subsidies to local government for wastewater infrastructure (Chéret, 1993). Quality standards are developed by the Ministry of Environment.

8.5.3   Germany

Wastewater management is the responsibility of the municipalities in Germany. If they are too small to address the financial and technical complexity of this task, the municipalities form Verbände (inter-municipal joint-venture autonomous enterprises) or, in the case of cities, the various utilities are amalgamated into one Stadtwerke (City Enterprise) encompassing water supply, power distribution, district heating, (often) sewerage and wastewater treatment and, importantly, public transport. The shares of such municipal enterprises are in the hands of the municipalities. The management has a large degree of autonomy, although critical decisions need approval by the board in which the representatives of the municipal enterprises have a majority. The enterprise is subject to taxation on any profits. However, because public transport and sewerage typically lose money, whereas power distribution and water supply commonly yield a benefit, the net profit is zero and taxation is avoided.

Depending on the local topography and pollution load, joint-ventures may be created, based on river basins, to manage water and wastewater, including the operation of treatment works. The Emscher Genossenschaft (Treatment Association for the Ems River) in the industrial heartland of the Ruhr region has an unusual arrangement, insofar as local municipalities (in proportion to their population), industries and other partners form a fully autonomous "water parliament". This "water parliament" undertakes to collect all domestic, and part of the industrial, sewage in the basin and, after pre-treatment, to treat it centrally near the mouth of the Ems in the Rhine. This arrangement has operated for almost a century although, currently, environmental quality is considered to be better served by providing more specialised decentralised treatment. Regulation and part of the water quality management are carried out by the Land's (State) Environment Department and in the Federal Ministry of Environment.

8.5.4 The Netherlands
Historically, The Netherlands has been very much influenced by the need to safeguard its low-lying lands from flooding from the sea or large rivers (Rhine, Meuse and Scheldt). Seventy per cent of the territory needs infrastructure to protect against floods, and the large areas of polders require continuous drainage and meticulous water management. Since the 12th century Polder Boards have been operational. These were unusual because they represented a separate line of local government; the councils of these boards were, and still are, composed of representatives elected by ballot by all those with a commercial or residential interest within the confines of the polder area. In return, all these groups pay a substantial contribution for dike maintenance and water management. After the 1950s, the task of water quality management and wastewater management, with a few exceptions, automatically became a new mandate of the newly-named Water Boards. The local sewerage remained the responsibility of the technical departments of municipalities. The boards cover an area of half to one full province, typically with half a million inhabitants. A move towards an increase in scale (mergers) started recently, in order to pool technical expertise and financial strength, and to allow a more integrated approach for complete water systems (e.g. inter-related canals, lakes).

The present water boards are not owned by local or national government, but have built up their own financial resources and institutional position. All polluting units in the country (households, industries and farms) pay a waste-water conveyance and treatment contribution which is added to the water supply bill and allows full cost recovery of all wastewater infrastructure. The boards also serve as water quality managers and, as such, report to the Ministry of Transportation and Water Management. Regulations are issued by this Ministry as well as by the Ministry of Environment.

8.5.5 Belgium, Flanders

Since 1986, Belgium has been a federal country, of which Flanders is the northern region. Flanders consists of five provinces with approximately five million inhabitants. In the early 1950s a comprehensive pollution control law was adopted investing the municipalities with the responsibility to treat sewage. However, although most industries gradually installed treatment works, reduced their pollution production or closed down, most domestic wastewater remained untreated due to the lack of institutional mechanisms to make municipalities co-operate, and due to the lack of financial means and political will. In the 1970s two regional governmental agencies were set up by national and provincial authorities to combine water quality management and wastewater management. This attempt again failed to produce more than a small proportion of the badly needed investments, partly because the country as a whole was in a state of re-organisation (with devolution of power to the regions) and partly because the government agencies could not generate the required finance. In 1989 the two agencies were reorganised into a "mixed" autonomous investment organisation, known as Aquafin, in which the regional government (responsible for 51 per cent) and a private partner co-operate, and into a Regional Wastewater Corporation (which became the Flemish Environmental Agency after 1992) for water quality management and operation of infrastructure. The private partner is one of the English private water companies which contributes technical expertise and substantial finance, for which it is compensated through tariffs. National and regional Ministries of Environment are responsible for regulation.

8.5.6 India

India must address the deficient sanitary conditions of the poor rural areas and urban squatter zones simultaneously with the industrialised and urbanised regions. Institutional analysis shows an allocation of mandates as illustrated in Figure 8.3.

	Regulation	Integrated planning	Construction	Operation of cost recovery
Rural and peri-urban	—	—	State Water Corp./Board	State Water Corp./Board; Local Govt
Urban	State PCB; CPCB	Min. Urb.Constr.; Min. Water Res.; State Water Corp./Board	State Water Corp./Board	Local Govt
Industrial	State PCB; CPCB	—	Industry	Industry

Figure 8.3 Typical mandate allocation amongst organisations for sanitation and wastewater management in India. The shaded area indicates the fields with comparatively weak effectiveness due to sub-optimal mandate definition and/or inappropriate organisational capacity. PCB: Pollution Control Board; CPCB: Central Pollution Control Board

Regulation and standard setting have achieved much progress and can be considered well organised. The Central and the State Pollution Control Boards were already functional by the 1960s. In the 1970s a basic comprehensive water quality standards system (MInimimal NAtional Standards — MINAS) was established which, among other things, specifies quality standards depending on the intended use of the water, and sets discharge standards that are specific for each industrial sector. These boards also regulate air and soil quality and monitor quality trends. The boards have been instrumental in forcing large factories to install primary or more advanced treatment, although they will not take any responsibility for the execution of the treatment programmes. Their effectiveness can be attributed, in part, to their clear, simple focus and well demarcated tasks, and to the relatively small size and high degree of professionalism which facilitate their management.

In the large cities, such as New Delhi, Bombay, Madras and Calcutta, city departments or corporations are responsible for drainage, sewerage, sanitation and sewage treatment. In the rest of the territory this responsibility falls with the state water boards or corporations, such as the Jal Nigam in Uttar Pradesh, and the Panchayat Raj Engineering Department in Andra Pradesh. However, these state organisations are primarily structured and equipped to develop and execute new construction schemes. Water supply and wastewater infrastructure for the larger towns, once built, are handed over to local government for O&M (local government is also supposed to take care of cost recovery). In the rural areas the state agencies retain responsibility for O&M.

Implementation has proved to be more difficult than regulation. The state boards and corporations were effective in the planning and construction of water supply and drainage, but progress has been below expectation for collecting and treating urban sewage and for providing sustainable water supplies and sanitation to rural communities. A key reason for the first deficiency is the very weak technological and managerial capacity at the level of local government, especially the capacity to recover (high) costs from the city population. Local water supply and sewerage corporations have a weak financial basis, poor personnel management and suffer from continuing political interference. In most cities and towns they resort to continuous crisis management. In the rural areas, these boards and corporations are ill-equipped to communicate with the local communities, decide on the service level for which the communities are willing to pay, involve them in the planning of the scheme and, importantly, organise and train them to assume responsibility for some of the local management and collection of fees. Some state boards are now experimenting with schemes to delegate more power to the district level.

The Indian Government has followed an alternative path in order to by-pass the institutional weaknesses. In 1986 the then Prime Minister, Rajiv Gandhi, launched a separate, high-profile and devoted programme to "clean up the Holy River Ganges" which would involve the construction of numerous municipal and industrial sewage treatment plants in the river basin (see Case Study I). In the wake of the programme several integrated urban environmental sanitation programmes were developed, made up of sewerage infrastructure as well as water supply, and assistance by government agencies to industry to advise them on the options for minimisation and prevention of waste discharges. This Ganga Action Plan (GAP) has a limited-time mandate and is centrally financed and guided by a special Project Directorate in the Ministry of Environment and Forests, although it is executed by the state and local authorities. One of its components, focusing on one of India's largest and most polluted cities, Kanpur, includes substantial institutional development. The success of the GAP has led to the development, in 1993, of the Yamuna and Gumti Action Plans, and will be expanded into a National Rivers Action Plan (see Case Study I). Operation and maintenance cost recovery is claimed to be complete, although these figures often hide an underestimation of the true costs, such as for major repairs, warehouse stocks, and for qualified and well-paid staff. Plans are being developed for improving cost recovery while at the same time spending more funds on better O&M (Box 8.2).

Box 8.2   Achieving cost recovery and operation and maintenance

Weak organisations may recover part of their costs but may be too political to resist the temptation to use the funds for other purposes. The only escape from the "poor O&M–poor cost recovery" trap is to improve on service incrementally by improving O&M in part of the water pollution network. In this way a better service is delivered and more income is earned, that can be re-invested exclusively in further O&M improvement. To ensure institutional sustainability of the planned, large sewage infrastructure of the city of Kanpur (Uttar Pradesh), a phased programme with set targets was devised (Anon, 1993). At present the infrastructure suffers from poor, if any, maintenance and low technical standards and, because of the low service levels and frequent breakdowns, consumers are dissatisfied and unwilling to pay fees. The city corporation lacks professional capacity, despite being overstaffed, and is highly political. The programme for the city of Kanpur comprises five steps to improve gradually the operational efficiency, consumer satisfaction and, hence, cost recovery (see table below). The increased financial means will allow further quality improvement.

Step	Targets	Time-frame
1	Sub-standard O&M with poor service delivery for basic services. Partial cost recovery of O&M and substantial state subsidies. State pays for investment and O&M of sewage treatment	Present
2	Sub-standard O&M but with marginally improved service delivery (water supply and sewerage) to a target area. Full cost recovery for O&M. State pays for sewage treatment	Feasible in short term: 3–5 years
3	Systematically improved O&M with better service delivery of basic services. Full O&M cost recovery. State pays for sewage treatment	Feasible in medium term: 4–10 years
4	As for step 3. Assets partially, to completely, depreciated and debt for investment serviced. State pays for sewage treatment	Feasible in longer term: 8–15 years
5	As for step 3. Complete depreciation of all assets and debt servicing, including for major expenditure on pumping stations and wastewater treatment	Not feasible in foreseeable future; to remain centrally subsidised

The fact that full, local cost recovery of wastewater treatment may not be feasible in the foreseeable future is not surprising because in some rich Western European countries this expensive part of the infrastructure is also still subsidised from central funds.

In the mean time, on-site sanitation retains a low priority in Urban Development Departments. The understanding of water management, and also of community management, remains poor. Nonetheless, several promising initiatives are being taken, particularly those involving the local urban communities in planning and operational phases. In addition, the tendering of concessions to private companies and non-governmental organisations (NGOs) for the installation and operation of blocks with lavatories and bathing facilities are being relatively successful.

8.5.7 South Korea: towards institutions for sustainable management
South Korea went through rapid changes in its institutional arrangement between 1985 and 1995. This was spurred by the country's rapid economic development and the associated pollution pressure. In addition, the country is comparatively poorly endowed with freshwater resources, all of which are intensively used. The development process led to increasing scale and scope within the water pollution control organisations and necessitated an integral water management concept.

In 1985, urban wastewater collection and treatment were mandated exclusively to the municipalities. These were faced with the need for major investments. The typical sub-sectoral approach (with limited vision on long-term sustainability) taken at that time is illustrated by, for example, the hydraulic design guidelines for sewers and sewage works. These were based on a projected linear increase of water consumption from 100–440 litres per capita per day. However, it was not recognised that the available water resources would not be able to sustain this level of consumption beyond the foreseeable future. Similarly, the ensuing treatment works would be so costly that, at best, only secondary sewage treatment would be possible, followed by discharge to coastal waters (because most cities lie close to the coast). However, the coastal ecosystems which supported the harvesting of sea kelp (an important economic activity) would be badly affected by the nutrient-rich effluents from the secondary treatment plants.

To integrate water and wastewater planning and management more effectively, a National Water Improvement Program was developed at national level in 1990. In 1992, region-specific Catchment Water Quality Master Plans were drafted by the Ministry of Public Works and in co-ordination with other ministries. The plans attempted to avoid resource losses and minimise expenditure. This regional planning and co-financing of infrastructure works is administered by Catchment Authorities that direct and complement municipal initiatives. As a consequence, as of 1994, the cities of Kwangju and Seoul envisaged the application of more modest hydraulic design guidelines, with the full reuse of sewage in nearby agriculture, the avoidance of any nutrient disposal in coastal waters, and with much lower investments in wastewater infrastructure.

8.5.8 Sri Lanka: turning an organisation around

Between 1985 and 1991 the United States Agency for International Development (USAID) assisted a major institutional development programme with the Water Supply and Drainage Board (NWSDB) (Edwards, 1988; Wickremage, 1991). This Board was functioning reasonably well in terms of construction of new schemes, but performance was less than satisfactory in operation and financial viability. In 1983, for example, collections covered only 12 per cent of O&M costs. The basic problem with NWSDB was that it had not been able to adjust to the significant differences brought about by its change from a government department to a public corporation. The new role demanded that its attention be changed from capital projects to O&M and the consumers. Deficiencies included minimal commitment to financial viability, negligible budget discipline, lack of corporate planning, little attention to communities and users, and over-sensitivity to political pressures. These deficiencies could not be overcome without a change in staff attitude supported by new staff skills and organisation procedures. Major objectives of the institutional development programme were:

- Decentralisation of management to regional offices in order to put it closer to the consumers.
- Change of organisational structure and attitudes in order to make O&M the most important mission of NWSDB.
- Close co-operation with Ministry of Health, NGOs and communities to provide co-ordinated support to public health programmes.

The process consisted of consultations, practical and formal training sessions, organisational analysis, and changes in the administrative organisation and procedures. In doing this, a large degree of "ownership" of the staff was created. The most notable changes were decentralisation of financial responsibilities (including setting up an accountability and Management Information System), management skill development, corporate planning (including setting up a Corporate Planning Division), financial viability (including tariff reform and collection efficiency improvement), human resources development (especially in basic management and accounting skills, and exposure programmes abroad), and community participation. The incentive structure for engineers was also revised.

At a cost of US$ 14 million the whole organisation was restructured in six years. After the programme, the performance of NWSDB was vastly improved on all accounts, and it showed a high degree of commitment to public water and health services. Importantly, its managerial system now ensured "institutional sustainability".

8.6   Capacity building

Capacity building in the water sector is a new concept that starts from three premises (Alaerts and Hartvelt, 1996):

- Water is a finite resource, for which numerous users compete, most notably the waste dischargers (who lower the usefulness of the water).
- Water is essential for a healthy economy as well as for the environment and, therefore, it is a resource that should be managed in a sustainable way.
- Institutional rather than technical factors cause weakness in the sector.

Capacity building, therefore, takes a comprehensive look at the sector, analyses its physical and institutional characteristics in detail, defines opportunities and key constraints for sustainable development, and then selects a set of short- and long-term action programmes. Very often the water sector performs poorly because of inappropriate or rigid institutional arrangements. If these can be improved, structural constraints are removed. Water is a finite resource and, therefore, demand management rather than new development is necessary because any additional supply created from a new water development is soon fully used and creates even more demand, which can no longer be fulfilled.

Countries must build "capacities" in order to achieve the goal of good sector development, which is effective in service delivery, efficient in resource use and sustainable. Through the Delft Declaration, the United Nations Development Programme (UNDP) developed the following definitions of the aims of capacity building which are applicable for the water sector (Alaerts *et al.*, 1991):

- Creating an enabling environment with appropriate policy and legal frameworks.
- Institutional development, including community participation.
- Human resources development and strengthening of managerial systems.

Experience, especially in developing countries and in economies in transition, shows that the main tasks ahead can be formulated as follows:

- Price setting, cost recovery and the enforcement of rules, are more difficult to implement than regulation (of water quality, for example) and, therefore, strategies to achieve these deserve priority.
- Many inefficiencies can be improved by allocating the right mandates and by reviewing the performance of the arrangement regularly. This will render organisations more alert and target-orientated.

- In rich as well as in poor countries, organisations must be orientated to the consumers of their "environmental services". In poor countries especially, engineers must be willing and able to co-operate with the community to facilitate O&M and cost recovery.
- Organisations must develop the right expertise profile.

A number of instruments can be applied in capacity building. These are:

- Technical assistance for sector analysis and programme development. Since 1992, UNDP has developed "water sector assessments" which analyse comprehensively national water sectors and which develop a priority action programme. Other agencies, such as The World Bank and the Asian and European Development Banks, are also engaged in similar exercises. Such analyses need to be performed by an interdisciplinary team.
- Technical assistance for institutional change. The expertise for this will differ depending on the institution that is under consideration and it may relate to policy, micro- or macro-economic structures, management systems, and administrative arrangements.
- Training for change at different levels, including decision-makers, senior staff and engineers with managerial assignments, junior staff and engineers with primarily executive tasks, technicians and operators, and other stake-holders (such as care-takers and people in local communities who have undertaken to operate or to manage community-based systems).
- Education of prospective experts who will play a role in the sector. This encompasses physical and technological sciences, as well as financial and administrative management, and behavioural sciences. The water pollution control sub-sector is so complex and develops so fast that in most developing countries not more than 10 per cent of the required technical expertise (as university graduates) is available. Many graduates are inadequately prepared for the tasks in their country (Alaerts, 1991).

8.7 Conclusions

Water pollution control comprises four main functions: water quality management, regulation and standard setting, on-site sanitation, and collection and treatment of domestic and industrial wastewater. Each function needs an appropriate institutional arrangement in order to make the whole sub-sector work effectively. In many instances the regulatory function has proved to be a comparatively easy part of the overall task.

The types of institutional arrangements for water pollution control often differ, but not always, from those for water supply. The "optimal" arrangement depends on the political and institutional environment, the economic policy, the roles and values of water in the country, the local topography and hydrogeology, and the natural environment.

Many types of arrangement exist and could fulfil the necessary requirements. No "ideal" type exists that could be prescribed to any country, at any moment, in the world. A prerequisite is that an appropriate match exists between the organisational mandates and structures and the institutional environment. Depending on local conditions, the preferred organisations may have a particular scale and scope. Typically, however, water pollution control requires a relationship with water management and hence large scales (10–100 km, covering a river or drainage basin or an agglomeration of municipalities). Usually, single municipalities are unable to generate the required vision, finance and technical knowledge. Where it is possible to enhance particular functions, mergers with other sub-sectors or utilities may be advisable.

As wastewater infrastructure is so expensive, the generation of finance is a key consideration for investment, and for operation and maintenance. Consequently, institutions must be designed to allow cost recovery. This necessitates devolution of decision making and operation and maintenance to lower administrative levels, i.e. closer to the consumer and citizen.

In order to render the organisations flexible, task and performance orientated, and financially well managed, they require a large degree of autonomy. For this purpose, the conventional command-and-control must be deregulated and replaced by measures that ensure self-regulation. This may include arrangements for competition (for service contracts, for example), avoidance or control of monopolies, or the prevention of executive organisations from regulating themselves. Delegated management and privatisation may be useful components in a deregulation strategy. However, the institutional environment must be equally developed to ensure adequate control of the private partners and to avoid monopoly and cartel formation.

8.8   References

Alaerts G.J. 1991 Training and education for capacity building in the water sector. In: G.J. Alaerts, T.L. Blair and F.J.A. Hartvelt [Eds] *A Strategy for Water Sector Capacity Building.* IHE Report 14. United Nations Development Programme, New York and International Institute for Hydraulic and Environmental Engineering, Delft.

Alaerts G.J., Blair T.L. and Hartvelt F.J.A. [Eds] 1991 *A Strategy for Water Sector Capacity Building.* IHE Report 14. United Nations Development Programme, New York and International Institute for Hydraulic and Environmental Engineering, Delft.

Alaerts G. and Hartvelt F.J.A. 1996 *Water Sector Capacity Building — Models and Instruments*. Capacity Building Monographs. United Nations Development Programme, New York.

Anon. 1993 *Programme Support for the Ganga Action Plan in Kanpur*. DGIS-Ministry of International Cooperation, The Hague and Ministry of Environment and Forests, New Delhi.

Chéret I. 1993 Managing water: the French model. In: I. Serageldin and A. Steer [Eds] *Valuing The Environment*. The World Bank, Washington, D.C.

de Capitani A. and North D.C. 1994 *Institutional Development in Third World Countries: The Role of the World Bank*. HRO Working Papers 42. The World Bank, Washington D.C.

Edwards D.B. 1988 *Managing Institutional Development Projects: Water and Sanitation Sector*. WASH Water and Sanitation for Health Project, Technical Report 49, Washington D.C.

ICWE (International Conference on Water and the Environment) 1992 *The Dublin Statement and Report of the Conference*, World Meteorological Organization, Geneva.

Israel, A. 1987 *Institutional Development*. Johns Hopkins University Press, Baltimore.

Lorrain, D. [Ed.] 1995 *Gestions Urbaines de l'Eau*. Ed. Economica, Paris.

Okun, D.A. 1977 *Regionalization of Water Management — A Revolution in England and Wales*. Applied Science Publishers, London.

Uphoff, N. 1986 *Local Institutional Development: An Analytical Sourcebook with Cases*. Kumarian Press, West Hartford.

Wickremage, M. 1991 Organisational Development — A Sri Lankan Experience. In: G.J. Alaerts, T.L. Blair and F.J.A. Hartvelt [Eds] *A Strategy for Water Sector Capacity Building*. IHE Report 14. United Nations Development Programme, New York and International Institute for Hydraulic and Environmental Engineering, Delft.

World Bank 1993 *Water Resources Management. Policy Paper*. IBRD/The World Bank, Washington, D.C.

Chapter 9[*]

INFORMATION SYSTEMS

9.1   Introduction

In the last decade of this age of information, a shift in awareness of the role of monitoring and information has become apparent. In the past, monitoring originated from the greater scientific ideal that underpins our quest for knowledge. The consequence, especially in advanced countries, is that monitoring is frequently, if not implicitly, linked to scientific investigation. Water quality monitoring, world-wide, tends to suffer from a chronic failure to establish meaningful programme objectives. In addition, it has become recognised that many western countries suffer from a "data rich, but information poor" syndrome. The responsible organisations acknowledge that they have collected many data, but are unable to answer the basic questions of those using the water. As a consequence, in many countries, data gathering programmes are considered expendable, and are being reduced or even eliminated because there is no clear view of the information product and of the cost-efficiency of monitoring (Ward *et al.*, 1986; Ongley, 1995; Ward, 1995a). In recent years there has been an increasing consensus of opinion that information is meant for action, decision-making and use. Data that do not lead to management action, or for which a use cannot be stated explicitly, are being labelled increasingly as "not needed" (Adriaanse *et al.*, 1995).

Regardless of the purpose of monitoring water, one theme runs constantly through all discussions about monitoring system design (Adriaanse *et al.*, 1995), i.e. how can monitoring be more cost effective? Typical issues to be addressed are, for example (Ongley, 1995): is a 10 per cent improvement in data reliability worth the 30–40 per cent increase in cost of the data-gathering programme and would it actually change or enhance managerial decisions? Or, can 90 per cent of the management decisions be made with only 50 per cent of the existing data programme?

[*]   *This chapter was prepared by M. Adriaanse and P. Lindgaard-Jørgensen*

Table 9.1   Different categories of uses of water resources

Category	Major uses
Category 1: Uses without quality standards	Transport system (water, wastewater, shipping) Mineral extraction (sand, gravel, natural gas, oil) Power generation (hydropower dams)
Category 2: Uses with defined quality standards	Process/cooling water in industry Irrigation in agriculture Fisheries Recreation and tourism Domestic water supply
Category 3: "Use" with "undisturbed" quality	Ecosystem functioning

Source: Dogterom and Buijs, 1995

In general, information is the basis for any management and control. Water management activities are not excluded from this general statement. Management measures not based on adequate and reliable information are, principally, unaccountable. There is, therefore, a profound need for effective information that is suitable for such use. As a consequence the development of accountable information systems is receiving much emphasis. Effective monitoring programmes are, increasingly, "tailor-made".

9.2   The importance of integration
Information needs for water pollution control can only be defined from within the overall context of water resources management. By considering the various influences and aspects involved in water resources management today, it is possible to identify some fundamental information needs. Some relevant aspects of water resources management are highlighted briefly below.

Functions and use
Various functions and uses of water bodies, whether in relation to human activities or ecological functioning (Table 9.1), can be identified from existing policy frameworks, international and regional conventions and strategic action plans for river basins and seas (Dogterom and Buijs, 1995). These specify divers requirements for water quality. Uses may compete or even conflict, especially in situations of water scarcity and deteriorating quality. In addition, functions and uses can be affected by human activities in both positive and negative ways (Figure 9.1). Chemical water quality issues that have given rise to conflicts between water uses in industrialised countries are summarised in Figure 9.2.

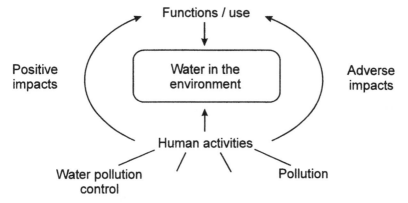

Figure 9.1 Interactions between human activities and functions and uses of water resources

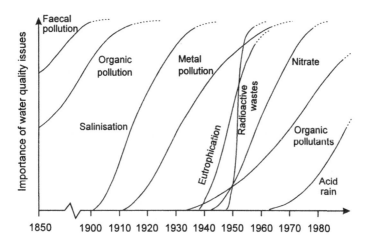

Figure 9.2 The sequence of water quality issues arising in industrialised countries (After Meybeck and Helmer, 1989)

Multi-functional approach

An integrated approach tries to find the balance between all desired uses, including ecosystem functioning. A multi-functional approach allows a hierarchy to be introduced to the uses. It allows flexibility in the application of water resources management policies at different levels of development and allows for prioritisation in time. This could be important for those countries where basic needs, such as supply of healthy drinking water, are so urgent that other uses must take a lower priority, or for countries where water resources have become deteriorated to such an extent that uses with stricter water quality needs can only be restored gradually over a long period of time and according to their priority (Niederländer *et al.*, 1995; Ongley, 1995).

The concept of integrated water management became widely adopted in the 1980s, and as a result the functions and uses of water bodies, their problems and threats, and the effects of water management measures, as well as the information needs to manage this complexity, are being viewed increasingly in an ecosystem context. The focus is now on the behaviour of water in the environment. Instead of breaking the environment into manageable parts, managers are leaving their restricted, traditional disciplines and taking a broad "systems" perspective of water quality management and monitoring (Ward, 1995b).

Various disciplines
Knowledge on various disciplines is required because the functions and uses of water resources may be related to physico-chemical, biological, morphological, hydrological and ecological features. The nature of water pollution issues and the effects of controlling measures do not allow a divided approach; they have to be characterised in an integrated way. For the same reason, information needs also require an integrated approach.

Appropriate media
Various media, such as the water itself, suspended matter, sediments and biota are integrated elements of a water body. Information needs are also concerned with appropriate media, wherever these media provide information that is considered to be characteristic for functions, problems and control measures. Interactions of water resources with air and soil demand the same approach (Laane and Lindgaard-Jørgensen, 1992).

Multiple sources
Multiple sources of water pollution require an integrated, balanced and site specific approach. If water pollution is dominated by well-defined point sources, monitoring of the discharged effluents may be the best approach. Generally, however, point sources are numerous and not well defined. In addition, diffuse sources are forming a substantial and growing aspect of water pollution problems. Knowledge of the relative contribution of different sources (agriculture, households, industries, aerial deposits) is often important to verify the effectiveness of control measures.

Table 9.2   Differences in the emission-based and the water quality-based
approaches to water pollution control

Management aspect	Emission-based approaches	Water quality based approaches
Effluent limits	No site-specific load	Site-specific concentrations
Required treatment techniques	Based on intrinsic (toxic) properties of chemicals in effluent; or technology based	Based on water quality criteria or preventing toxic effects in the effluent receiving water
Data requirements	Basic chemical and ecotoxicological data	Basic chemical and ecotoxicological data. Physical, chemical and biological characteristics for the receiving water and the fate of discharged chemicals
Monitoring	Effluent	Receiving water
Competition	Equality for the law	Inequality
Practice	May tend to worst case approach in general, but may underestimate effects of discharges in specific situations	May tend to dilution as a solution in general, but stricter standards are possible when effects are intolerable in specific situations

Source: Stortelder and Van de Guchte, 1995

Approaches in water pollution control
There are two approaches to water pollution control: the emission-based
approach and the water quality-based approach (Stortelder and Van de
Guchte, 1995). The differences between these approaches result from the
systems applied for limiting discharge and in the charging mechanisms.
However, these differences are also reflected in the strategies taken for
hazard assessment and the monitoring of discharges to water, i.e. whether it is
focused on the effluents or on the receiving water; both have their advantages
and disadvantages (Table 9.2). A combined approach can make optimal use
of the advantages.

Watershed management
Ecosystems are not restricted to boundaries defined by humans, such as
between local governments or countries. Consequently, integrated watershed
management is becoming more common. The Convention on Protection and
Use of Transboundary Watercourses and International Lakes, Helsinki
(UNECE, 1992) underlines the need for an integrated watershed approach in
water management and for adequate monitoring and assessment of transboundary waters.

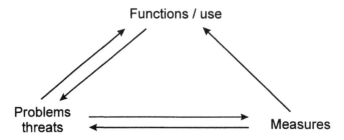

Figure 9.3 Core elements in water management and water pollution control

Institutional collaboration

In many countries the responsibility for collecting water information is divided between, for example, different ministries, executive boards, agencies. This approach risks duplication and a lack of harmonisation, and prevents an integrated approach. Often, responsibilities for water resources management and water pollution control rest with different ministries and with different governmental levels (federal, regional, local). The establishment of collaborative partnerships and the co-ordination of monitoring efforts between competing ministries or institutions can greatly enhance the quality of the information obtained and make better use of available resources.

9.3   Specifying information needs

Information needs are focused on the three core elements in water management and water pollution control, namely the functions and use of water bodies, the actual problems and threats for future functioning, and the measures undertaken (with their intended responses) to benefit the functions and uses (Figure 9.3).

Monitoring is the principle activity that meets information needs for water pollution control. Models and decision support systems, which are often used in combination with monitoring, are also useful information tools to support decision making. Figure 9.4 illustrates some of the key components of the environmental management system.

Monitoring objectives are set according to the focus of water management and water pollution control activities and according to the issues that are capturing public attention. Monitoring objectives may be of many kinds, but fall mainly within five basic categories:

- Assessment of water bodies by regular testing for compliance with standards that have been set to define requirements for various functions and uses of the water body concerned.
- Testing for compliance with discharge permits or for setting of levies.

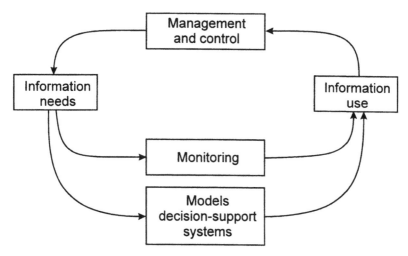

Figure 9.4 Components of environmental management information systems

- Verification of the effectiveness of pollution control strategies, i.e. by obtaining information on the degree of implementation of measures and by detection of long-term trends in concentrations and loads.
- Early warning of adverse impact for intended water uses, e.g. in case of accidental pollution.
- Increasing awareness of water quality issues by in-depth investigations, for example by surveys investigating the occurrence of substances that are potentially harmful. Surveys provide insight into many information needs for operational water management.

A monitoring objective, once defined, identifies the target audience. It makes clear who will be the users of the information and why the information is needed. It also identifies the field of management and the nature of the decision-making for which the information will be needed. It should be recognised that the detection of trends, in itself, is not a monitoring objective but a type of monitoring. Only when the intended use of the trend information is specified can it be considered to be an objective.

Once objectives have been set it is important to identify the information that is needed to support the specified objective. The content and level of detail of the information required depends upon the phase of the policy life cycle (Figure 9.5). In the first phase, research and surveys may identify priority pollution problems and the elements of the ecosystem that are appropriate indicators. Policies will be implemented for these. In the second and third phases, feedback on the effectiveness of the measures taken is obtained by assessing spatial distributions and temporal trends. Contaminants may

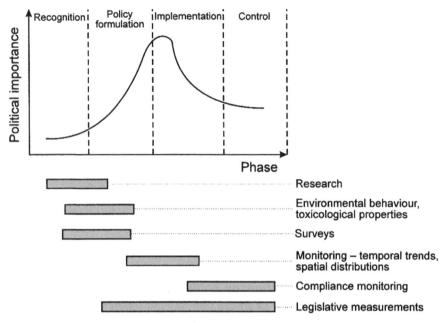

Figure 9.5 The policy life cycle and typical measurement activities applied in the respective phases

endanger human health by affecting aquatic resources, such as drinking water, and therefore specific monitoring programmes may be initiated to check, on a regular basis, the suitability of such resources. Legislation may also prescribe measurements required for certain decision-making processes, such as the disposal of contaminated dredged material. In the last phase, monitoring may be continued, although with a different design, to verify that control is maintained. The associated information needs change with the respective policy phases (Winsemius, 1986; Cofino, 1995).

Decision-makers have to decide upon the contents and performance of their desired information products. They are the users of the information (for management and control action) and they have to account for their activities to the public. Specification of information needs is a challenging task which requires that the decision-making processes of information users are formulated in advance. Various aspects of the information product must be specified, such as:

- The water quality assessment needs and the methods to be applied have to be defined, putting an emphasis on the development of a strategy of assessment rather than on a simple inventory of arbitrary needs for the measurement of substances.

- The methods for reporting and presenting the information product must be considered; these are closely related to the assessment methods applied. Visualised, aggregated information (such as indexes) is often much more effective (and therefore more appreciated) than bulky reports.
- Appropriate monitoring variables have to be selected. Selected variables should be indicators that characterise, adequately, the polluting effluent discharge or that are representative for the functions and uses of water bodies, for water quality issues or for testing the effectiveness of pollution control measures.
- Relevant margins of information have to be considered. To assess the effectiveness of the information product, the information needs have to be quantified; for example, what level of detail is relevant for decision-making? Such margins have to be specified for each monitoring variable. A relevant margin can be defined as "the information margin that the information-user considers important".

Information needs must be specified such that they enable design criteria for the various elements of the information system to be derived. Specified, relevant margins are a strong tool for network design. With these, sampling frequencies and the density of the network can be optimised, especially if reliable time-series of measurements are available. Relevant margins highlight the detail required in the presentation. Decisions on the development of more accurate analytical methods should be related to relevant margins or threshold values in water quality. However, the latter should be related critically to cost-effectiveness.

In general, a monitoring and information system can be considered as a chain of activities (Figure 9.6). Essentially, the chain is closed with the management and control action of the decision-maker, whereas past schemes have shown a more top-down sequence of a restricted number of activities, starting with a sampling network chosen arbitrarily and ending up with the production of a set of data. Building an accountable information system requires that the activities in the chain are designed sequentially, starting from the specified information needs.

While monitoring is continuing, information needs are also evolving. This has already been illustrated by the policy life cycle in Figure 9.5. In time, there will be developments in management and control, and targets may be reached or policies may change, implying that the monitoring strategy may need to be adapted. Dynamic information needs require a regular reappraisal of the information system; it is essential to add, to cancel, to revise and to bring the concept up to date. In order to visualise this the circle of Figure 9.6

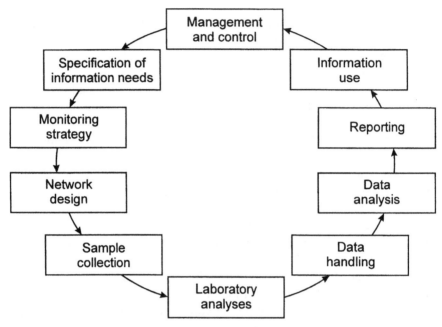

Figure 9.6 Chain of activities in an information system

may be modified to a spiral (Cofino, 1994), reflecting the ongoing nature of the monitoring and incorporating the feedback mechanism.

9.4 Information gathering and dissemination

9.4.1 System organisation and information flow

The objective of an information system for water pollution control is to provide and to disseminate information about water quality conditions and pollution loads in order to fulfil the user-defined information needs. Information systems can be based either on paper reports circulated in defined pathways, or on a purely computerised form in which all information and data are stored and retrieved electronically. In practice, most information systems are a combination of these. However, given the availability of powerful and inexpensive hardware and software, it is now almost unthinkable to design an information system without making use of computers for data management and analysis. The main types of data to be processed in an information system are:

- Data on the nature of the water bodies (size and availability of water resources, water quality and function, and structure of the ecosystem).
- Data on human activities polluting the water bodies (primarily domestic wastewater and solid waste, industrial activities, agriculture and transport).

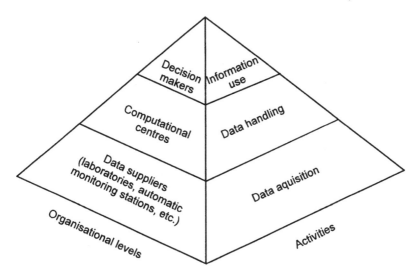

Figure 9.7 Information "pyramid" showing information system activities and their corresponding organisational levels

- Data on the physical environment (e.g. topography, geology, climate, hydrology).

Such data must be drawn from networks of national, regional and local monitoring stations on water quality and on pollution sources. Guidance for the establishment of such networks is given in section 9.6.

The flow of data in information systems must be well defined in order to fulfil the requirements of users and the overall demand for reliability. Data flow is considered in three directions, upwards, downwards and horizontally. Upward flow of information from lower to higher organisational structures reduces the amount of detail but enhances the information value through the interpretation of the data. Downward flow is important for the purpose of communicating decisions in relation to national standards and policies, and also to make a feedback to those involved in data acquisition and data-handling within the information system. Horizontal flow, through data sharing between organisations, is essential for developing an integrated approach to environmental monitoring and management and to make efficient use of data that are often collected and stored in a large number of institutions.

The vertical flow of information can often be described as a three-tiered system with respect to the organisational levels and the activities performed at each level. This is illustrated in the "information pyramid" (Figure 9.7) which reflects the large number of data at the lowest level which, as they reach higher levels of the triangle, become less detailed but of greater information value. The first level is responsible for primary data acquisition through monitoring, data validation and storage of data. Often the data will be dynamic, such as measurements and analyses and, typically, will be used locally (such as for compliance control). It is very important to implement basic quality assurance and control systems for all procedures generating primary data because the data generated at this level will influence the result of data analysis, reports and decisions also taken at other levels.

Data handling (the second level) is typically carried out at computational centres and can be organised thematically, such as on water quality in rivers, lakes or groundwaters or by pollution source, for example municipal and industrial wastewater, non-point pollution from agriculture. Computational centres can also be divided geographically according to river basins or to administrative boundaries, i.e. to local or regional level. These centres have the primary task of converting data into information. They are, therefore, the users of primary data from the data acquisition level as well as being the service centres producing the required information. Typically these centres use and maintain adequate graphical and statistical tools, forecasting tools (e.g. models) and presentation and reporting tools. In addition, they often maintain data of a more static nature, such as geographical data, and they may also be responsible for primary data acquisition within their specific area of responsibility.

The third level (information use) is made up of the decision-making authorities who are the end-users of the information produced. At this level, information is used for checking and correcting the policies and management procedures applied. However, this level is also responsible for the final generation of the information disseminated to the public and to other interested parties, such as private sector and international bodies and organisations. As such, this level may have its own tools for integrating the information on the water environment with information from other media and sectors.

9.4.2   Data acquisition

Data acquisition deals with the generation and storage of data from monitoring activities. Data should be stored to ensure that they maintain accuracy and to allow easy access, retrieval and manipulation. The volume of data to be acquired and stored is dictated by the size and level of ambition of the monitoring network. For small volumes of data, manual systems may be used efficiently to store and retrieve data, produce time series plots and to perform simple statistical analysis. Nevertheless, a system based on microcomputers, and using simple systems like spreadsheets, may substantially improve data handling capacity, simultaneously enabling basic statistical and graphical analyses that are straightforward and easy to perform. For larger volumes of data, a generalised data storage system, based on a relational database, will provide more powerful data management capabilities. In addition to being used for storage and retrieval of data, special programmes can be written for such systems to automate data entry, analysis and generation of reports.

The following general requirements for storing data in databases can be identified (Ward *et al.*, 1990):

- Data must be stored and retrieved unambiguously.
- Software must be portable.
- Software must be easy to use.
- Protection against wilful or accidental damage must be assured.
- Unambiguous output must be assured.
- Flexible enquiry and reporting should be possible.

9.4.3   Data handling

Data handling covers the analysis and transformation of data into information. Tools for this are described in more detail in section 9.5. The preparation of reports and the dissemination of information is another important aspect of an information system. Issues, such as for whom the reports are intended, at what frequencies should they be generated, and the level of detail of each report, should be clarified and the reporting systems should be planned as an integral part of the information system.

Reports containing results from routine analyses of data collected from a monitoring programme (i.e. daily, weekly, monthly, quarterly or yearly), and that present developments in water quality or pollution load since the preceding monitoring period should be prepared using a fixed format. The reporting can then be automated using a customised data management system. Other types of report present information generated on the basis of data from various pollution sources and locations and analysed by means of advanced tools such as models and geographical information systems (GIS). These types of report are particularly useful in water pollution control because they focus on water quality as well as on pollution sources. Some examples are:

- *State of the environment (SOE) reports*. These are environmental summary assessments used to inform decision makers, environmental organisations, scientists and the public about the quality of the environment. Such reports normally include the state of the environment; changes and trends in the state of the environment; links between human and environmental health and human activities, including the economy; and the actions taken by society to protect and to restore environmental quality.
- *Environmental indicator reports*. These are considered to be an effective way of communicating with the public, amongst others, and of presenting

information about the development of a number of indicators over time and space. Environmental indicators are sets of data selected and derived from the monitoring programme and other sources, as well as from data bases containing statistical information, for example, on economy, demography, socio-economics. For pollution control in rivers, examples of useful indicators are dissolved oxygen, biochemical oxygen demand (BOD), nitrate, uses and extent of available water resources, degree of wastewater treatment, use of nitrogenous fertilisers and land-use changes, accidents with environmental consequences. An example of an indicator report for the state of Danish rivers is given in Figure 9.8.

9.4.4   Use and dissemination of information

Use of information is the third and highest level of the information system. At this level the information, mostly in the form of reports, can be used to support decision makers. New approaches to water pollution control put much emphasis on the active participation of the public, as well as industries. It will, therefore, be increasingly important to disseminate to these parties relevant and easily understandable information about the state of the environment, as well as the extent to which environmental policies and private and public environmental investments are improving the state of the environment.

Other activities can be used in addition to the dissemination of reports and may help to raise the environmental awareness of governments, sectoral ministries and administration, as well as the private and public sector. Examples of these activities include seminars, meetings and public hearings held in connection with the launching of significant reports, such as the state of the environment report or environmental indicator reports.

9.5   From data to information tools

To avoid the "data rich but information poor" syndrome, data analysis, information generation and reporting should be given the same attention as the generation of the data themselves. Water pollution control requires access to statistical, graphical and modelling tools for analysis and interpretation of data. Theoretically, most of these analyses can be performed manually, although this approach is often so time consuming that for large data sets and complex data treatment methods it excludes the generation of the type of information required (Ward *et al.*, 1990; Demayo and Steel, 1996).

9.5.1   Graphical information

Data analysed and presented using graphical methods is probably the most useful approach for conveying information to a wide variety of information users, both technical and non-technical. Graphical analyses are easy to

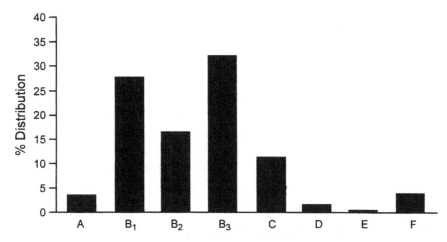

Figure 9.8 Percentage distribution of the types of quality objectives adopted for Danish water courses (according to the regional plan maps of the countries) (see Table 9.2 for definition of quality objectives) (After DEPA, 1991)

perform, the graphs are easy to construct and the information value is high when graphs are properly presented. The types of information that can be presented most effectively by graphical methods are:

- Time series (temporal variation).
- Seasonal data (temporal variation).
- Water quality at geographic locations (spatial variations).
- Pollution loads at geographic locations.
- Statistical summaries of water quality characteristics.
- Correlations between variables.
- Spatial and temporal comparisons of water quality variables.

Widely used methods include time series graphs and graphs which may be used to give a visual indication of data distribution (e.g. box and whisker plots) and to indicate how distribution changes over time or between locations (Ward *et al.*, 1990; Demayo and Steel, 1996; Steel *et al.*, 1996).

9.5.2 Statistical information
Statistical information is the most useful treatment of data for making quantitative decisions, such as whether water quality is improving or getting worse over time, or whether the installation of a wastewater treatment plant has been effective, or whether water quality criteria or emission standards are being complied with. Statistics can also be used to summarise water quality and emission data into simpler and more understandable forms, such as the mean and median (Demayo and Steel, 1996).

Another important application of statistics, in relation to water pollution control, is the transformation of data to give an understanding of the average and extremes of water quality conditions, and also the changes or trends that may be occurring. Statistical methods to provide this kind of information can be classified as graphical (as described above in section 9.5.1), estimation or testing-of-hypothesis methods (Ward *et al.*, 1990; Demayo and Steel, 1996). The classical method of trend analysis, for example, is estimation of a linear trend slope using least square regression, followed by a *t*-test of the statistical significance of the slope parameters. Standard software packages exist for most statistical methods. An explanation of the use of statistical methods, together with some examples, is available in Demayo and Steel (1996).

9.5.3 Water quality indices and classes

A water quality index is obtained by aggregating several water quality measurements into a single number (NRA, 1991). Indices are, therefore, simplified expressions of a complex set of variables. They have proved to be very efficient in communicating water quality information to decisions makers and to the public. Different water quality indices are in use around the world and among the best known are biological indices, such as the Saprobic Index (NRA, 1991; Friedrich *et al.*, 1996).

Many countries world-wide use a classification system for the water quality of rivers, dividing the rivers into four (or more) classes of quality, ranging from bad to good. Such systems are mostly based on the use of biological indices, sometime in combination with chemical indices (DEPA, 1992; Friedrich *et al.*, 1996). In Denmark, for example, quality objectives for the condition of Danish water courses have been adopted and approved as binding directives in the regional plans of the county councils. These quality objectives for water courses are laid down according to the physical and flow conditions of the water course and to the water quality conditions accepted by the authorities responsible for the quality of the water bodies. Table 9.3 shows these quality objectives and Figure 9.8 shows the percentage distribution of the types of quality objectives adopted for Danish water courses. Objectives A and B, which apply to more than 75 per cent of the lengths of all water courses, include biological criteria for areas with strengthened objectives or high scientific interest (A) or general objectives for areas sustaining a fish population (B) (DEPA, 1991).

Water quality indices and classifications should not be the only method used for analysing and reporting data from a water quality monitoring system, because it may not be possible to determine less obvious trends in

Table 9.3   Types of quality objectives for Danish water courses

Quality objectives		Maximum Saprobic Index
A	Area with specific scientific interests	II
B₁	Spawning and fry	II
B₂	Salmonid water	II
B₃	Carponides water	II (II–III)

Source: Based on information from the National Agency of Environmental Protection, Denmark

water quality and some water quality variables may change dramatically without affecting the overall classification.

9.5.4   Models

Water quality models can be a valuable tool for water management because they can simulate the potential response of the aquatic system to such changes as the addition of organic pollution or nutrients, the increase or decrease in nutrient levels, or water abstraction rates and changes in sewage treatment operations. The potential effects of toxic chemicals can also be estimated using models (SAST, 1992; Vieira and Lindgaard-Jørgensen, 1994). Mathematical models are, therefore, useful tools for water quality management because they enable:

- The forecasting of impacts of the development of water bodies.
- The linking of data on pollution loads with data on water quality.
- The provision of information for policy analysis and testing.
- The prediction of propagation of peaks of pollution for early warning purposes.
- The enhancement of network design.

In addition, and equally important, they enable a better understanding of complex water quality processes and the identification of important variables in particular aquatic systems.

Obtaining the data necessary for construction or verification of models may require additional surveys together with data from the monitoring programme. If models are to be used routinely in the management of water quality, it is also important to verify them and for the model user to be aware of the limitations of the models.

The development of models into combined systems linking physical, chemical and biological processes has enabled a better understanding and modelling of chemical and biochemical processes and behavioural reactions. It has also shown how such processes interact with basic physical processes (i.e. flow, advection and dispersion). These types of models are gradually being used for water quality management. Several models have been dedicated for specific water quality management purposes such as environmental impact assessment, pre-investment planning of wastewater treatment facilities, emergency modelling and real-time modelling (SAST, 1992; Vieira and Lindgaard-Jørgensen, 1994).

Knowledge-based systems (also called decision support systems) are computer programmes that are potentially capable of identifying unexpected links and relationships based on the knowledge of experts. Knowledge-based systems can be used for network design, data validation and interpretation of spatial data. Knowledge-based systems are also applicable for managing the complex rules of legislation, regulations or guidelines. In recent years, knowledge-based systems have been introduced for environmental applications (Hushon, 1990). Most of these systems have focused on data-interpretation, although systems have also been developed for sampling strategy; for example, Olivero and Bottrell (1990) developed a sampling strategy for soils and Wehrens et al. (1993) reported the design of a decision support system for the sampling of aquatic sediments.

Simple knowledge-based systems can provide, for example, the necessary information to decide if, and what, action should be taken when specific pollutant concentrations exceed certain standards. One of the advantages of decision support systems is that they can make the knowledge of a few experts available to many non-experts. Furthermore, developing knowledge-based systems forces experts to make their knowledge explicit and, in this way, new knowledge may be discovered. Knowledge-based systems can also work with incomplete knowledge and uncertainty.

The development of knowledge-based systems has only begun recently. Therefore, the lack of experience with their use suggests caution is necessary when first implementing such systems. Possible problems to be considered are:

- The development of knowledge-based systems is time-consuming and, often, expensive.
- The acquisition of knowledge is difficult because the number of experts is small and many experts may never have conceptualised the process by which they reach particular conclusions.
- The adaptation of knowledge-based systems to new situations often requires the assistance of the persons who built the system.

Knowledge-based systems can be considered as a branch of artificial intelligence (Walley, 1993). Another promising branch (recently gaining increased interest) is artificial neural networks. Artificial neural networks are very powerful at pattern recognition in data sets and at dealing with uncertainties in the input data. They are, therefore, especially applicable in situations where expert knowledge cannot easily be made explicit or where considerable variability in input data can occur. The standardisation provided by the application of artificial neural networks will lead to improved data interpretation, particularly for biological assessments. Most applications of artificial neural networks are still in an experimental stage although some interesting examples can be found for biological classification of river water quality (Ruck *et al.*, 1993) and the automatic identification of phytoplankton (Dubelaar *et al.*, 1990).

9.5.5 Geographical information systems

Data used for water pollution control, such as water quality, hydrology, climate, pollution load, land use and fertiliser application, are often measured in different units and at different temporal and spatial scales. In addition, the data sources are often very diverse (Demayo and Steel, 1996).

To obtain information about, for example, spatial extent and causes of water quality problems (such as the effects of land-use practices), computer-based GISs are valuable tools. They can be used for data presentation, analysis and interpretation. Geographical information systems allow the georeferencing of data, analysis and display of multiple layers of geographically referenced information and have proven their value in many aspects of water pollution control. For example, they have been used to provide information on:

- Location, spatial distribution and area affected by point-source and non-point source pollution.
- Correlations between land cover and topographic data with environmental variables, such as surface run-off, drainage and drainage basin size.
- Presentation of monitoring and modelling results at a geographic scale.

A typical GIS system consists of:

- A data input system which collects and processes spatial data from, for example, digitised map information, coded aerial photographs and geographically referenced data, such as water quality data.
- A data storage and retrieval system.
- A data manipulation and analysis system which transforms the data into a common form allowing for spatial analysis.
- A data reporting system which displays the data in graphs or maps.

9.5.6 Environmental management support systems

Advanced systems combining databases, GIS and modelling systems into one application are sometime called environmental management support systems. These systems are designed to fulfil a specific purpose, such as the management of water resources and they allow integrated assessments of the effectiveness of environmental policies and planning, such as good agricultural practice and application of best available technology (Vieira and Lindgaard-Jørgensen, 1994). Such systems require a substantial effort in monitoring and system design, implementation and updating. However, because they may serve as a basis for policy development and assessment for a long period of time, they can be a cost-effective tool for controlling high priority water quality problems.

A system integrating monitoring and modelling of water resources (groundwater as well as surface water) has the following elements:

- A GIS-based database of all relevant spatial data, such as topography, river systems (including drainage), soil types, present water resources and land use, plans and restrictions for future water resources and land use (including, for example, forest planting, quantities and distribution of animal manure, livestock watering permits, water reclamation, wells and permitted abstractions), waste disposals and other point sources, and administrative limits.
- A geological database with all relevant geological and hydrogeological data.
- A time series database including data on climate, run-off, pressure level of groundwater, water quality (surface water as well as groundwater), water reclamation and water abstraction.
- Hydrological and water quality models set up and calibrated to different levels of detail with respect to the type of data and the density of monitoring and modelling network.

9.6   Design of monitoring networks and selection of variables

To obtain the necessary focus within a monitoring network for water pollution control, network design should be initiated by surveys to identify potential water quality problems and water uses, and by inventories of pollution sources in order to identify major pollution loads. The objectives of any monitoring activities (see also section 9.3) should first be identified by analysis of the requirements of the users of the data. Examples of specific monitoring objectives are:

- To follow changes (trends) in the input of pollutants to the aquatic environment and in compliance with standards.
- To follow changes (trends) in the quality of the aquatic environment (rivers, lakes and reservoirs) and in the development of water uses.
- To evaluate possible relationships between changes in the quality of the environment and changes in the loads of pollutants and human behaviour, particularly changes in land-use patterns.

- To give overall prognoses of the future quality of water resources and to give assessments of the adequacy of water pollution control measures.

The key function of network design is to translate monitoring objectives into guidance as to where, what and when to measure. Network design, therefore, deals with the location of sampling, with sampling frequency and with the selection of water quality variables (Ward *et al.*, 1990). Obtaining the necessary information for water pollution control may require the following types of monitoring stations:

- *Baseline stations:* monitoring water quality in rivers and lakes where there is likely to be little or no effect from diffuse or point sources of pollution and that will provide natural, or near-natural, effects and trends.
- *Impact stations:* monitoring both water quality and the transport of pollutants. These are located downstream of present and possible future areas of urbanisation, industry, agriculture and forests, for example. To protect water intakes, additional monitoring stations can be placed upstream of the intakes.
- *Source monitoring stations:* monitoring water quality and enabling calculation of pollution loads. These are located at major point sources and also in catchments which are primarily influenced by non-point source pollution.

An additional requirement for selecting the geographic location of stations for baseline and impact monitoring is that they should be at, or close to, current hydrological recording stations or where the necessary hydrological information can be computed reliably. This is because no meaningful interpretation of analytical results for the assessment of water quality is possible without the corresponding hydrometric data base. All field observations and samples should be associated with appropriate hydrological measurements. Other requirements for selecting station locations include accessibility and ease of sampling, safety for operators and transit time for samples going to the laboratory.

If possible, source monitoring stations should be placed at the outlet of major municipal and industrial wastewater discharges (Nordic Fund for Technology and Industrial Development, 1993). Point source monitoring, which requires substantial personnel resources, should be based preferably on self-monitoring performed by municipalities and industries, in combination with public inspection and control systems. The frequency of monitoring should reflect the variability, as well as the magnitude, of the pollution load, i.e. large volume sources should be monitored more frequently than small volume sources.

If monitoring at an outlet is not possible or the discharge is very small, the pollution load from industries may be calculated from information on the type of production and the actual production capacity using standard emission rates. For discharges from urban areas, loads can be calculated using person equivalents. The validity of the calculated information should be checked against values of pollution transport based on results from impact monitoring stations upstream and downstream of the discharges.

Direct monitoring of pollution loads from non-point sources to the water bodies is not possible. However, an impact monitoring station, located downstream of a catchment dominated primarily by non-point sources, such as agriculture, may be used for the evaluation of trends in loads from these sources (DEPA, 1992). If this is not possible because the catchment contains both point and non-point sources, some evaluation of trends in non-point loads may be achieved by subtracting the load from the point sources (monitored at the relevant point source monitoring station in the catchment) from the values obtained at the downstream impact station.

Additional evaluation of the pollution load from diffuse sources can be obtained from data on land-use, including land-use for agriculture, forestry, urban areas, landfills and waste dumps. The information required in relation to agriculture and forestry includes animal and livestock production, types of crops, soil types, use of fertiliser (by type and amount), and use of pesticides. Data on population size is appropriate for the evaluation of pollution loads from smaller urban and rural areas where there is no infrastructure for wastewater collection and treatment. To transform this type of data into usable information, tools such as models and GIS are necessary (see section 9.5).

Where monitoring stations are located in lakes with long retention times, the evaluation of pollution loads may require information from the monitoring of atmospheric deposition of nitrogen, phosphorus and heavy metals, especially in more industrialised areas.

The selection of sampling frequencies and variables is usually based on a compromise between average station densities, average sampling frequencies and a restricted number of variables (depending on the character of the industrial and agricultural activities in the catchment together with the financial resources of the monitoring agency). Table 9.4 gives some guidance for the development of a water pollution control programme with different levels of complexity. It should also be recognised that sampling frequency and the number of samples required may have to be adapted in order to allow the necessary statistical analysis (Ward et al., 1990; Demayo and Steel, 1996).

An advanced monitoring programme in areas with major industrial and agricultural sources of pollution, including the use of pesticides and chemical fertilisers, requires additional media, such as sediment and biological material in which heavy metals and some hazardous chemicals accumulate,

Table 9.4   Selection of analyses and resources for different levels of water
pollution control monitoring programmes

Monitoring level	Sampling freq. (a^{-1})	Water analysis	Sediment analysis	Biological monitoring	Source monitoring	Required resources
Simple	6	°C, pH, O_2, TSS, major ions, visual observation			°C, pH, O_2, TSS, COD, BOD	Small sampling team, general chemistry laboratory
Inter-mediate	6–12	As above plus PO_4, NH_4, NO_2, BOD, COD	Trace elements	Biological indices	As above plus PO_4, NH_4, NO_2, and trace elements	Specialised chemical laboratory, team of hydro-biologists
Advanced	> 12	As above plus soluble organic pol-lutants, DOC, POC and some trace elements	As above plus organic micro-pollutants	As above plus chemical analysis of target organisms	As above plus toxicity tests and organic micro-pollutants	Major central-ised laboratory, ecotoxicol-ogists, national research institute

Source: Adapted from Chapman, 1996

and variables, particularly some heavy metals and specific organic
compounds, when compared with pollution control monitoring of municipal
wastes or traditional agricultural methods. Some industrial discharges may
contain toxic chemicals that can affect aquatic life. The introduction of
aquatic toxicity tests, using the effluents from industrial sources, may be an
effective way of giving information on toxicity (OECD, 1987).

9.7   Monitoring technology

This section gives only a brief summary of types of monitoring technology for
water pollution control. The main emphasis is on any additional requirements
compared with more basic water quality monitoring, i.e. requirements such as
technology for monitoring pollution sources, sampling sediment, biological
monitoring and laboratory equipment necessary for advanced analysis of some
heavy metals and specific organic chemicals. Further guidance on monitoring
technology and laboratory methods is given in the *GEMS/WATER Operational
Guide* (WHO, 1992) and Bartram and Ballance (1996).

9.7.1   Source monitoring

The volumetric flow rate is particularly important for the determination of
pollution loads coming from point sources. Flow should preferably be

recorded continuously or, if this is not possible, at least during the period of sampling (Nordic Fund for Technology and Industrial Development, 1993). Suitable manually-operated equipment for monitoring flow includes a meter linked to a propeller, electromagnetic sensors or even a system using buckets and time recording (the latter can provide a good estimate).

Water or effluent samples can be taken manually, using simple equipment such as buckets and bottles, or automatically using vacuum or high speed pumps. Spot-samples, giving the concentration just at the time of sampling, should only be used if there is no other alternative. Instead, time-proportional or flow-proportional samples should be taken over a period of time (e.g. 24 hours) to give a better estimation of the variation of loads over time.

Variables such as temperature, pH, redox potential, turbidity and concentration of dissolved oxygen may be monitored *in situ*, using hand-held portable meters. For other variables, such as chemical oxygen demand (COD), BOD or nutrients or advanced variables such as heavy metals and specific organic chemicals, the samples have to be transported to and analysed at a laboratory. Such variables are often specified in discharge permits.

Discharges from some industrial processes may have an adverse effect on aquatic organisms, as a result of toxic components. This toxicity can be evaluated by different types of biological tests in which the organisms are exposed to the effluent (OECD, 1987). An example of such a method is Microtox, which is an off-line method for measuring acute toxicity using bioluminescent bacteria. The principle of the test, which is standardised in some European countries, is to measure the light production of the bacteria before and after exposure to the wastewater for a defined period of time. The result can be used to estimate if the discharge is likely to affect aquatic life in the water body receiving the discharge. Other tests, which may be more relevant, but also more laborious, are based on the exposure of fish or other organisms known to be abundant in the receiving water body (OECD, 1987; Chapman and Jackson, 1996; Friedrich *et al.*, 1996).

9.7.2   Particulate matter sampling and biological monitoring

Monitoring programmes for particulate matter and biological material need careful design. In general, the frequency of sampling is low compared with water sampling. However, the analysis of samples is often more time consuming (Bartram and Ballance 1996; Chapman, 1996). Monitoring of particulate matter (suspended or deposited on the bottom) is particularly important because heavy metals and some hazardous organic industrial chemicals and pesticides are associated with the particulate matter and accumulate in deposited sediments; therefore, water samples do not give an accurate representation of the pollution load from such substances (Thomas and Meybeck, 1996). Sampling can be performed with inexpensive grab or core samplers (for bottom sediment) or by filtration or centrifugation of water samples (for suspended material). Chemical analyses can be performed on extracts of the samples (Ongley, 1996).

Whereas water quality monitoring provides a picture of the quality of the water at the time of sampling, biological monitoring can give an integrated picture of water quality over the life time of the selected fauna and flora. It is impossible to monitor separately the thousands of chemicals often occurring simultaneously in the environment, but biological methods provide an indication of their combined effects (Tørsløv and Lindgaard-Jørgensen, 1993; Chapman and Jackson, 1996; Friedrich *et al.*, 1996). Consequently, biological monitoring has been introduced into many water quality monitoring systems.

9.7.3 Advanced analysis

Water pollution control of industrial chemicals and pesticides needs more advanced and expensive equipment, and better laboratory infrastructure, than may be found in many ordinary water quality laboratories (Suess, 1982). Appropriate equipment includes atomic absorption spectrophotometers (AAS) for heavy metals analysis, gas chromatographs (GC) and liquid chromatographs for organic pollutants (in combination with effective preconcentration (Ballance, 1996).

9.7.4 Automation of monitoring and information systems

Over the last decade, much has been achieved in the automation of monitoring and automatic transfer of data from the monitoring system into the information system. New developments using sensor technology and telemetry, for example, will probably speed up this process. The following presents a short summary of the main approaches to sampling and analysis (SAST, 1992; Griffiths and Reeder, 1992):

- Manual or automatic on-site water sampling with subsequent analysis using portable analytic equipment. This approach is primarily of importance for physical and chemical variables, such as pH, temperature, redox potential, conductivity and turbidity, as well as for variables which have to be monitored *in situ* (e.g. dissolved oxygen). New developments in monitoring kits and hand-held instruments for chemical variables will increase the number of variables that can be monitored on-site.
- Manual or automatic on-site water sampling with subsequent transport to central facilities for analysis and further processing. At present this is the most common approach. In some areas, where the transportation time to a

laboratory is very long or the road infrastructure is not sufficiently developed, analysis using a mobile laboratory may be feasible.

- On-site measurement (using sensors) and simultaneous on-site analysis. Such methods reduce the operational cost by limiting personnel require-ments although they are presently not developed to a sufficient level for widespread use.
- Remote sensing of regional characteristics, such as land use, by satellites or airborne sensors. Such methods have gained much interest in recent years, particularly for applications using GIS.

Early warning is important for cases of accidental pollution of surface water (surface water early warning) and for cases where there is a direct danger from accidental pollution of surface water (effluent early warning). Early warning has two objectives; providing an alarm and detection. Alarms may be used to alert water users and to trigger operation management. They mainly inform water supply undertakings that are treating surface water for potable water supplies. To a lesser extent they may inform all other direct users of the water body, e.g. for animal husbandry, arable farming and industry. Detection systems may be used to trace discharges or to identify operation failures. As a result of timely warnings, intakes and uses of water can be suspended, the spread of the pollutant can sometimes be limited to certain less vulnerable areas by water management measures (e.g. control of locks/weirs, water distribution), and the continued, perhaps calamitous, discharge can be prevented (specifically for effluent early warning).

In addition to the measurements made by an early warning monitoring system other components play an important role. These components include:

- A communication system, in which warning procedures are defined and through which all those involved in the river basin can be informed quickly.
- A model for the calculation of the transit time of a confirmed accidental pollution from a warning centre or a monitoring station to the place where the water is used or abstracted.
- A toxic substances inventory providing information on the deleterious properties of substances.

An adequate early warning system integrates all these components. There have been major developments in early warning systems in the last 20 years. Integrated, early warning systems have been developed for the river basins of the Rhine (Spreafico, 1994), the Ile de France region (Mousty et al., 1990) and the Elbe (IKSE, 1992), among others. An integrated system is now under development for the Danube river basin (EPDRB, 1994), one of the largest river basins in Europe.

9.8 References

Adriaanse, M., Van de Kraats, J., Stoks, P.G. and Ward, R.C. 1995a Conclusions monitoring tailor-made. In: M. Adriaanse, J. Van de Kraats, P.G. Stoks and R.C. Ward [Eds] *Proceedings of the International Workshop Monitoring Tailor-made.* Institute for Inland Water Management and Waste Water Treatment (RIZA), Lelystad, The Netherlands.

Bartram, J. and Ballance, R. [Eds] 1996 *Water Quality Monitoring. A Practical Guide to the Design and Implementation of Freshwater Quality Studies and Monitoring Programmes.* Published on behalf of UNEP and WHO by Chapman & Hall, London.

Chapman, D. [Ed.] 1996 *Water Quality Assessments. A Guide to the Use of Biota, Sediments and Water in Environmental Monitoring.* Second Edition. Published on behalf of UNESCO, WHO and UNEP by Chapman & Hall, London.

Chapman D. and Jackson, J. 1996 Biological monitoring. In: J. Bartram and R. Ballance [Eds] *Water Quality Monitoring. A Practical Guide to the Design and Implementation of Freshwater Quality Studies and Monitoring Programmes.* Published on behalf of UNEP and WHO by Chapman & Hall, London, 263–302.

Cofino, W.P. 1995 Quality management of monitoring programmes. In: M. Adriaanse, J. Van de Kraats, P.G. Stoks and R.C. Ward [Eds] *Proceedings of the International Workshop Monitoring Tailor-made.* Institute for Inland Water Management and Waste Water Treatment (RIZA), Lelystad, The Netherlands.

DEPA 1991 *Environmental Impact of Nutrient Emissions in Denmark.* Published on behalf of Danish Ministry of the Environment by Danish Environmental Protection Agency.

DEPA 1992 *Redegørelse fra Miljøstyrelsen — Aquatic Environment Nationwide Monitoring Programme 1993–1997.* No. 3. Published on behalf of Danish Ministry of the Environment by Danish Environmental Protection Agency.

Demayo, A. and Steel, A. 1996 Data handling and presentation. In: D. Chapman [Ed.] *Water Quality Assessments. A Guide to the Use of Biota, Sediments and Water in Environmental Monitoring.* Second Edition. Published on behalf of UNESCO, WHO and UNEP by Chapman & Hall, London, 511–612.

Dogterom, J. and Buijs P.H.L. 1995 *Concepts for Indicator Application in River Basin Management.* Report 95.01. International Centre of Water Studies (ICWS), Amsterdam.

Dubelaar, G.B.J., Balfoort, H.W. and Hofstraat, H.W. 1990 Automatic identification of phytoplankton. In: *North Sea Pollution: Technical Strategies for Improvement.* N.V.A. Rijswijk, The Netherlands, 539–542.

EPDRB (EPDRB) 1994 *Strategic Action Plan (SAP) for the Danube River Basin 1995–2005*. Task Force for the Environmental Programme for the Danube River Basin, Brussels.

Friedrich, G., Chapman, D. and Beim, A. 1996 The use of biological material. In: D. Chapman [Ed.] *Water Quality Assessments. A Guide to the Use of Biota, Sediments and Water in Environmental Monitoring*. Second Edition. Published on behalf of UNESCO, WHO and UNEP by Chapman & Hall, London, 175–242.

Griffiths, I.M. and Reeder, T.N. 1992 Automatic river quality monitoring systems operated by the National Rivers Authority — Thames Region, UK. *Eur. Wat. Poll.Cont.*, **2** (2), 523–30.

Hushon, J.M. 1990 *Expert Systems for Environmental Applications*. American Chemical Society Symposium Series 431. American Chemical Society, Washington, D.C.

IKSE 1992 *Internationales Warn- und Alarmplan Elbe*. Magdenburg, July 1992, aktualisiert September 1994, International Kommission zum Schutz der Elbe, Magdeburg, Germany.

Laane, W. and Lindgaard-Jørgensen, P. 1992 Ecosystem approach to the integrated management of river water quality. In: P.J. Newman, M.A. Piavaux and R.A. Sweeting [Eds] *River Water Quality. Ecological Assessment and Control*. EUR 14606 EN-FR. Commission of the European Communities, Luxembourg.

Meybeck, M. and Helmer, R. 1989 The quality of rivers: from pristine state to global pollution. *Paleogeog. Paleoclimat. Paleoecol.* (Global Planet. Change Sect.) **75**, 283–309.

Mousty, P., Morvan, J.-P. and Grimaud, A. 1990 Automatic warning stations, recent serious industrial river pollution incidents, and prediction models for pollutants propagation - some European examples. *Wat. Sci. Tech.* **22**, 259–264.

Niederländer, H.A.G., Dogterom, J., Buijs, P.H.L. and Hupkes, R. and Adriaanse, M. 1996 *State of the Art in Monitoring and Assessment*. UNECE Task Force on Monitoring and Assessment, Working Programme 1994/95, Volume No. 5. Institute for Inland Water Management and Waste Water Treatment (RIZA), Lelystad, The Netherlands.

Nordic Fund for Technology and Industrial Development 1993 *Handbook on Processing Data in Municipal and Industrial Waste Water Systems*. Nordic Fund for Technology and Industrial Development, Copenhagen.

NRA 1991 *Proposals for Statutory Water Quality Objectives*. National Rivers Authority Water Quality Series No. 5. HMSO, London.

OECD 1987 *The Use of Biological Tests for Water Pollution Assessment and Control*. Environment Monographs No. 11. Organisation for Economic Co-operation and Development, Paris.

Olivero, R.A. and Bottrell, D.W. 1990 Expert systems to support environmental sampling, analysis and data validation. In J.M. Hudson [Ed.] *Expert Systems for Environmental Applications*. ACS Symp. Series 431. American Chemical Society, Washington, D.C.

Ongley, E.D. 1995 The global water quality programme. In: M. Adriaanse, J. Van de Kraats, P.G. Stoks and R.C. Ward [Eds] *Proceedings of the International Workshop Monitoring Tailor-made*. Institute for Inland Water Management and Waste Water Treatment (RIZA), Lelystad, The Netherlands.

Ongley, E.D. 1996 Sediment measurements. In: J. Bartram and R. Ballance [Eds] *Water Quality Monitoring. A Practical Guide to the Design and Implementation of Freshwater Quality Studies and Monitoring Programmes*. Published on behalf of UNEP and WHO by Chapman & Hall, London, 315–33.

Ruck, B.M., Walley, W.J. and Hawkes, H.A. 1993 Biological classification of river water quality using neural networks. In: *Proceedings of 8th International Conference on Artificial Intelligence in Engineering*. Toulouse, France.

SAST 1992 Research and Technological Development for the Supply and Use of Freshwater Resources I. Krüger Consult AS and Danish Hydraulic Institute. Prepared for the Strategic Analysis in Science and Technology (SAST) Monitoring Programme, Commission of the European Communities, Luxembourg.

Spreafico, M. 1994 Early warning system of the river Rhine. In: *Advances in Water Quality Monitoring*. Report of a WMO regional workshop in Vienna (7–11 March 1994). World Meteorological Organization, Geneva.

Steel, A. Clarke, M. and Whitfield, P. 1996 Use and reporting of monitoring data. In: J. Bartram and R. Ballance [Eds] *Water Quality Monitoring. A Practical Guide to the Design and Implementation of Freshwater Quality Studies and Monitoring Programmes*. Published on behalf of UNEP and WHO by Chapman & Hall, London, 335–62.

Stortelder, P.B.M. and Van de Guchte, C. 1995 Hazard assessment and Monitoring of discharges to water: concepts and trends. *Eur. Wat. Poll. Cont.*, **5**(5).

Suess, M.J. [Ed.] 1982 *Examination of Water for Pollution Control. A Reference Handbook*. Volume 1, Sampling Data Analysis and Laboratory Equipment. Published on behalf of WHO by Pergamon Press, Oxford.

Thomas, R. and Meybeck, M. 1996 The use of particulate material. In: D. Chapman [Ed.] *Water Quality Assessments. A Guide to the Use of Biota,*

Sediments and Water in Environmental Monitoring. Second Edition. Published on behalf of UNESCO, WHO and UNEP by Chapman & Hall, London, 127–74.

Tørsløv, J. and Lindgaard-Jørgensen, P. 1993. Effect of mixtures of chlorophenols, surfactants and aniline on growth of *Pseudomonas florescence. Ecotox. Env. Safe.*

UNECE 1992 *Convention on the Protection and Use of Transboundary Watercourses and International Lakes.* United Nations Economic Commission for Europe, Geneva.

Vieira, J.R. and Lindgaard-Jørgensen, P. 1994 Management support systems for the aquatic environment — concepts and technologies. *J. Hydr. Res. - Hydroinf.,* **32**, 163–83.

Walley, W.J. 1993 Artificial intelligence in river water monitoring and control. In: W.J. Walley and S. Judd [Eds.] *Proceedings of the Freshwater Europe Symposium on River Water Quality, Monitoring and Control.* 22–23 February 1993, Birmingham. Published by Aston University, Aston, UK, 179–94.

Ward, R.C., Loftis, J.C. and McBride, G.B. 1986 The "data-rich but information poor" syndrome in water quality monitoring". *Envi. Manag.,* **10**(3), 291–7.

Ward, R.C., Loftis, J.C. and McBride, G.B. 1990 *Design of Water Quality Monitoring Systems.* Van Nostrand Reinhold, New York.

Ward, R.C. 1995a Monitoring tailor-made: what do you want to know? In: M. Adriaanse, J. Van de Kraats, P.G. Stoks and R.C. Ward [Eds] *Proceedings of the International Workshop Monitoring Tailor-made.* Institute for Inland Water Management and Waste Water Treatment (RIZA), Lelystad, The Netherlands.

Ward, R.C. 1995b Water quality monitoring as an information system. In: M. Adriaanse, J. Van de Kraats, P.G. Stoks and R.C. Ward [Eds] *Proceedings of the International Workshop Monitoring Tailor-made.* Institute for Inland Water Management and Waste Water Treatment (RIZA), Lelystad, The Netherlands.

Wehrens, R., van Hoof, P., Buydens, L., Kateman, G., Vossen, M., Mulder, W.H. and Bakker, T. 1993 Sampling of aquatic sediments. The design of a decision support system and a case study. *Anal. Chim. Acta,* **271**, 11–24.

WHO 1992 GEMS/WATER Operational Guide. Third edition, Unpublished WHO document GEMS/W.92.1. World Health Organization, Geneva.

Winsemius, P. 1986 *Guest in Own House, Considerations about Environmental Management.* Samson H.D. Tjeenk Willink, Alphen aan de Rijn.

Chapter 10[*]

FRAMEWORK FOR WATER POLLUTION CONTROL

10.1   Introduction

This chapter synthesises the aspects of water pollution control presented in Chapters 1–9 and brings their main themes together in order to recommend an approach for comprehensive water resources management. There is, inevitably, some repetition of key messages from the preceding chapters. However, for a more detailed treatment of the specific aspects of water pollution control presented below, readers are advised to study the appropriate chapters. Examples of the different approaches to water pollution control can be found in the case studies indicated.

10.1.1 Background: Agenda 21

In recent years water quality problems have attracted increasing attention from authorities and communities throughout the world, especially in developing countries but also in countries in transition from centrally planned economies to market economies. In the latter, previously neglected aspects of environmental protection are now becoming a major obstacle for further and sustainable economic and social development.

Degradation of surface and groundwater sources has previously been an inherent consequence of economic development and remedial action to compensate for, or to reduce, environmental impacts have always been a lesser priority. Consequently, when the impacts of pollution and the costs of remedial actions are finally acknowledged, the cost of preventive precautionary measures is higher than if they had been implemented at the appropriate time. Thus, negligence of water quality problems often leads to a waste of (economic) resources, resources that might have been used for other purposes if the water quality problems had been given proper attention in the first place.

The international community has now acknowledged the severity of the problems incurred by deteriorating water quality and agreed formally to take

[*]   *This chapter was prepared by H. Larsen and N.H. Ipsen*

action to protect the quality of freshwater resources. The most recent demonstration of this was provided by the United Nations Conference on Environment and Development (UNCED) in Rio de Janeiro in 1992, from which came "Agenda 21". In Chapter 18 of this document (UNCED, 1992), on protection of the quality and supply of freshwater resources, key principles and recommendations for sound water resources management are laid down. These were crystallised, matured and elaborated through a series of preparatory meetings, including the Copenhagen Informal Consultation (CIC) in 1991 and the International Conference on Water and the Environment (ICWE) in Dublin in 1992.

The principles for water resources management that have formed the basis for the guidelines presented here are derived from the conclusions reached in Dublin and Rio de Janeiro and are:

- Freshwater is a finite and vulnerable resource, essential to sustain life, development and the environment.
- Land and water resources should be managed at the lowest appropriate levels.
- The government has an essential role as enabler in a participatory, demand-driven approach to development.
- Water should be considered a social and economic good, with a value reflecting its most valuable potential use.
- Water and land-use management should be integrated.
- Women play a central part in the provision, management and safeguarding of water.
- The private sector has an important role in water management.

10.1.2 Scope of guidelines

The recommendations and principles from Agenda 21 cover water resources management in general, i.e. including availability of water, demand regulation, supply and tariffs, whereas water pollution control should be considered as a subset of water resources management. Water resources management entails two closely related elements, that is the maintenance and development of adequate *quantities* of water of adequate *quality* (see Case Study V, South Africa). Thus, water resources management cannot be conducted properly without paying due attention to water quality aspects. It is very important to take note of this integrated relationship between water resources management and water pollution control because past failures to implement water management schemes successfully may be attributed to a lack of consideration of this relationship. All management of water pollution should ensure integration with general water resources management and vice versa.

The approach presented in this chapter concentrate specifically on aspects that relate to water quality, with special emphasis on the conditions typically prevailing in developing countries and countries in economic transition (e.g. eastern European countries). The intention is to demonstrate an approach to water pollution control, focusing on processes that will support effective management of water pollution. A step-wise approach is proposed, comprising the following elements:

- Identification and initial analysis of water pollution problems.
- Definition of long- and short-term management objectives.
- Derivation of management interventions, tools and instruments needed to fulfil the management objectives.
- Establishment of an action plan, including an action programme and procedures for implementation, monitoring and updating of the plan.

The suggested approach may be applied at various levels; from the catchment or river basin level to the level of international co-operation. The Danube case study (Case Study IX) is an example of the latter. This chapter demonstrates the approach by taking the national level as an example.

10.2   Initial analysis of water quality problems

Management of water pollution requires a concise definition of the problem to be managed. The first task is recognition of an alleged water quality problem as being "a problem". This assumes an ability to identify all relevant water quality problems. The next task is to make sure that useful information is acquired that enables identification and assessment of existing and potential future water quality problems. Thus managers must be able to identify problem areas that require intervention within the water quality sector or the sector for which they are responsible. Nevertheless, even if all existing and potential water quality problems could be identified it may not be feasible to attempt to solve them all at once. All managers are limited by budgetary constraints imposed by political decision makers. Therefore, tools for analysis and prioritisation of water quality problems are indispensable and help make the best possible use of the available resources allocated to water pollution control.

10.2.1 Identification of water quality problems

On a national scale, or regional scale depending on the size of the country, the initial step should be to conduct a water resources assessment. In this context, a water resources assessment is an integrated activity, taking into account water pollution control as well as more general water resources issues. At this very early stage it may be difficult to determine whether a certain problem is purely one of water quality or whether it also relates to the availability of

---

Box 10.1   Summary of water resources assessment

Objective
- To establish a basis for rational water resources management and water pollution control

Action
- To estimate the spatial and temporal occurrence of quantities and qualities of water resources.
- To assess water requirements and development trends, and associated requirements for water quality.
- To assess whether the available resources meet the present and projected demands and requirements in terms of both quantity and quality.

Result
- An overview of the current and expected status and problems of general water resources and water quality.

---

water resources. For example, an identified problem of supplying clean water to a local community may be a problem of scarcity of freshwater resources but may also be caused by inadequate treatment of wastewater discharged into the existing water supply source, thereby rendering the water unfit for the intended use. The water resources assessment should constitute the practical basis for management of water pollution as well as for management of water resources. The recommendation of preparing water resources assessments is fully in line with that given in Agenda 21 (UNCED, 1992), according to which water resources assessments should be carried out with the objective *"... of ensuring the assessment and forecasting of the quantity and quality of water resources, in order to estimate the total quantity of water resources available and their future supply potential, to determine their current quality status, to predict possible conflicts between supply and demand and to provide a scientific database for rational water resources utilization".*

More specifically, the recommended assessment should identify the occurrence (in space and time) of both surface and groundwater quantity and their associated water quality, together with a tentative assessment of trends in water requirements and water resources development (see Box 10.1). The assessment should be based, as far as possible, on existing data and knowledge in order to avoid unnecessary delays in the process of management improvement. The objective of the assessment is not to solve the problems but to identify and list the problems, and to identify priority areas within which more detailed investigations should be carried out. As stated by WMO/UNESCO (1991), *"Water Resources Assessment is the determination of the sources, extent, dependability, and quality of water resources, on which is based an evaluation of the possibilities for their utilization and control"*. An example of implementation of water resources assessments is given in Case Study IV, Nigeria.

10.2.2 Categorisation of water quality problems
Identified water quality problems may fall into different categories requiring application of different management tools and interventions for optimal resolution of the problems. For example, it is important to know whether a certain water quality problem pertains only to a local community or whether it is a national problem. If a problem exists at the national scale it might be necessary to consider imposing general effluent standards, regulations or other relevant measures. By contrast, if the problem is limited to a small geographic region it might only be necessary to consider issuing a local by-law or to intervene to settle a dispute through mediation.

It may also be useful to categorise water quality problems as either "impact issues" or "user-requirement issues". Impact issues are those derived from human activities that negatively affect water quality or that result in environmental degradation. User-requirement issues are those which derive from an inadequate matching of user-specified water quality requirements (demand) and the actual quality of the available resources (supply). Both types of issues require intervention from a structure or institution with powers that can resolve the issue in as rational a manner as possible, taking into consideration the prevailing circumstances.

According to the traditional water pollution control approach, user-requirement issues would often be overlooked because the identification of such problems is not based on objectively verifiable indicators. Whereas an impact issue can be identified by the presence of, for example, a pollution source or a human activity causing deterioration of the aquatic resources (e.g. deforestation), user-requirement issues are identified by a lack of water of adequate quality for a specific, intended use.

10.2.3 Prioritisation of water quality problems
In most cases the resources (financial, human, and others) required for addressing all identified water quality problems significantly exceed the resources allocated to the water pollution control sector. Priorities, therefore,

need to be assigned to all problems in order to concentrate the available resources on solving the most urgent and important problems. If this is not done the effect may be an uncoordinated and scattered management effort, resulting in a waste of scarce resources on less important problems. Ultimately, the process of assigning priority to problems requires a political decision, based on environmental, economic, social and other considerations, and therefore it is not possible to give objective guidelines for this. Nevertheless, some aspects to be considered when assigning priority to water quality problems can be identified as follows:

- Economic impact.
- Human health impact.
- Impact on ecosystem.
- Geographical extent of impact.
- Duration of impact.

As an example, the uncontrolled proliferation of the water hyacinth, *Eichhornia*, in some water bodies may lead to a deterioration in water quality, for example due to oxygen depletion caused by the decay of dead plants, but may also hamper navigation and transport, perhaps with considerable economic consequences. Thus, based on this simple analysis, combating the proliferation of water hyacinth should be given a higher priority than might be indicated by purely environmental considerations.

Another aspect to take into account in assigning priority is the geographical extent of the impact, i.e. whether a particular problem, for example caused by a discharge of wastewater, has only a local impact in an area of a few hundred meters along the river or whether there is an impact in the entire river system downstream of the discharge. The likely answer depends, for example, on the size of the discharge and the retention time in the receiving water bodies, the degradability of the pollutant, and the occurrence of sensitive species in the receiving water body. In addition, the duration of impact should be considered. A discharge of easily degradable organic material may cause considerable deterioration in water quality but only for the duration of the discharge. When the discharge ceases the impact also disappears, although there is often a time lag between the discharge ceasing and no further effects being detected. By contrast, the discharge of a persistent pollutant that is bioaccumulated in the aquatic environment can have an effect long after the discharge has ceased.

10.3 Establishing objectives for water pollution control

When establishing objectives for water pollution control, an essential task is the definition of the ultimate aim. An ultimate aim of effective water pollution control might only be achievable after some considerable time due to financial, educational or other constraints. The further the aims are from the initial situation the more difficult it is to put strategy into practice because a lot of assumptions and uncertainties need to be included. To overcome this problem the following step-wise strategy should be considered:

- Identification of required management interventions.
- Definition of long-term objectives.
- Analysis of present capacity.
- Definition of realistic short-term objectives.

10.3.1 Required management interventions

Having identified and classified relevant water pollution problems, and having assigned priority to them, the next step is to identify appropriate interventions to cope with the problems. For every problem identified, therefore, an assessment should be made of the most appropriate means for intervention. Furthermore, an indication should be given of the relevant administrative level(s) to be involved. The proposed interventions may vary significantly in detail and scope. Depending on the problem in question and the existing institutional framework for management of water pollution, they may range from formulation of a national policy for a hitherto unregulated issue to the establishment of a database containing water quality monitoring results in a local monitoring unit. Examples of typical, required management interventions are:

- Policy making, planning and co-ordination.
- Preparation/adjustment of regulations.
- Monitoring.
- Enforcement of legislation.
- Training and information dissemination.

In many countries, no comprehensive and coherent policy and legislation exists for water pollution control or for environmental protection (see Case Study XIII, Yemen). This does not prevent water pollution control from taking place before such policies have been formulated and adopted, but the most efficient and effective outcome of water pollution control is obtained within a framework of defined policies, plans and co-ordinating activities. There may be obvious shortcomings in the existing situation that need urgent attention and for which remedial actions may be required independently of the overall general policy and planning. Such interventions and remedial actions should be taken whether or not an overall policy exists. A lack of policy should not delay the implementation of identified possibilities for obvious improvements in water pollution control. In many developed countries, regulations supporting legislation are also lacking, inadequate or outdated (see

Case Study X, Russia). Adjustment of regulations is an ongoing process that has to adapt continuously to the socio-economic development of society.

A typical weakness in legislation, which should be avoided, is the tendency to state explicitly within the act economic sanctions for non-compliance (such as fees, tariffs or fines). It is much more complicated and time consuming to change or to amend an act than to amend the supporting regulations and management procedures. Hence, stating economic sanctions within an act entails an associated risk that enforcement of the legislation could become ineffective and outdated due to economic inflation. Examples of inadequate, or lack of enforcement of, existing legislation are widespread and can be illustrated by Case Studies III, IX, X and VI (Philippines, Danube, Russia and Brazil).

Improvement in water quality monitoring systems is an intervention required world-wide, not only in developing countries. There are, however, huge differences from country to country in the shortcomings induced by inadequate, existing monitoring systems. In most developing countries the problem is one of too little monitoring due to a lack of allocated resources for this activity. In several central and eastern European countries the problem is different. Extensive monitoring programmes have been functioning for many years and many raw data have been collected. What has been missing in a number of cases is an ongoing analysis and interpretation of the data, i.e. transformation of the data into useful information, followed by a subsequent adjustment of the monitoring programmes.

10.3.2 Long-term objectives
Definition of long-term objectives includes the identification of key functions that will have to be performed in order to achieve reasonably effective water pollution control at all administrative levels. This evaluation and description of necessary management functions and levels should be made without giving too much consideration to the existing administrative capacity at various administrative levels. It may be assumed, for example, that there is a reasonable capacity to carry out the necessary tasks designated at each level in the long-term strategy. However, a reasonable assessment of the full potential for development of the general level of management should form the basis for the long-term objectives. If the present situation is characterised by extremely scarce financial and human resources and major obstacles to economic and social development, it would not be appropriate to define very high standards of water pollution control in the long-term objective, simply because this situation would most likely never occur. The situation obtained by fulfilling the long-term objectives for water pollution control, should be

one that is satisfactory to society (considering the anticipated general level of development at that future moment).

The guiding principles for water resources management (see section 10.1) should be reflected in the long-term strategy. For example, management at the lowest appropriate level should be pursued through the identification of the lowest appropriate level for all identified key functions, irrespective of the present level of management. For some functions, the lowest appropriate level is a local authority or unit, while for other functions it is a central authority (e.g. Case Study I, India). The case study for China (Case Study II), however, provides an example of the opposite approach, i.e. centralised control of pollution. Table 10.1 gives an example of how elements of a long-term strategy for water pollution control could be described.

10.3.3 Analysis of present capacity
Having defined long-term objectives it is necessary to assess how the present situation matches the desired situation. The key issue is identification of the potential of, and constraints upon, the present management capacity and capability in relation to carrying out the management functions defined in the long-term objectives. Such aspects as suitability of institutional framework, number of staff, recruitability of relevant new staff, educational background, and availability of financial resources should be considered. The needs for training staff and for human resources development to enhance management performance should also be identified and plans made for initiation of this development.

In many countries, problems associated with an absence of clear responsibilities, with the overlapping of institutional boundaries, duplication of work and a lack of co-ordination between involved institutions, are common obstacles to effective water pollution control (see Case Studies V, III, XIII, X and IV for South Africa, Philippines, Yemen, Russia and Nigeria).

The analysis must include all relevant administrative levels, for example through intensive studies at the central level combined with visits and studies in selected regions at lower administrative levels. The regions or districts should not be selected randomly but with a view to selecting a representative cross-section of diversity in water quality problems and their management. An example of such an analysis is given in Table 10.2.

10.3.4 Short-term strategy
In relation to short-term strategy, the duration of the "short-term" has to be defined. A period of approximately five years is suggested, because this is

Table 10.1   Summary of long-term strategy for water pollution control

Function	National level	Intermediate level	Local level
Formulation of international policies	Defining the country's position with regard to cross-border issues of water pollution. Providing information for negotiations with upstream and down-stream riparian states	None	None
Wastewater discharge regulation	Processing waste-water discharge applications and issuing discharge permits	Commenting on applications in relation to district development planning. Organising public hearings. Assisting in checking that permissions are adhered to. Disseminating information on national standards through public health authorities	Assisting in the monitoring of potentially harmful discharges; framing and enforcing local rules and maintaining structures to avoid contamination of domestic water sources through sub-district water and sanitation committees and water user groups

Source: Directorate of Water Development/Danida, 1994

roughly the planning horizon that can be controlled reasonably well and foreseen without too much dependency on future development scenarios.

The output of the capacity analysis provides the basis for establishing a short-term strategy, taking into account the identified potential for, and constraints associated with, achieving the long-term objectives. For example, a long-term objective might be to decentralise water quality monitoring activities. However, if the current manpower skills and analytical capabilities at the lower administrative levels do not allow implementation of this strategy (see Case Study VII, Mexico), a short term strategy might be defined, maintaining monitoring activities at a central level but simultaneously upgrading the skills at the lower levels by means of training activities and orientation programmes. Alternatively, monitoring could be restricted in the short-term to those activities that can currently be carried out by the lower levels, and additional monitoring activities could be gradually included along with upgrading of manpower skills and analytical facilities.

In general, when defining the short-term strategy it should be ensured that the fulfilment of the short-term objectives will significantly contribute to

Table 10.2   Example of an analysis of present management capacity

Functions	Potentials	Constraints
Formulation of international policies	Establishment of a Water Policy Committee has been agreed	Lack of formal agreements between upstream and downstream riparian countries. Lack of reliable information on the quantity and quality of shared water resources
Wastewater discharge regulation	Staff with necessary knowledge available at national level. Required administrative structures and procedures at national level are relatively uncomplicated. District Water Officers can assist in monitoring activities	Lack of qualified staff at district local level to deploy for discharge control. Lack of monitoring equipment. Very limited access to laboratory facilities

Source: Directorate of Water Development/Danida, 1994

Table 10.3   Example of a short-term strategy for water pollution control

Functions	National level	Lower levels
Formulation of international policies	Establish Water Policy Committee, its secretariat and its international subcommittees	None
Wastewater discharge regulation	Establish unit for administering wastewater discharge permits as per regulations	Identify wastewater dischargers requiring licensing. Establish procedures for administering the licensing system as per regulations. Local authorities to report on pollution problems and to comment on wastewater discharge applications

Source: Directorate of Water Development/Danida, 1994

achieving the long-term objectives. An example of definition of a short-term strategy for water pollution control, based on the above example of a long-term strategy with identified potentials and constraints, is given in Table 10.3.

10.4   Management tools and instruments

This section discusses a number of management tools and instruments together with principles for their application and for the combination of different tools (for a more thorough description of tools and instruments see preceding chapters). The range of tools and instruments should be considered as an input to the overall process of achieving effective water pollution control, that is a toolbox for the water pollution manager. They are necessary means to address the identified problems. The manager's task is to decide which tool(s) will most adequately solve the present water pollution problem and to ensure that the selected tool(s) are made available and operational within the appropriate institutions.

10.4.1 Regulations, management procedures and by-laws

Regulations are the supporting rules of the relevant legislation. Regulations can be made and amended at short notice, and in most cases need only the approval of the minister to become binding. In specific cases, approval by the cabinet may be necessary. Regulations specify the current policies, priorities, standards and procedures that apply nationally.

Management procedures are a set of guidelines and codes of practice that ensure consistent responses in problem solving and decision making. Such procedures contain a further level of detail supporting the legislation and the regulations and specifying the steps to be taken in implementing particular provisions, such as regulation of wastewater discharge. Regulations and procedures pertaining to wastewater discharge would typically include, for example, descriptions of procedures for applying and granting a permit to discharge wastewater to a recipient, procedures for monitoring compliance with the permit, fees and tariffs to be paid by the polluter, and fines for non-compliance.

As a general rule it should be ensured that only regulations that are enforceable are actually implemented. If the existing enforcement capacity is deemed insufficient, regulations should be simplified or abandoned. Regulations and management procedures made at the national level need not necessarily apply uniform conditions for the entire country, but can take account of regional variations in water pollution and socio-economic conditions.

By-laws (that are binding on local residents) can be made by a legally established corporate body, such as a district or province government and can, for example, determine the regulation and pollution of local water resources. By-laws made by lower level institutions cannot contradict those made by higher level institutions (see Chapter 5).

10.4.2 Water quality standards

Water quality standards are, in fact, part of regulations but are discussed separately here because some important aspects relating specifically to the use of standards should be noted (see Chapters 2 and 5). Numerous sets of water quality standards, or guidelines for water quality standards, have been issued during the course of time by various agencies and authorities (e.g.

United States Environmental Protection Agency (EPA), World Health Organization (WHO), European Union (EU)) intending to define the maximum acceptable limit of water pollution by various pollutants. Standards for ambient water quality (quality objectives) are commonly designated according to the intended use of the water resource (e.g. drinking water, fishing water, spawning grounds), while effluent standards are usually based on either of the following two principles, or a combination of both (see Case Study II, China):

- Fixed emission standard approach, requiring a certain level of treatment of all wastewater, regardless of the conditions and intended use of the receiving water body.
- Environmental quality standard approach, defining the effluent standards in order to enable compliance with the quality objectives for the receiving water body.

Standards or guidelines developed according to the first approach must be very restrictive in order to protect the environment effectively, because they must take into account the most critical situations and locations. Thus, this approach might lead to unnecessary treatment costs in some situations. In other cases, it may lead to inappropriate treatment and excessive pollution, depending on the applied emission standards and the assimilative capacity of the receiving water body (see Case Study V, South Africa). The major advantage of this approach is its rather simple administrative implications.

The second approach allows for a more flexible administration of environmental management, and optimisation of treatment efforts and costs because the level of treatment may be tuned to the actual assimilation capacity of the receiving waters (which must be assessed on an individual basis). The problem with this approach is the difficulty in practical application; knowledge of the assimilative capacity requires studies of the hydraulic, dispersive, physico-chemical and biological conditions prevailing in the water body. In addition, plans for future development in the area should be taken into account. The above factors suggest that a strategy based on the fixed emission standard approach may be the most appropriate, at least as a starting point in many developing countries because of their often limited administrative capacities. However, the dangers associated with automatically adopting water quality standards from western industrialised countries must be emphasised. The definition of water quality standards should, to a large extent, be a function of the level of economic and social development of a society. For example, a number of water quality standards applied in western countries are based on the best available technology (BAT) and generally achievable technology (GAT) principles. These require organisations to treat their wastewater according to

BAT for hazardous substances and according to GAT for other substances. Whereas the economic costs of applying these principles may be affordable in a highly industrialised country, they may be prohibitive for further industrial and economic development in developing countries.

In central and eastern European countries, water quality standards and emission standards are often more stringent. In some cases they are too stringent to be met and in other cases they are even too stringent to be measured (see Case Study IX, Danube). As a result the standards have often been ignored by both polluters and managers. In addition, the necessary administrative capacity to enforce very high water quality standards may exceed that available. As mentioned previously, it is highly recommended that only regulations that can be enforced are implemented.

Water quality standards applied in developing countries should, therefore, be adjusted to reflect the local (achievable) economic and technological level. The implication of this approach is that standards may be tightened along with the rise in economic capability to comply with higher standards. Furthermore, since a high level of wastewater treatment is often easier and cheaper to achieve when considered during the planning and design phase of any industrial production, more strict effluent standards (when compared with existing discharges) may be imposed on new discharges of wastewater. These measures would allow for both economic development and the gradual increase in environmental protection.

10.4.3 Economic instruments

The use of economic instruments is on the increase in many countries but is far from reaching its full potential. Until now, most governments have relied primarily on regulatory measures to control water pollution. However, application of economic instruments in water pollution control may offer several advantages, such as providing incentives for environmentally sound behaviour, raising revenue to help finance pollution control activities and ensuring that water quality objectives are achieved at the least possible (overall) cost to society.

The main types of economic instruments applicable in a water pollution context include (Warford, 1994; see Chapter 6):

- Resource pricing.
- Effluent charges.
- Product charges.
- Subsidies or removal of subsidies.
- Non-compliance fees (fines).

Prerequisites for the successful implementation of most economic instruments are appropriate standards, effective administrative, monitoring and enforcement capacities, institutional co-ordination and economic stability. Various degrees of administration are associated with the application of different economic instruments. Effluent charges, for example, require a well-established enabling environment and large institutional capacity and co-ordination. By contrast, product charges are relatively simple to administer (Warford, 1994).

Among the key factors in the successful implementation of economic instruments is the appropriate setting of prices and tariffs. If prices are set too low, polluters may opt to pollute and pay, as seen in some eastern and central European countries (see Case Study IX, Danube). Moreover, artificially low prices will not generate adequate revenues for system operation and maintenance (see Case Study VII, Mexico). Setting appropriate prices is very difficult because, ideally, prices should cover direct costs, opportunity costs and environmental costs (externalities) (Nordic Freshwater Initiative, 1991).

Economic instruments incorporate the polluter-pays-principle to various degrees. Subsidies, for example, clearly counteract the polluter-pays-principle but may, in some cases, be applied for political or social reasons. By contrast, effluent charges go hand-in-hand with the polluter-pays-principle. In the case of resource pricing, progressive charging scales may be used to allow large-scale users to subsidise the consumption of small-scale users, and thereby balance considerations of social needs and sustainable use of the resource.

10.4.4 Monitoring systems

There are a number of important elements to consider in relation to the implementation and functioning of a monitoring system (see Chapter 9):

- Identification of decision and management information needs.
- Assessment of capacity (economic and human) to maintain the monitoring system.
- Proper design of the monitoring programme and implementation of routines according to defined objectives.
- Data collection.
- Data handling, registration and presentation.
- Data interpretation for management.

Traditionally, monitoring programmes collect data either from chemical and biological analysis of water samples or from on-line field equipment. However, depending on available laboratory facilities, instruments, transport and human resources, for example, all monitoring programmes are restricted

---

Box 10.2   An example of indirect estimation of pollution load

Load estimates can be based on, for example, measurements available from a monitoring system. However, very often it is only possible to cover part of a lake or river catchment with monitoring stations, and hence only some of the major contributors to pollution load, due to the limited resources available. The rest of the catchment has to be taken into consideration using experience and representative measurements from elements of a similar catchment. Furthermore, it is possible to give recommendations of unit loads from personal equivalents (p.e.) in relatio to economic status. Unit loads from different types of industry and run-off of pollutants from, for example, agricultural land and forests can also be deduced according to the farming or forestry practised.

---

in some way and may collect data primarily by direct sampling. A number of information gaps often have to be filled, therefore, before a rational decision about monitoring system design can be taken with respect to a specific water quality problem. Although they are less accurate, indirect techniques for obtaining the necessary information exist for a variety of water quality-related factors. It is possible, for example, to obtain reasonable estimates of pollution quantities from various sources from a knowledge of the activities causing the pollution (see Box 10.2).

Another frequent problem associated with traditional monitoring programmes is the lack of coupling between measured concentrations and water flow or discharge measurements, thereby rendering quantification of pollution transport difficult. Estimation techniques also exist for these situations, where hydrometric networks are not established or functioning, or where instruments are not available for measuring flow, such as in wastewater discharges.

The actual design of a fully operational and adequate national monitoring system must, from the beginning, take account of the requirements of the additional management tools which are being considered for use (see Case Study III, Philippines). The complexity and size of the area to be monitored, the number of pollutants monitored, and the frequency of monitoring, have to be balanced against the resources available for monitoring. To a large extent the data that become available determine the level of complexity of the management tools that can be supported by the monitoring system. An example of the kind of support needed for other management tools is the requirement for reliable and frequent data to support the enforcement of effluent standards (see Case Study XII, Jordan). In this situation the monitoring programme needs to be tailored to suit the detailed requirements for enforcement, as defined in the supporting regulations.

10.4.5 Water quality modelling tools

Modelling tools are treated here as any set of instructions based on a deterministic theory of cause-effect relationships which are able to quantify a specific water quality problem and thereby support rational management decisions. This can be done at different levels of complexity, some of which are discussed below:

- *Loadings*. Preliminary decisions can be taken with respect to reduction of loadings from a ranking of the size of actual pollution loadings to a particular receiving water body. The rationale is to assess where the greatest reduction in pollution can be obtained in relation to the costs involved.
- *Mass balances*. Mass balances can be established using load estimates from pollution sources in combination with the water flow or residence time in the water body. The significance of the different loadings can be evaluated by comparing their magnitude to their contribution to the resulting concentration of the pollutant in the receiving waters. The significance of the different loadings for the pollution level of the receiving water body provides the rational basis for decisions on effective reduction of the pollution level in those waters.
- *Effect evaluation*. Assessment of changes in the identified pollution sources and their resulting concentration in the receiving waters can be made at various levels, from using simple, empirical relations to long-term mass balance models. An example of a well known empirical relation is the Vollenweider method for estimating eutrophication effects in lakes (Vollenweider, 1968, 1975, 1976). Based on experience from measurements in a large number of lakes, the method relates pollution discharges and static lake characteristics (such as water depth and retention time) to expected effects on the Secchi depth and algal concentrations. Effect evaluation may also combine considerations about cost effective pollution reduction at the source, the resulting pollution concentration in receiving waters and the resulting effects in the ecosystem.
- *Simple mathematical mass balance models*. Application of this tool allows consideration of the possible changes over time in relation to any reductions proposed in pollution load. Many types of these biogeochemical models have been developed over the years and some are available in the public domain.
- *Advanced ecological models*. If higher level effects of pollution loadings on an ecosystem are to be determined, more sophisticated ecological models are available. Such models may create the basis for a refined level of

prediction (see Case Study III, Philippines) and should be used in cases of receiving waters with high complexity and importance, provided sufficient resources (financial, human or institutional) exist or can be allocated.

The above examples serve to illustrate that quantitative assessments of pollution problems can be performed at various levels of complexity, from hand calculations to advanced state-of-the-art ecological modelling.

10.4.6 Environmental impact assessment and cross-sectoral co-ordination

Impact assessment plays a central role in the process of providing information on the implications for water quality arising from development programmes and projects. However, in addition to impacts on the physical environment, impacts on the water resources often imply impacts on the biological and socio-economic environment. Assessments of impacts on water quality should, therefore, often be seen as an integral part of an environmental impact assessment (EIA). Environmental impact assessments are being used increasingly as environmental management tools in numerous countries (see Case Studies II and IV, China and Nigeria).

The main objectives of impact assessments used for the purposes of water quality management are to identify potential impact on water quality arising from proposed plans, programmes and projects. They therefore serve:

- To assist decision makers in making informed decisions on project developments and final project prioritisation.
- To provide, where possible, relevant and quantitative water quality information so that potential impacts can be avoided or reduced at the project and programme design stage.
- To provide a basis for development of management measures to avoid or reduce negative impacts under, and/or after, project implementation.

The impact assessment should form an integral part of multiple resource development planning and feasibility studies for the projects. It should provide for a quantified assessment of the physical, biological and related economic and social impacts of proposed projects as well of the likelihood of such impacts occurring. Thus, the impact assessment should accomplish its purpose by providing decision makers with the best quantitative information available regarding intended, as well as unintended, consequences of particular investments and alternatives, the means and costs to manage undesirable effects, and the consequences of taking no action.

An important element in any impact assessment is the encouragement of public participation in the process. The general public should be given an opportunity to express their views on proposed projects and programmes, and procedures should be established for considering these views during the decision making process. In many cases, non-governmental organisations (NGOs) with considerable insight in environmental issues can be identified and may provide valuable contributions to the impact assessment. Public participation can often ease the implementation of projects and programmes as a result of the increased feeling of ownership and influence that it produces amongst directly-involved users (see Case Studies III, V, VI and IX for the Philippines, South Africa, Brazil and Danube).

In addition to identifying and describing water quality impacts that a proposed programme or project would cause if no management measures were taken, an impact assessment should:

- Specify the necessary measures to protect water quality.
- Ensure that these are included in the project implementation plan.

Finally, evaluations of water quality impacts and technical and economic feasibility should be linked so that effective project modification and water quality management can be developed. Water quality aspects and economic evaluations should be linked to ensure that both water quality benefits and drawbacks of the project, as well as the costs of water quality management, can be accounted for in a subsequent cost-benefit analysis.

The operational functions of the water quality impact assessment should be to provide the necessary background for:

- Approval or rejection of wastewater discharge permit applications.
- Inclusion of operation conditions in wastewater discharge permits.
- Input to EIAs.
- Inclusion of water quality consequences in the final prioritisation of development projects (made by authorities at different levels).
- Developing modifications in the technical design of development projects with the aim of protecting water resources.

Capacity for making and overseeing water quality impact assessments should be developed within the relevant water or environment authorities, although the actual assessments should not necessarily always be made by the authority itself, for example line ministries, local authorities or private companies may undertake the task. However, detailed procedures and guidelines should be developed and co-ordinated with the development of general EIA procedures within the country.

The integrated water resources management approach implies that sectoral developments are evaluated for possible impacts on, or requirements for, the water resources and that such evaluations are considered when designing and allocating priority to development projects. Consequently, the water resources management systems must include cross-sectoral information exchange and co-ordination procedures, techniques for evaluation of individual projects with respect to their implications for water resources, and procedures ensuring that water resources aspects are included in the final design and prioritisation of projects.

As a general rule a rapid screening of the project for possible water resources implications, regarding water quality as well as other aspects, should be carried out and if the project is likely to cause water related problems it should be subject to:

- Impact assessment (possibly EIA).
- An evaluation of possible specific requirements affecting the involved water resource and recommendations for project design to fulfil such requirements.
- Identification of possible interaction with, or competition from, other planned or ongoing projects in relation to use of the same water resource.
- Recommendations on possible improvements in project design to provide optimal exploitation of water resources.

Finally, the evaluations and recommendations should be included in the prioritisation process of the project emphasising both environmental and economic implications arising from the water resources issues.

The integration of water pollution issues in the prioritisation process makes it necessary that tools and procedures exist for securing adequate exchange of information between bodies preparing the project, the water pollution authorities and the final decision makers. These requirements are:

- That information about new proposals for projects which may impact or imply specific requirements for water quality should reach the water pollution authorities in good time for the elaboration of impact assessments and recommendations before final decisions are taken (including consideration of potential alternative exploitation of the involved water resources).
- That the same authorities should possess rapid access to relevant information about registered, planned and ongoing water-related projects through, for example, adequate database tools.

10.4.7 Principles for selecting and combining management tools

When deciding on which management tools and instruments to apply in order to improve water pollution control in a given situation, some underlying principles should be considered to help achieve effective management. The principles are:

- Balance the input of resources against the severity of problem and available resources.
- Ensure sustainability.

- Seek "win-win" solutions, whereby environmental as well as other objectives are met.

Balance the input of resources

This principle entails a reasonable input of financial, human or other resources to handle a specific problem, according to the priority and severity previously assigned to that problem. For example, if the discharge of waste-water is concentrated at a few locations in a country, leaving most regions or districts unaffected by wastewater discharge, and if this situation is antici-pated to continue, there would be no need to build technical and administrative capacities to handle the problem in all regions or districts. Similarly, the treatment requirements and the threshold size for activities requiring a wastewater discharge permit might be more lenient if only a few dischargers exist and if the receiving waters show no symptoms of pollution.

Ensure sustainability

This principle has a bearing upon the methods and technical solutions that should be considered for the purposes of water pollution control. In most developing countries possibilities for the operation and maintenance of advanced technical equipment are very scarce or non-existent. Among donors and recipients of projects there has been a tendency to favour quite advanced and sensitive technical solutions, even in situations where more simple and durable equipment would have been sufficient and adequate (see Case Study VII, Mexico). This can result in entire development programmes failing to be implemented successfully. Thus, as a general rule in many devel-oping countries, it is best to keep technical solutions simple. The recommendation to use simple stabilisation ponds for wastewater treatments is one such example (as in Case Study VII, Mexico).

Sustainability also entails building on existing structures, where appropriate, instead of building new structures. Existing institutions or methods have, to some extent, proved their viability. It is more likely that the allocation of resources for existing institutions would be continued rather than additional resources would be allocated for new institutions.

Seek "win-win" solutions

"Win-win" situations (Bartone *et al.*, 1994; Warford, 1994; see also Chapter 6) are created by applying instruments that lead to improvement in water pollu-tion control as well as in other sectors (e.g. improved health or improvement in economy). This means that the difficult balancing between environmental benefits and other drawbacks is avoided. Economic instruments are often in the "win-win" category.

Regulatory versus economic instruments
Compared with economic instruments, the advantages of the regulatory approach to water pollution control is that it offers a reasonable degree of predictability about the reduction of pollution, i.e. it offers control to authorities over what environmental goals can be achieved and when they can be achieved (Bartone *et al.*, 1994). A major disadvantage of the regulatory approach is its economic inefficiency (see also Chapter 6). Economic instruments have the advantages of providing incentives to modify the behaviour of polluters in support of pollution control and of providing revenue to finance pollution control activities. In addition they are much better suited to deal with non-point sources of pollution. However, setting of appropriate prices and charges is crucial to the success of economic instruments and is often difficult to achieve.

Against this background, it seems appropriate for most countries to apply a mixture of regulatory and economic instruments for controlling water pollution. In developing countries, where financial resources and institutional capacity are very limited, the most important criteria for balancing economic and regulatory instruments should be cost-effectiveness (those that achieve the objectives at the least cost) and administrative feasibility.

Finally, in cases of highly toxic discharges, or when a drastic reduction or complete halt in the discharge is required, regulatory instruments (e.g. a ban) rather than economic instruments should be applied.

Levels of water pollution control
According to Soliman and Ward (1994), the various management tools available may be applied and combined at five categories (levels) of water pollution control, reflecting an increasing level of development and economic and administrative capacity:
- *Crisis management.* Non-proactive mode; doing very little management (e.g. no regulation); action is taken only in response to disasters or emergencies, where a group of specialists is assigned to handle the problem; no efforts made to prevent the problem in the future. This approach is adequate in only a very few cases today.
- *The criteria/standard only strategy.* At this stage, the risk of environmental problems occurring justifies a more proactive approach to water pollution management; water quality criteria and standards may be formulated;

monitoring of compliance with standards; still a passive mode of management in which no attempts are made to modify the system.

- *Controlling strategy.* If the results of monitoring using the previous strategy showed that water quality standards have been violated, additional management tools are applied; effluent standards and wastewater discharge permits may be introduced in combination with enforcement and penalty procedures to handle violations. Management has entered the proactive mode.

- *Compliance assistance strategy.* In many developing countries, widespread violations of permits may still occur because the treatment costs needed to meet the effluent standards are higher than many industries can afford. In this situation, decision makers may decide to offer financial aid to firms and municipalities in order to treat their effluents adequately, rather than closing down the installations, which would often be the only alternative to accepting continued violations. Setting priorities for financial and technical assistance is a vital component at this stage, where management has reached a supportive mode.

- *Enhancement of the science/policy of management.* Management designing the future; grants for research in water pollution control and for application of modern techniques; forecasting future potential problems and preparing to prevent the occurrence of such problems; management in an interactive mode.

10.5   Action plan for water pollution control

10.5.1 Components of and processes within an action plan

The preceding sections have described various elements and aspects of what could be considered as an action plan for water pollution control. Some elements are identical to elements from traditional master plans but, contrary to prescriptive and rather rigid master plans, the action plan concept provides a flexible and dynamic framework for development and management of water resources. It is very important to recognise the dynamic nature of the action plan concept because a significant value of the concept lies in its flexibility. The action plan should be continuously monitored and adjusted in order to take account of recent development trends. Only a flexible and non-prescriptive approach will allow for such changes.

An overview of the components and the processes within the action plan concept are given in Figure 10.1. One of the main results of the action plan is a list of actions proposed for implementation in order to achieve the goal of effective and sustainable water quality management. For easy

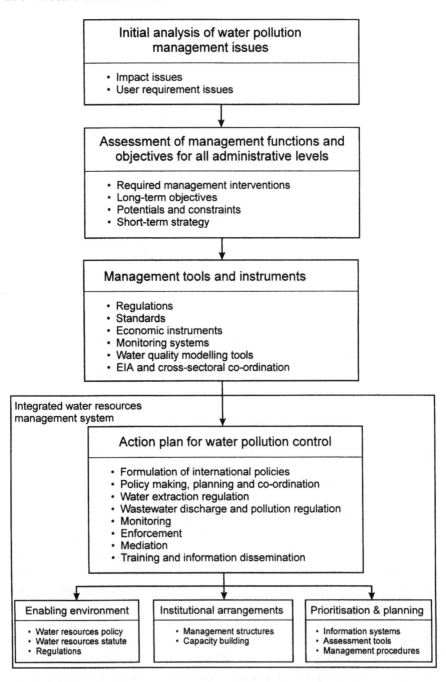

Figure 10.1 Elements and processes of an action plan for water pollution control

implementation and updating, the action list should preferably be prepared using a common format for each identified necessary action. For example, each action could be accompanied by information on the background (justification) for inclusion, objective and expected output, and the tasks necessary to be carried out. This information will facilitate easy transformation of the relevant actions into projects, if appropriate. The actions can typically be organised according to the following categories (Figure 10.1):

- Actions supporting the development of an enabling environment, i.e. a framework of national legislation, regulations and local by-laws for encouraging sound management of water pollution and constraining potentially harmful practices.
- Actions supporting development of an institutional framework which allows for close interaction between national, intermediate and local levels.
- Actions enhancing planning and prioritisation capabilities that will enable decision makers to make choices (based on agreed policies, available resources, environmental impacts and the social and economic consequences) between alternative actions.

Training and capacity development are an integrated element of the proposed actions that apply to all categories. In addition to skill-based training related to developing assessment capabilities, there may be a need for different training, education and information activities at various levels (such as orientation programmes, curriculum development and extension training) in order to carry out the functions described in the short term strategy.

In accordance with the underlying principles of the government as an enabler in a demand-driven approach but with management occurring at the lowest appropriate levels, it is necessary to create a structure that facilitates decentralisation of management (see Case Study IX, Danube). National agencies should be concerned with essential functions that are not dealt with at other levels and they should act as enablers that review and revise the overall structure so that it responds to current needs and priorities.

The recommended framework should be one that attempts to reach a balance between national and local levels carrying out the identified management functions previously outlined. The envisaged organisational framework should, as far as possible, build on existing structures.

10.5.2 Implementation, monitoring and updating of the action plan

Depending of the number of proposed actions contained in the action plan, a phased implementation of the actions may be desirable. For example, the actions could be scheduled according to the following criteria:

- *Cohesion.* Some actions may cluster together.

- *Conditionality*. The pattern of actions may largely follow the overall pattern of the action plan, i.e. creating the legislative framework which establishes the enabling environment, building the appropriate institutional structures, and producing the required water quality management procedures and tools.
- *Dependency*. Some actions cannot be started until others are completed; for example, training related to developing an integrated extension service cannot take place until agreement has been reached to establish such a service.
- *Urgency*. Some actions are started in the initial phase because they are ranked as high priority.

A feasible, overall concept for phased implementation that might be considered is:

- Creating/adjusting the enabling environment, e.g. policies, legal procedures, regulations.
- Building/shaping the institutional structures.
- Producing/applying the required management tools and instruments.

It is very important to recognise that the action plan will have no significance if the action programme is not implemented, and unless all concerned parties are aware of the principles and procedures of the plan and are prepared to co-operate in its implementation. The action programme is the backbone of the action plan. Therefore, procedures for monitoring the progress of implementation should form part of the plan. Key indicators should be identified illustrating the progress, as well as the associated success criteria.

As indicated above, an obvious key indicator for monitoring the progress of the action plan would be the progress of setting up key institutional structures. Other useful indicators, depending on the actions listed, could be attendance at training courses and workshops, whether or not a permit system for wastewater discharges is implemented, number of analyses performed as part of a water quality monitoring programme. To document the progress of the action plan (or lack of it), a regular system for reporting on the monitoring activities should be instituted.

The action plan as a continuous process calls for frequent updating (see Case Study III, Philippines) and the addition of new actions as contexts change, requirements develop, or as progress falls below expectations or schedules. Modifications of earlier proposed actions may also be relevant. Regular monitoring reports should be accompanied by updated project/action lists.

10.6   References

Bartone, C., Bernstein, J., Leitmann, J. and Eigen, J. 1994 *Toward Environmental Strategies for Cities: Policy considerations for Urban Development*

Management in Developing Countries. UNDP/UNCHS/World Bank, Urban Management Programme, Washington, D.C.

Directorate of Water Development/Danida, 1994 *Uganda Water Action Plan.* Directorate of Water Development, Uganda and Danida, Denmark

Nordic Freshwater Initiative 1991 *Copenhagen Report. Implementation Mechanisms for Integrated Water Resources Development and Management.* Background document for the UN Conference on Environment and Development, Nordic Freshwater Initiative, Copenhagen.

Soliman, W.R. and Ward R.C. 1994 The evolving interface between water quality management and monitoring. *Wat. Int.*, **19**, 138–44.

UNCED 1992 Chapter 18 Protection of the quality and supply of freshwater resources. In: *Agenda 21.* United Nations Conference on Environment and Development, Geneva.

Vollenweider, R.A. 1968 *Scientific Fundamentals of the Eutrophication of Lakes and Flowing Waters, with Particular Reference to Nitrogen and Phosphorus as Factors in Eutrophication.* Organisation for Economic Co-operation and Development, Paris.

Vollenweider, R.A. 1975 Input-output models. With special reference to the phosphorus loading concept in limnology. *Schw. Z. Hydrolog.* **27**, 53–84.

Vollenweider, R.A. 1976 Advances in defining critical load levels for phosphorus in lake eutrophication. *Mem. dell'Inst. Ital. di Idrobiol.*, **33**, 53–83.

Warford, J.J. 1994 Environment, health, and sustainable development: the role of economic instruments and policies. Discussion paper. Director General's Council on the Earth Summit Action Programme for Health and Environment, World Health Organization, Geneva.

WMO/UNESCO 1991 *Report on Water Resources Assessment: Progress in the Implementation of the Mar del Plata Action Plan and a Strategy for the 1990s.* World Meteorological Organization, Geneva and United Nations Educational, Scientific and Cultural Organization, Paris.

Case Study I[*]

THE GANGA, INDIA

I.1   Introduction

There is a universal reverence to water in almost all of the major religions of the world. Most religious beliefs involve some ceremonial use of "holy" water. The purity of such water, the belief in its known historical and unknown mythological origins, and the inaccessibility of remote sources, elevate its importance even further. In India, the water of the river Ganga is treated with such reverence.

The river Ganga occupies a unique position in the cultural ethos of India. Legend says that the river has descended from Heaven on earth as a result of the long and arduous prayers of King Bhagirathi for the salvation of his deceased ancestors. From times immemorial, the Ganga has been India's river of faith, devotion and worship. Millions of Hindus accept its water as sacred. Even today, people carry treasured Ganga water all over India and abroad because it is "holy" water and known for its "curative" properties. However, the river is not just a legend, it is also a life-support system for the people of India. It is important because:

- The densely populated Ganga basin is inhabited by 37 per cent of India's population.
- The entire Ganga basin system effectively drains eight states of India.
- About 47 per cent of the total irrigated area in India is located in the Ganga basin alone.
- It has been a major source of navigation and communication since ancient times.
- The Indo-Gangetic plain has witnessed the blossoming of India's great creative talent.

I.2   The Ganga river

The Ganga rises on the southern slopes of the Himalayan ranges (Figure I.1) from the Gangotri glacier at 4,000 m above mean sea level. It flows swiftly

---

[*]   *This case study was prepared by Y. Sharma*

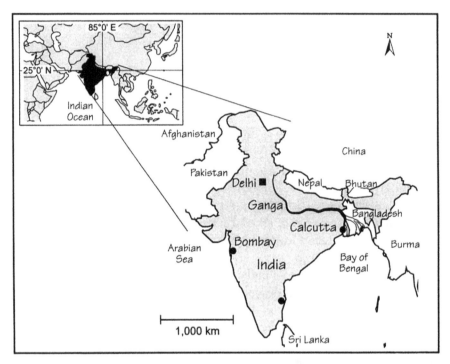

Figure I.1 Location map of India showing the Ganga river

for 250 km in the mountains, descending steeply to an elevation of 288 m above mean sea level. In the Himalayan region the Bhagirathi is joined by the tributaries Alaknanda and Mandakini to form the Ganga. After entering the plains at Hardiwar, it winds its way to the Bay of Bengal, covering 2,500 km through the provinces of Uttar Pradesh, Bihar and West Bengal (Figure I.2). In the plains it is joined by Ramganga, Yamuna, Sai, Gomti, Ghaghara, Sone, Gandak, Kosi and Damodar along with many other smaller rivers.

The purity of the water depends on the velocity and the dilution capacity of the river. A large part of the flow of the Ganga is abstracted for irrigation just as it enters the plains at Hardiwar. From there it flows as a trickle for a few hundred kilometres until Allahabad, from where it is recharged by its tributaries. The Ganga receives over 60 per cent of its discharge from its tributaries. The contribution of most of the tributaries to the pollution load is small, except from the Gomti, Damador and Yamuna rivers, for which separate action programmes have already started under Phase II of "The National Rivers Conservation Plan".

The Ganga river carries the highest silt load of any river in the world and the deposition of this material in the delta region results in the largest river

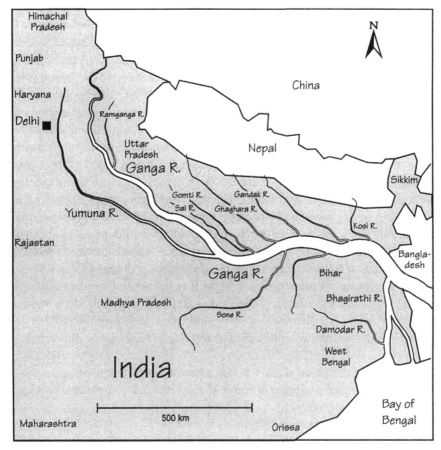

Figure I.2 Map of India showing the route of the Ganga river

delta in the world (400 km from north to south and 320 km from east to west). The rich mangrove forests of the Gangetic delta contain very rare and valuable species of plants and animals and are unparalleled among many forest ecosystems.

I.2.1 Exploitation

In the recent past, due to rapid progress in communications and commerce, there has been a swift increase in the urban areas along the river Ganga. As a result the river is no longer only a source of water but is also a channel, receiving and transporting urban wastes away from the towns. Today, one third of the country's urban population lives in the towns of the Ganga basin. Out of the 2,300 towns in the country, 692 are located in this basin, and of these, 100 are located along the river bank itself.

The belief the Ganga river is "holy" has not, however, prevented over-use, abuse and pollution of the river. All the towns along its length contribute to the pollution load. It has been assessed that more than 80 per cent of the total pollution load (in terms of organic pollution expressed as biochemical oxygen demand (BOD)) arises from domestic sources, i.e. from the settlements along the river course. Due to over-abstraction of water for irrigation in the upper regions of the river, the dry weather flow has been reduced to a trickle. Rampant deforestation in the last few decades, resulting in topsoil erosion in the catchment area, has increased silt deposits which, in turn, raise the river bed and lead to devastating floods in the rainy season and stagnant flow in the dry season. Along the main river course there are 25 towns with a population of more than 100,000 and about another 23 towns with populations above 50,000. In addition there are 50 smaller towns with populations above 20,000. There are also about 100 identified major industries located directly on the river, of which 68 are considered as grossly polluting. Fifty-five of these industrial units have complied with the regulations and installed effluent treatment plants (ETPs) and legal proceedings are in progress for the remaining units. The natural assimilative capacity of the river is severely stressed.

The principal sources of pollution of the Ganga river can be characterised as follows:

- Domestic and industrial wastes. It has been estimated that about 1.4×10^6 $m^3 d^{-1}$ of domestic wastewater and $0.26 \times 10^6 m^3 d^{-1}$ of industrial sewage are going into the river.
- Solid garbage thrown directly into the river.
- Non-point sources of pollution from agricultural run-off containing residues of harmful pesticides and fertilisers.
- Animal carcasses and half-burned and unburned human corpses thrown into the river.
- Defecation on the banks by the low-income people.
- Mass bathing and ritualistic practices.

I.3   The Ganga Action Plan

I.3.1   Scientific awareness

There are 14 major river basins in India with natural waters that are being used for human and developmental activities. These activities contribute significantly to the pollution loads of these river basins. Of these river basins the Ganga sustains the largest population. The Central Pollution Control Board (CPCB), which is India's national body for monitoring environmental pollution, undertook a comprehensive scientific survey in 1981–82 in order to classify river waters according to their designated best uses. This report was the first systematic document that formed the basis of the Ganga Action Plan (GAP). It detailed land-use patterns, domestic and industrial pollution loads, fertiliser and pesticide use, hydrological aspects and river classifications. This inventory of pollution was used by the Department of Environment in 1984 when formulating a policy document. Realising the need for urgent intervention the Central Ganga Authority (CGA) was set up in 1985 under the chairmanship of the Prime Minister.

The Ganga Project Directorate (GPD) was established in June 1985 as a national body operating within the National Ministry of Environment and Forest. The GPD was intended to serve as the secretariat to the CGA and also as the Apex Nodal Agency for implementation. It was set up to co-ordinate the different ministries involved and to administer funds for this 100 per cent centrally-sponsored plan. The programme was perceived as a once-off investment providing demonstrable effects on river water quality. The execution of the works and the subsequent operation and management (O&M) were the responsibility of the state governments, under the supervision of the GPD. The GPD was to remain in place until the GAP was completed. The plan was formally launched on 14 June 1986. The main thrust was to intercept and divert the wastes from urban settlements away from the river. Treatment and economical use of waste, as a means of assisting resource recovery, were made an integral part of the plan.

It was realised that comprehensive co-ordinated research would have to be conducted on the following aspects of Ganga:

- The sources and nature of the pollution.
- A more rational plan for the use of the resources of the Ganga for agriculture, animal husbandry, fisheries, forests, etc.
- The demographic, cultural and human settlements on the banks of the river.
- The possible revival of the inland water transport facilities of the Ganga, together with the tributaries and distributaries.

One outcome of this initiative was a multi-disciplinary study of the river in which the 14 universities located in the basin participated in a well co-ordinated, integrated research programme. This was one of largest endeavours, involving several hundred scientists, ever undertaken in the country and was funded under the GAP. The resultant report is a unique, integrated profile of the river.

The GAP was only the first step in river water quality management. Its mandate was limited to quick and effective, but sustainable, interventions to contain the damage. The studies carried out by the CPCB in 1981–82 revealed that pollution of the Ganga was increasing but had not assumed serious proportions, except at certain main towns on the river such as industrial Kanpur and

Calcutta on the Hoogly, together with a few other towns. These locations were identified and designated as the "hot-spots" where urgent interventions were warranted. The causative factors responsible for these situations were targeted for swift and effective control measures. This strategy was adopted for urgent implementation during the first phase of the plan under which only 25 towns identified on the main river were to be included. The studies had revealed that:

- 75 per cent of the pollution load was from untreated municipal sewage.
- 88 per cent of the municipal sewage was from the 25 Class I towns on the main river.
- Only a few of these cities had sewage treatment facilities (these were very inadequate and were often not functional).
- All the industries accounted for only 25 per cent of the total pollution (in some areas, such as Calcutta and Kanpur, the industrial waste was very toxic and hard to treat).

I.3.2 Attainable objectives

The broad aim of the GAP was to reduce pollution and to clean the river and to restore water quality at least to Class B (i.e. bathing quality: 3 mg l^{-1} BOD and 5 mg l^{-1} dissolved oxygen). This was considered as a feasible objective and because a unique and distinguishing feature of the Ganga was its widespread use for ritualistic mass bathing. The other environmental benefits envisaged were improvements in, for example, fisheries, aquatic flora and fauna, aesthetic quality, health issues and levels of contamination.

The multi-pronged objectives were to improve the water quality, as an immediate short-term measure, by controlling municipal and industrial wastes. The long-term objectives were to improve the environmental conditions along the river by suitably reducing all the polluting influences at source. These included not only the creation of waste treatment facilities but also invoking remedial legislation to control such non-point sources as agricultural run-off containing residues of fertilisers and pesticides, which are harmful for the aquatic flora and fauna. Prior to the creation of the GAP, the responsibilities for pollution of the river were not clearly demarcated between the various government agencies. The pollutants reaching the Ganga from most point sources did not mix well in the river, due to the sluggish water currents, and as a result such pollution often lingered along the embankments where people bathed and took water for domestic use.

I.3.3 The strategy

The GAP had a multi-pronged strategy to improve the river water quality. It was fully financed by the central Government, with the assets created by the central Government to be used and maintained by the state governments. The main thrust of the plan was targeted to control all municipal and industrial wastes. All possible point and non-point sources of pollution were identified. The control of point sources of urban municipal wastes for the 25 Class I towns on the main river was initiated from the 100 per cent centrally-invested project funds. The control of urban non-point sources was also tackled by direct interventions from project funds. The control of non-point source agricultural run-off was undertaken in a phased manner by the Ministry of Agriculture, principally by reducing use of fertiliser and pesticides. The control of point sources of industrial wastes was done by applying the polluter-pays-principle.

A total of 261 sub-projects were sought for implementation in 25 Class I (population above 100,000) river front towns. This would eventually involve a financial outlay of Rs 4,680 million (Indian Rupees), equivalent to about US$ 156 million. More than 95 per cent of the programme has been completed and the remaining sub-projects are in various stages of completion. The resultant improvement in the river water quality, although noticeable, is hotly debated in the media by certain non-governmental organisations (NGOs). The success of the programme can be gauged by the fact that Phase II of the plan, covering some of the tributaries, has already been launched by the Government. In addition, the earlier action plan has now evolved further to cover all the other major national river-basins in India, including a few lakes, and is known as the "National Rivers Conservation Plan".

Domestic waste

The major problem of pollution from domestic municipal sewage (1.34×10^6 $m^3 d^{-1}$) arising from the 25 selected towns was handled directly by financing the creation of facilities for interception, diversion and treatment of the wastewater, and also by preventing the other city wastes from entering the river. Out of the 1.34×10^6 $m^3 d^{-1}$ of sewage assessed to be generated, 0.873×10^6 $m^3 d^{-1}$ was intercepted by laying 370 km of trunk sewers with 129 pumping stations as part of 88 sub-projects. The laying of sewers and the renovation of old sewerage was restricted only to that required to trap the existing surface drains flowing into the river. Facilities for solid waste collection using mechanised equipment and sanitary landfill, low-cost toilet complexes (2,760 complexes), partly-subsidised individual pour flush toilets (48,000), 28 electric crematoriums for human corpses, and 35 schemes of river front development for safer ritualistic bathing, were also included. A total of 261 such projects were carried out in the 25 towns. The programme also included 35 modern sewage treatment plants. The activities of the various sub-projects can be summarised as follows:

Approach to river water quality improvement	Number of schemes
Interception and diversion of municipal wastewater	88
Sewage treatment plants	35
Low-cost sanitation complexes	43
Electric crematoriums	28
River front facilities for bathing	35
Others (e.g. biological conservation of aquatic species, river quality monitoring)	32
Total	261

A total of 248 of these schemes have already been commissioned and those remaining are due to be completed by 1998.

Industrial waste
About 100 industries were identified on the main river itself. Sixty-eight of these were considered grossly polluting and were discharging 260×10^3 $m^3 \ d^{-1}$ of wastewater into the river. Under the Water (Prevention and Control of Pollution) Act 1974 and Environment (Protection) Act 1986, 55 industrial units (generating $232 \times 10^3 \ m^3 \ d^{-1}$) out of the total of 68 (identified) grossly polluting industrial units complied and installed effluent treatment plants. In addition, two others have treatment plants under construction and currently one unit does not have a treatment plant. Legal proceedings have been taken against the remaining 12 industrial units which were closed down for non-compliance.

Integrated improvements of urban environments
Apart from the above, the GAP also covered very wide and diverse activities, such as conservation of aquatic species (gangetic dolphin), protection of natural habitats (scavenger turtles) and creating riverine sanctuaries (fisheries). It also included components for landscaping river frontage (35 schemes), building stepped terraces on the sloped river banks for ritualistic mass-bathing (128 locations), improving sanitation along the river frontage (2,760 complexes), development of public facilities, improved approach roads and lighting on the river frontage.

Applied research

The Action Plan stressed the importance of applied research projects and many universities and reputable organisations were supported with grants for projects carrying out studies and observations which would have a direct bearing on the Action Plan. Some of the prominent subjects were PC-based software modelling, sewage-fed pisciculture, conservation of fish in upper river reaches, bioconservation in Bihar, monitoring of pesticides, using treated sewage for irrigation, and rehabilitation of turtles.

Some of the ongoing research projects include land application of untreated sewage for tree plantations, aquaculture for sewage treatment, disinfection of treated sewage by ultra violet radiation, and disinfection of treated sewage by Gamma radiation. Expert advice is constantly sought by involving regional universities in project formulation and as consultants to the implementing agencies to keep them in touch with the latest technologies. Eight research projects have been completed and 17 are ongoing. All the presently available research results are being consolidated for easy access by creation of a data base by the Indian National Scientific Documentation Centre (INSDOC).

Public participation

The pollution of the river, although classified as environmental, was the direct outcome of a deeper social problem emerging from long-term public indifference, diffidence and apathy, and a lack of public awareness, education and social values, and above all from poverty.

In recognition of the necessity of the involvement of the people for the sustainability and success of the Action Plan, due importance was given to generating awareness through intensive publicity campaigns using the press and electronic media, audio visual approaches, leaflets and hoardings, as well as organising public programmes for spreading the message effectively. In spite of full financial support from the project, and in spite of a heavy involvement of about 39 well known NGOs to organise these activities, the programme had only limited public impact and even received some criticism. Other similar awareness-generating programmes involving school children from many schools in the project towns were received with greater enthusiasm. These efforts to induce a change in social behaviour are meandering sluggishly like the Ganga itself.

Technology options

The choice of technology for the GAP was largely conventional, based on available options and local considerations. Consequently, the sewers and pumping stations and all similar municipal and conservancy works were executed in each province by its own implementing agencies, according to their customary practices but within the commonly prescribed specifications, fiscal controls and time frames. The choice of technology for most of the large domestic wastewater treatment plants was carefully decided by a panel of experts, in close consultation with those external aid agencies which were supporting that particular project. A parallel procedure was adopted in-house for all other similar projects. For all the larger sewage treatment plants the unanimous choice was to adopt the well-accepted activated sludge process. For other plants trickling filters were considered more appropriate. In smaller towns where land was available and the quantity of wastewater was small, other options such as oxidation ponds were chosen. However, unconventional technologies like the rope bound rotating biological contactors (RBRC), sewage irrigated afforestation, upflow anaerobic sludge blanket (UASB) technology and plants for chromium recovery from tannery wastewater were tried out with a fair degree of success. Some of these new and simpler technologies, with their low-cost advantages, will emerge as the large-scale future solution to India's sanitation problems.

Operation and maintenance
The enduring success of the pollution abatement works under the GAP is essential for sustainability. Most of these works were carried out by the same agencies which were eventually responsible for maintaining them as part of their primary functions, such as the city development authority, the municipality, or the irrigation and flood control department. The responsibility for subsequent O&M of these works automatically passed to these agencies. The most crucial components for preventing river pollution were the main pumping stations which were intercepting the sewage and diverting it to the treatment plants. These large capacity pumping stations, operating at the city level, had been built for the first time in India, and it was considered unlikely that the municipalities would have adequate resources and skilled personnel to be able to manage them. An integral part of the earlier planning of these sewage treatment works had been self sufficiency from resource recovery by the sale of treated effluent as irrigation water for agriculture, by the sale of dried sludge as manure (because it was rich in nutrients) and from the generation of electricity from the bio-gas production in the plant. It was considered that the generation of bio-electricity would be sufficient to offset much of the cost of the huge energy inputs required. In time it was realised that all these assumptions were only partly true. The state governments took over the responsibility of O&M through the same agencies that had built the plants by

providing the funds to cover the deficit of the O&M expenditures. The central Government shared half of this deficit until 1997. In the broader interest of pollution control, future policies will also be similar, where the state governments undertake the responsibility for pollution control works because the local bodies are unable to bear the cost of the O&M expenditures with such limited resources.

I.4   Implementation problems

The implementation of a project of this magnitude over the entire 2,500 km stretch of the river, covering 25 towns and crossing three different provinces, could only be achieved by delegating the actual implementation to the state government agencies which had the appropriate capabilities. The state governments also undertook the responsibility of subsequently operating and maintaining the assets being created under the programme. The overall inter-agency co-ordination was done by the GPD through the state governments. The defined project objectives were ensured by the GPD through appraisal of each project component submitted by the implementing agency. The overall fiscal control was exercised by the GPD by close professional monitoring of the physical progress through independent agencies.

The progress in the first four years was satisfactory. The swift commissioning of the interception and diversion works as an immediate priority, ensured that most of the city wastes were collected and re-released to the river downstream of the city, thus earning public approval for the remarkably clean city waterfronts. However, some of the major sewage treatment plants (STPs) could not be completed in the original time frame. The delays in the completion of these major plants were unavoidable because treatment plants of such large capacity for domestic wastewater were being built for the first time in the country. The involvement of the external aid agencies was initially useful in introducing new technologies, such as chrome recovery plants for tannery wastewaters, low energy input technologies like the UASB and *in situ* sewer rehabilitation technology. However, the involvement of aid agencies, with their associated mandatory procedures, also added to the complexities of decision-making, especially in the large STP projects. The aid was awarded on a turn-key basis by inviting global bids. On account of the huge capital outlay, the final approvals were a multi-stage process and sometimes quite removed from the actual execution level. The collective wisdom of many experts was at times at odds with the opinions of the executing agency officials, who had to take the final responsibility. The procedural delays experienced with mid-project decisions on some issues of these turn-key contracts gave the contractors grounds to justify their own shortcomings in causing the original delays. Therefore, project schedules had to be relaxed several times. Of the original 261 sub-projects, 95 per cent are now complete and functioning satisfactorily. The remaining projects are mainly STPs and are in progress, due to be completed by 1998.

I.5   River water quality monitoring

Right from its inception in 1986, the GAP started a very comprehensive water quality monitoring programme by obtaining data from 27 monitoring stations. Most of these river water quality monitoring stations already existed under other programmes and only required strengthening. Technical help was also received for a small part of this programme from the Overseas Development Agency (ODA) of the UK in the form of some automatic water quality monitoring stations, the associated modelling software, training and some hardware. The monitoring programme is being run on a permanent basis using the infrastructure of other agencies such as the CPCB and the Central Water Commission (CWC) to monitor data from 16 stations. Some research institutions like the Industrial Toxicology Research Centre (ITRC) are also included for specialised monitoring of toxic substances. The success of the programme is noticeable through this record of the water quality over the years, considered in proportion to the number of improvement schemes commissioned. To evaluate the results of this programme an independent study of water quality has also been awarded to separate universities for different regional stretches of the river.

I.6   The future

Apart from the visible improvement in the water quality, the awareness generated by the project is an indicator of its success. It has resulted in the expansion of the programme over the entire Ganga basin to cover the other polluted tributaries. The GAP has further evolved to cover all the polluted stretches of the major national rivers, and including a few lakes. Considering the huge costs involved the central and state governments have agreed in principle to each share half of the costs of the projects under the "National Rivers Action Plan". The state governments are also required to organise funds for sustainable O&M in perpetuity. Initially, the plan was fully sponsored by the central Government.

I.7   Conclusions and lessons learned

The GAP is a successful example of timely action due to environmental awareness at the governmental level. Even more than this, it exhibits the achievement potential which is attainable by "political will". It is a model which is constantly being upgraded and improved in other river pollution prevention projects. Nevertheless, some very important lessons have been learned which are being incorporated into further projects. These include lessons learned about poor resource recovery due to poor resource generation, because of the lower organic content of Indian sewage. This may be due to less nutritious dietary habits, higher water consumption, fewer sewer connections, higher grit loads, insufficient flows and stagnation leading to bio-degradation of the volatile fractions in the pipes themselves. The assumed BOD design load of the plants were, in some cases, considered much higher than the actual BOD loading. This was due to a lack of practical experience within India and the fact that western experiences were not entirely appropriate.

There were also many lessons learned associated with the project objectives, which overlapped in many areas with urban infrastructure development, especially when the GAP was mistakenly assumed to be a city improvement plan. This led to an initial rise in general expectations followed by disappointments when the GAP was found to limit itself only to river pollution abatement without pursuing popular measures. This could have been one of the main reasons why it attracted some sharp criticism. In spite of close co-ordination with the Ministry of Urban Development at the central and state government levels, this communication gap still remains because future planning is still based on narrow considerations and short-term objectives (solely due to resource constraints), without addressing the root causes, which were also being overlooked earlier for precisely the same reasons. Thus the river pollution plan being "action" orientated, avoids involvement in long-term town planning, which continues to remain deficient with respect to environmental sanitation. This is due to a lack of overview by any stakeholding agency and to the blinkered foresight by the already beleaguered city authorities who remain perpetually short of funds for their daily crisis-management.

The most important lesson learned was the need for control of pathogenic contamination in treated effluent. This could not be tackled before because of a lack of safe and suitable technology but is now being attempted through research and by developing a suitable indigenous technology, which should not impart traces of any harmful residues in the treated effluent detrimental to the aquatic life. This is an aspect difficult to control in surface waters in tropical areas, but it is very important for the Ganga because the river water is used directly by millions of devout individuals for drinking and bathing.

I.8   Recommendations

The Action Plan started as a "cleanliness drive" and continues in the same noble spirit with the same zeal and enthusiasm on other major rivers and freshwater bodies. Its effectiveness could however be enhanced if these efforts could be integrated and well-accepted within the long-term objectives and master plans of the cities, which are constantly under preparation without adequate attention to the disposal of wastes. More information on polluted groundwater resources in the respective river basins will prove useful, because the existing levels of depletion and contamination of groundwater resources, which are already overexploited and fairly contaminated, will increase the dependency in the future on the rivers, as the only economical source of drinking water. This aspect has not been seriously considered in any long-term planning.

I.9   Source literature

This chapter was prepared from publicity material issued by the Ganga Project Directorate, New Delhi.

Case Study II[*]

SHANGHAI HUANGPU RIVER, CHINA

II.1 Introduction

The Huangpu River flows through the heart of Shanghai (Figure II.1). It supplies water to the 13 million people in the metropolis and is also important for navigation, fishery, tourism and receiving wastewater.

Around the mid-1980s, about 70 per cent of the 5.5×10^6 m^3 of industrial wastewater and domestic sewage, mostly untreated or partially treated, was being discharged directly, or through urban sewers, to the Huangpu River and its branches. As the result, the Huangpu River became very seriously polluted. The urban section of the Huangpu River turned black and anoxic for about 100 days in the early 1980s and this increased to more than 200 days in the 1990s.

Since 1979, the Shanghai Municipal Government has given much attention to the integrated pollution control of the Huangpu River. In the late 1970s to the early 1980s, environmental legislation and standards were stipulated for ambient water quality and effluent, and institutions for enforcement were created. In 1982, an overall survey of pollution sources, ambient water quality and hydrology of the major water bodies was carried out. In the mid-1980s, the Huangpu River pollution control plan was drawn up, following which financial resources were pooled, locally and from abroad, for major investment projects, particularly for the development of an infrastructure for the new water supply intake and for wastewater pollution control. Progress in this plan is described below.

II.2 Background information

II.2.1 Urban, social and economic profile

The city of Shanghai is situated in the Yangtze River (Chiang Jiang) delta plain on the south side of the Yangtze River, within the Tai Lake (Taihu) Basin (Figure II.2). The total area of Greater Shanghai is 6,340.5 km^2, of which about 140 km^2 are classified as urban and consists of 10 central districts. The rest of

---

*   *This case study was prepared by Chonghua Zhang*

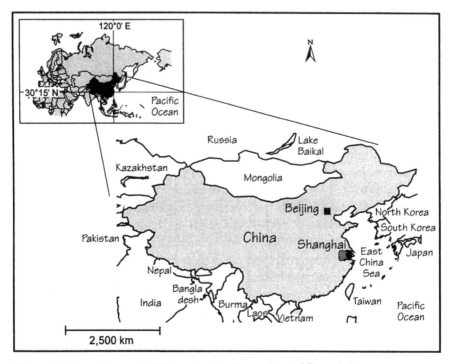

Figure II.1 Location map of China showing the position of Shanghai

the area includes two satellite towns and 10 rural counties. The Huangpu River runs through the city from south west to north east and finally enters the Yangtze River at Wusong Kou (Figure II.3).

Shanghai is a densely populated city. In 1992 its population was 12.9 million, including an urban population of about 8 million. Shanghai is one of the nation's major centres for economics, trading, finance, politics, communication, science, technology and culture. It is notably the largest industrial base in China, with 145 of the total 161 industrial sectors represented (the exceptions are mining related sectors). In 1993, Shanghai had about 39,000 industrial enterprises, of which the major sectors were textiles, machinery, automobiles, shipbuilding, chemicals, electronics, metallurgy and pharmaceutical chemicals. Although Shanghai has only 1.17 per cent of the country's population, it contributes about 11 per cent of the country's gross national industrial output. Being the most advanced city in the country, Shanghai is viewed by planners as a window to the outside world through which various approaches to modernisation can be introduced into China. In recent years, Shanghai has been attracting about 30 per cent of the total foreign investment to China.

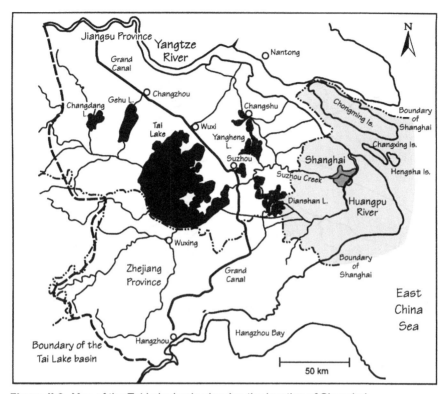

Figure II.2 Map of the Tai Lake basin showing the location of Shanghai

II.2.2 Water resources

Shanghai is very rich in water resources. The main rivers are the Yangtze River in the north and the Huangpu River, a tributary of the Yangtze, in the delta area. The Huangpu River also belongs to the Tai Lake Water System and is important for discharging flood water from the Tai Lake. The amount of flood water discharged from the Tai Lake area during the wet season, usually in the summer, strongly affects the flow rate of the Huangpu River and its water quality. The average annual flow rate of the Huangpu River is 315 m^3 s^{-1}. There are hundreds of man-made canals in Shanghai. They are inter-connected to form a web around the Huangpu River. About 80 per cent of Shanghai falls within this web of water networks. The major water bodies within the Huangpu River Basin are:

- The Yangtze River. This is the third largest river in the world, providing the greatest freshwater resource for Shanghai. Many inner, navigation rivers are connected to the Yangtze River, making it the largest continental

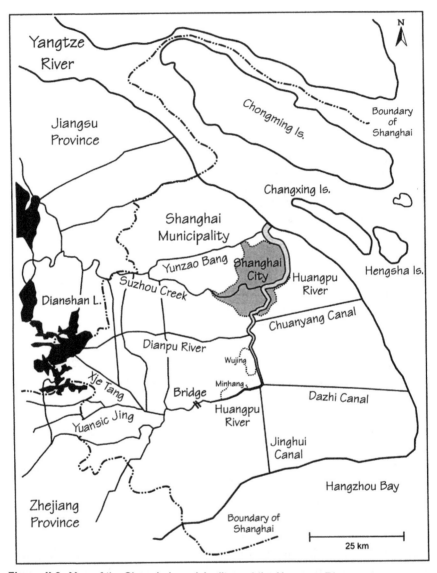

Figure II.3 Map of the Shanghai municipality and the Huangpu River system

navigation channel in Asia. The annual average flow rate is about 10,000 m³ s⁻¹.

- The Suzhou River (also called Suzhou Creek). This is the major river which connects Tai Lake and the Huangpu River. It has a total length of 125 km (including 54 km in Shanghai) with an average width of 58.6 m, an

Table II.1   Main branches of the Huangpu River

River name	Length (km)	Width (m)	Depth (m)
Longhuagang	3.4	22.8	3.2
Qiujiang	6.4	37.8	1.6
Yangpugang	4.3	11.7	1.9
Hongkougang	2.0	17.5	2.5
Yunzaobang	38.0	92.0	5.0
Damaogang	17.3	176.0	6.5
Xietang	23.2	170.0	6.0
Yuanxiejing	16.5	178.0	7.8
Taipuhe	16.5	150–180	3.5

average depth of 3.4 m and a water level gradient of 0.8 cm km^{-1}. The Suzhou River is the most important navigation channel, promoting commerce for towns and villages between Tai Lake and Shanghai City.

- Dianshan Lake. This lake has a surface area of 64 km^2. It is a rich freshwater fishery resource and has beautiful scenery and many historic relics, making it attractive for tourism.

Surface run-off in the Shanghai area varies significantly from year to year. In a very dry season the run-off can be only 40 per cent of that for an average year. The flow received from Tai Lake also varies significantly from year to year, ranging from 5.11×10^9 to 12.83×10^9 m^3 a^{-1}.

Groundwater is extracted and used mainly as cooling water in industry. Over-exploitation of groundwater in the past caused serious land subsidence in the area and in recent years, therefore, groundwater extraction has been controlled. Between 1981 and 1990 an average of about 88×10^6 m^3 a^{-1} of groundwater were extracted in Shanghai.

The Huangpu River is tidal. The tidal effect complicates the flow pattern of the river and also the water quality of the tidal sections. The Huangpu River receives about 40.9×10^9 m^3 of tidal water from the Yangtze River. The total tidal influx of the Huangpu River is about 47.47×10^6 m^3 a^{-1}, including all the other tidal water received by smaller rivers (about 6.57×10^6 m^3) (Table II.1).

II.2.3 Water pollution in the Huangpu River basin

In 1992, the piped water and groundwater consumption was 2.26×10^9 m^3 and the wastewater discharged was 2.03×10^9 m^3, or about 5.5×10^6 m^3 per day. About 25 per cent of the industrial wastewater was subject to primary and secondary treatment and about 14 per cent of the domestic wastewater received secondary treatment.

According to a pollution source survey in 1985, the water bodies that received the greatest industrial wastewater loads were:

- The Huangpu River and its minor tributaries: 71 per cent.
- The Suzhou River, the largest tributary: 10 per cent.
- The Yangtze River, Hangzhou Bay and East Sea: 19 per cent.

It is estimated that 58 per cent of the industrial wastewater was discharged directly to rivers and the rest was discharged to sewers. However, about 70 per cent of the sewage collected by sewerage systems was discharged indirectly to rivers and to the estuary of the Yangtze River.

The annual run-off from rural areas within the web of the Huangpu River is estimated to be 1.5×10^9 m^3, bringing 4,600 tonnes of nitrogen and 900 tonnes of phosphorus to the rivers and lakes each year. A new source of pollution is livestock manure. In 1992, 7.2×10^6 tonnes of livestock manure and other wastes were generated.

There are four attributes to the pollution of the Huangpu River. First, wastewater discharged to the Huangpu River contains large amount of organic substances, which create a significant demand for dissolved oxygen in the water. Second, about 81 per cent of the total waterways in the city are polluted. Third, the most serious pollution occurs in the urban section, particularly at the water intake points for the Nanshi Water Treatment Plant and the Yangpu Water Treatment Plant. Finally, the tidal nature of Huangpu River restricts the release of organic pollutants to downstream stretches.

II.3   Institutional development and industrial pollution control

II.3.1 Environmental regulations and organisations

The Environmental Protection Law of China was stipulated in 1978 by the National People's Congress and includes the authorisation for creating agencies for the management of environmental protection. Following on from that, the Chinese Government enacted laws for the control of water, air, noise, solid waste pollution and radioactive substances. Around the mid-1980s, environmental quality standards (EQSs) for surface water and effluent standards for industrial wastewater were promulgated. Shanghai has adopted all the national environmental regulations and standards but, in order to meet local requirements, the city has also established water quality objectives and the associated standards for rivers, canals and lakes.

nvironmental protection institutions in China were established at all levels of government agencies, including central, provincial, prefecture, municipal, district and county governments. A typical environmental protection system for a large city, such as Shanghai, comprises the municipal environmental protection bureau, several district environmental protection bureaux, a centre for environmental monitoring, a number of district monitoring stations, a research institute and several pollution levy collection offices in the districts (Figure II.4). The total number of staff employed varies from 300–700 depending on the size of the city. The Shanghai Environmental Protection Bureau (Shanghai EPB) employs about 700 people.

Government ministries in China, including industrial, agricultural, urban construction and military ministries, have also created functional departments or divisions of environmental protection to deal with pollution problems. These environmental units are mainly set up for self-monitoring and enforcement. They have also created pollution control divisions at the provincial and municipal level. In Shanghai, the textile bureau has about 100 full-time staff for environmental protection and who are responsible for environmental management and monitoring and pollution control technology development.

II.3.2 Old and new pollution control measures

Three very important environmental regulations were stipulated in the early 1980s by the national Environmental Protection Agency (Qu Geping, 1991a). These should be implemented in parallel to project design, construction and commissioning and are known as the "three simultaneous actions" system of environmental protection in China:

- Environmental impact assessment (EIA) system for new and expanding projects.
- Implementation of pollution control measures for new and expanding projects.
- Pollution fee charges (Table II.2).

In the late 1980s, five new regulations were stipulated for the further control of existing pollution (Qu Geping, 1991b):

- A system of objective responsibility in environmental protection, making the highest governmental official directly responsible for the needs of the environment and the associated specific improvements within his area of responsibility.
- A system of quantitative assessment for the integrated control of urban environments, with 20 specific environmental variables selected for monitoring and assessment in 32 provincial capitals.
- Pollution discharge permits.
- Setting a deadline for reaching the target of pollution control.
- Centralised control of pollution.

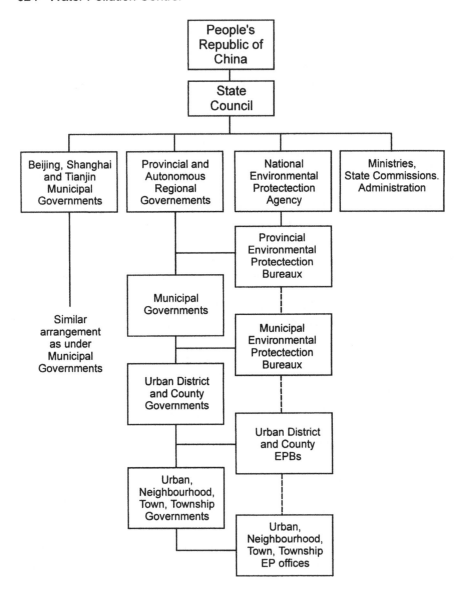

Figure II.4 Chart showing the organisation of Environmental Protection in China

II.3.3 Sources of finance

By means of legislation, the Chinese government created several funding channels for pollution control. The most important is that for new industrial and technology renovation projects which requires up to 7 per cent of the investment costs to be reserved for pollution control.

Table II.2   Wastewater discharge fees in Shanghai

Pollutant	GVOD (tonne.time)	Grade A (Yuan per tonne.time)	Grade B (Yuan per tonne.time)	Elementary Fee of Grade B (Yuan)
Total Hg	2,000	2.00	1.00	2,000
Total Cd	3,000	1.00	0.15	2,550
Total Cr	150,000	0.06	0.03	4,500
Cr^{+6}	150,000	0.09	0.02	10,500
Total As	150,000	0.09	0.02	10,500
Total Pb	150,000	0.08	0.03	7,500
Total Ni	150,000	0.08	0.03	7,500
Bap	3,000,000	0.06	0.03	90,000
pH	5,000	0.25	0.05	1,000
Colour	100,000	0.14	0.04	10,000
Suspended solids	800,000	0.03	0.01	16,000
BOD	30,000	0.18	0.05	3,900
COD	20,000	0.18	0.05	2,600
Petrols	25,000	0.20	0.06	3,500
Animal and plant material	25,000	0.12	0.04	2,000
Volatile phenols	250,000	0.06	0.03	7,500
Cyanide	250,000	0.07	0.04	7,500
Sulphide	250,000	0.05	0.02	7,500
NH_3-N	25,000	0.10	0.03	1,750
Fluoride	25,000	0.30	0.09	5,250
Phosphate (asp)	250,000	0.05	0.02	7,500
Methylaldehyde	200,000	0.12	0.06	12,000
Aniline	200,000	0.12	0.06	12,000
Nitrobenzene	200,000	0.10	0.04	12,000
Detergent (LAS)	25,000	0.30	0.09	5,250
Cu	250,000	0.04	0.02	5,000
Zn	100,000	0.06	0.02	4,000
Mn	100,000	0.06	0.02	4,000
Organophosphorus pesticides (as P)	250,000	0.07	0.04	7,500

GVOD   Grading value for over-standard discharge (tonne.time)

BOD   Biochemical oxygen demand

COD   Chemical oxygen demand

Amount charged = charge rate × total amount of discharge exceeding the pollutant standard (APDES) (tonne.time)

where: APDES (tonne.time) = amount wastewater discharge × time for which pollutant standard is exceeded;

GVOD (tonne wastewater × time exceeded) is the boundary value of APDES (tonne.time);

When APDES < GVOD: amount charged = charge rate A × APDES

When APDES > GVOD: amount charged = charge rate B × APDES + elementary fee of Grade B

Total amount of pH value exceeding the standard = (pH value of the wastewater − pH discharge standard) × amount of wastewater discharge

When implementing the "Maximum permitted discharge of wastewater or minimum permitted recycle rate of water" in the *Integrated Wastewater Discharge Standard*, the fee levied on the amount of discharge exceeding the standard is based on the minimum charge rate for water supply superimposed on the fee of pollutant discharge exceeding the standard.

The charge standard for wastewater with pathogen discharge exceeding the standard is 0.14 Yuan per tonne wastewater

The pollution control fund is another important source of pollution investment. Industries that do not meet discharge standards are required to pay pollution fines. Fines are collected by the municipalities and allocated for pollution control in the form of a fund. During the 1980s, the average annual levy collected in Shanghai was about 100 million RMB yuan (about US$ 15 million). The fund allocated for pollution control can be used for local enforcement (20 per cent), for pollution enforcement (e.g. for monitoring equipment) and for investment in industrial pollution control (80 per cent). In the 1980s, the funds were mainly used for end-of-pipe pollution control. In the 1990s, however, they were used for industrial plant relocation and centralised treatment. The pollution fund is responsible for financing about 30 per cent of existing industrial pollution control projects.

II.3.4 Accomplishments and limitations

The two main types of water-borne pollutants of concern are heavy metals and organic substances. Heavy metal pollution is toxic and irreversible in the environment. Shanghai EPB has recognised the control of heavy metal pollution to be a priority since the late 1970s. In the early 1980s, Shanghai EPB centralised all the scattered electroplating enterprises into just a few locations. Their wastes were treated on site with joint treatment methods. As a result, the reduction in heavy metal waste has exceeded 95 per cent since the mid-1980s.

Organic pollution is widely distributed amongst many industrial sectors in Shanghai. Developing a strategy for controlling industrial organic pollution is complicated and requires integrated planning with domestic wastewater control. Nevertheless, Shanghai took strong measures against the major polluters in the city. Several pulp mills, responsible for about 25 per cent of the biochemical oxygen demand (BOD) in the Huangpu River, were closed down in the 1980s. Pre-treatment is now widely practised by industries producing concentrated organic effluent, such as food and pharmaceutical industries. The relocation of scattered industrial units to industrial parks is very much encouraged in Shanghai.

Pollution control in new and expanding projects has been quite successful by state-owned enterprises in Shanghai. In the 1980s, the compliance rate of state enterprises with requirements for EIA and the three "simultaneous actions" reached 100 per cent in Shanghai. Due to the successful control of new pollution sources and some major polluters, the pollution load from industry in 1990 did not increase relative to pollution in the mid-1980s, although industrial productivity increased four-fold.

Table II.3 Water quality planning objectives for the Huangpu River system

Phased quality objectives	Upstream section				Estuary of Changjiang River
	Water source protection zone	Sub-water source protection zone	Downstream section	Urban section	
Present					
DO (mg l⁻¹)	> 5	> 4			> 6
BOD₅ (mg l⁻¹)	< 5	< 5			< 3
NH₃-N (mg l⁻¹)	< 1	< 1			< 0.5
	Maintain current situation	Maintain current situation	No further worsening	No further worsening	Class II
Short-term (1990)					
DO (mg l⁻¹)	> 5	> 4	> 2.5	> 2.5	> 6
BOD₅ (mg l⁻¹)	<5	< 5	< 10	< 10	< 3
NH₃-N (mg l⁻¹)	<1	< 1	< 3.5	< 3.5	< 0.5
	Attain Class II standard	Attain Class III standard	Eliminate anaerobic condition	Eliminate anaerobic condition	Attain Class II standard
End of the century (2000)					
DO (mg l⁻¹)	> 6	> 6	> 4	> 4	> 6
BOD₅ (mg l⁻¹)	< 3	< 1	< 5	< 5	< 3
NH₃-N (mg l⁻¹)	< 0.5	< 0.5	< 1	< 1	< 0.5
	Protect Class II standard	Attain Class II standard	Attain Class III standard	Attain Class III standard	Attain Class II standard

Despite the successes mentioned, the water quality of the Huangpu River remains very poor because a large amount of remaining organic substances are still left untreated. There remains much more to be done if the water quality is to be improved to an acceptable level.

II.4   Pollution control strategy for the Huangpu River

In Shanghai, there are two environmental problems related to the Huangpu River. First, the river is a source of water supply for the whole city which has been taking water from the most polluted section of the river for domestic use. Second, the Huangpu River has a very serious pollution problem to solve. These two problems are related although the former is more urgent. It is not possible to keep the existing water intake in service for drinking water supply purpose, even in the near future, because the risks from pollution are too great. Against this background, two separate projects were proposed under the Huangpu River Waste Water Integrated Prevention and Control Planning (Shanghai EPB, 1985):

- Moving the water supply intake point upstream of the Huangpu River project.
- A Shanghai sewerage collection and wastewater treatment project.

Water quality objectives (Table II.3) were set by taking into consideration:

- The requirements of the water body functions at each section of Huangpu River.
- The existing pollution status.
- The self-purification capability of the river.
- The medium- and long-term urban planning of Shanghai.
- The financing capability of the city.

The integrated pollution control of Huangpu River is a large system project composed of many sub-projects. The scope of the project covers the main stream of the Huangpu River, its main branches, the urban area, the old and new industrial zones, Dingshan Lake, the flood control plan of Tai Lake, the upstream canals, the estuary of Yangtze River, the East China Sea and Hangzhou Bay. During project implementation, several factors had to be considered, including financing the capital costs, local technical capability, drinking water quality improvement, urban sanitation improvement, demolition of houses, relocation of people, the impact to traffic and the costs of operating the new system. The whole project must be supported by a combination of engineering and other measures, such as laws, policies and management. The basic approaches were as follows:

- Moving the water intake further upstream in the Huangpu River immediately because it would bring a direct benefit for the health of the people.

- Pollution control of Suzhou River as a priority over the Huangpu River pollution control plan because the Suzhou River passes through the downtown area of Shanghai and is responsible for about 30 per cent of the pollution of Huangpu River.
- Taking advantage of the environmental assimilative capability of Yangtze River and East China Sea for discharges of sewage that has been properly pre-treated.
- Protecting the source water of the upstream Huangpu River (particularly from pollution from new, private rural industries) in order to guarantee the water quality for the new water supply intake and to avoid future pollution.
- The strengthening of the current pollution enforcement program of the Shanghai Government, including setting up a special regulatory system for industrial and domestic pollution control in the upper Huangpu River.

II.4.1 Moving water supply intake upstream of the Huangpu River Project

As mentioned above, due to the expansion of the city over many years, the water quality of the present water supply intake points does not, and probably will never, meet the water quality standards for the drinking water source for Shanghai. The City thus decided that moving the water supply intake locations upstream in the Huangpu River was the only viable long-term solution. A study was conducted from the late 1970s to the early 1980s to evaluate different options and to determine the most cost-effective approach.

In the study for the selection of the new intake point (Shanghai Municipal Urban Construction Design Institute, 1993) the main issues considered were:
- The impact of increasing sewage in the mid-section of the Huangpu River as a result of no further extraction by the water treatment plants in the present locations in the city section of Huangpu river.
- Pollution intrusion upstream under tidal influence, especially during the dry season.
- Options for pipe routes from existing water treatment plants to the selected intake points in relation to the costs associated with engineering construction and relocation of people.
- The financing of the project and the number of phases for implementation.

Based on hydrological conditions, the probability of four proposed intake points being affected by wastewater discharges from various points was calculated (Table II.4). The section between Minghang and the Bridge was found to be suitable for locating the new intake point (Figure II.5) The project was then divided into two phases: the relocation of the water intake to Linjiang and the location of the Bridge as the ultimate water intake.

Table II.4 Probability of the four sections of the Huangpu River being affected by wastewater discharge at different points

| | Section of Huangpu River | | | | | | | |
	Zhgang		Minhang		Daqiao		Mishidu	
Discharge period	June	Aug.	June	Aug.	June	Aug.	June	Aug.
Minhang			90	50	95	39	90	26
Wujing	96	43	92	30	64	7	26	0
Changqiao WTP	92	26	73	8	18	0	5	0
Nanshi WTP	45	4	9	0	0	0	0	0
Pudong WTP	26	0	3	0	0	0	0	0
Yangshupu WTP	2	0	0	0	0	0	0	0
Zhabei WTP	0	0	0	0	0	0	0	0

WTP   Water treatment plant

The water intake relocation project consists of the following three major components:

- A water diversion channel and steel transmission pipes with the total length of 70 km, in which the section of each hole of square concrete channel is 8–10 m^2 and bearing an inner pressure of 1.35 kg cm^{-2}.
- Three pipes crossing the Huangpu River to the Yangshupu Water Treatment Plant (WTP), Nanshi WTP and Lingjiang Pump Station with diameters of 3 and 4 m.
- Four large-size intake pump stations and booster pump stations equipped with 35 large water pumps.

The designed intake capacity is 5,000,000 m^3 d^{-1} serving 6 million citizens.

Project implementation and benefit
The project was divided into two phases for implementation. The first phase of the project, which was completed in July 1987, succeeded in drawing water from the Lingjiang Pump Station. This phase consisted of:

- The two large pump stations of Lingjiang and Yanqiao.
- A three hole concrete water transmission channel with a length of 17.5 km.
- Steel water transmission branches with a length of 16.68 km.
- River-crossing jacking pipes for Yangshupu and Nanshi WTPs with a diameter of 3,000 mm and a length of 2.63 km.
- Connection engineering between Yangshupu, Nanshi, Yangsi and Jujiajiao WTPs.
- Corresponding communication engineering.

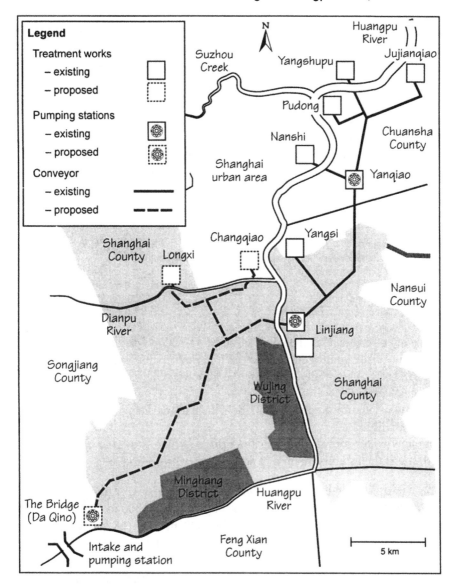

Figure II.5 Map showing the route of the new water conveyer between the City and the new water intake point at Bridge

The completion of the first phase of the new water intake project at Lingjiang, enabled Yangshupu and Nanshi WTPs to provide relatively clean water to 4 million people in the main city (i.e. compared with previously). The total investment for phase one was about US$ 70 million. The completion of this phase, however, will not meet the required quality for water supply because it is not free from the risk of pollution. This problem was experienced in the summer of 1988 when the water quality of Lingjiang deteriorated seriously. This was caused by a reduction in the flow from upstream (the Tai Lake flood release) by about 15 per cent compared with the average flow of a normal year, and the tidal intrusion carrying sewage from downstream to the upstream section of the Huangpu River.

In the second phase of the project, the intake will be moved further upstream to the neighbourhood of the Huangpu Bridge. The main investments associated with this are for the following components:

Item	No. required	Dimensions
Bridge pump station	1	$5,400,000 \, m^3 \, d^{-1}$
A reservoir with aeration facilities	1	$40,000 \, m^3$
Water transmission main channel	1	3.4×3.8 m; length 16.6 km
Water transmission channel	1	2.5×2.8 m; length 3.5 km
River-crossing jacking pipe	1	DN3700; length 0.88 km
Water transmission branch channel	2	2.5×3.0 m; length 6.3 km

The expected results after the completion of the second phase are:
- The raw water quality for the water treatment plants will be improved significantly, essentially meeting the requirements for drinking water sources (Table II.5).
- The raw water after treatment will meet the national standards for drinking water quality.
- The new water source area near the bridge (which is a large open space) will merit the establishment of a source water protection area.

II.4.2 Shanghai sewerage collection and wastewater treatment project
In 1992, the total sewage discharge of the city was $5,500,000 \, m^3 \, d^{-1}$, in which industrial wastewater accounted for $3,750,000 \, m^3 \, d^{-1}$ (68 per cent) and domestic wastewater accounted for $1,750,000 \, m^3 \, d^{-1}$ (32 per cent). Only 3 per cent (about $180,000 \, m^3 \, d^{-1}$), consisting mainly of domestic sewage, was collected and treated by municipal wastewater treatment plants. The West sewer main received $700,000 \, m^3 \, d^{-1}$ and the South sewer main received $300,000 \, m^3 \, d^{-1}$. Both sewers were built in the 1970s and discharge 18 per cent of their wastewater to the Yangtze River without any treatment. The remaining 79 per cent was discharged directly to the Huangpu River, of which about 30 per cent came from the tributary, i.e. the Suzhou River. About 25 per cent of the industrial wastewater received primary and/or secondary treatment (Table II.6).

Table II.5   Comparison of main water quality indicators obtained at different intake points on the Huangpu River

| Indicator | Water intake points | | | | Relative improvement at Daqiao intake[1] |
	Yangshupu WTP	Nanshi WTP	Changqiao WTP	Daqiao intake	
Ammonia-N (mg l^{-1})	2.10	1.68	1.00	0.35	Reduced to 6–2.7 times
Dissolved oxygen (mg l^{-1})	2.70	4.69	4.72	5.00	Increased to 1.9–1.1 times
Phenol (mg l^{-1})	0.007	0.004	0.004	0.001	Reduced to 7–4 times
Chloride (mg l^{-1})	50 (1,500)	45 (1,380)	44 (225)	32 (< 93)	Reduced to 1.56–1.3 times (16.1–2.4 times)

[1]   Improvement compared with Yanghupu, Nanshi and Changqiao WTPs

Table II.6   Nature and disposal of sewage in Shanghai

	Quantity of sewage (10^3 m^3 d^{-1})	Proportion (%)	Remark
Total quantity	5,500		
Industrial	3,750	68	
Domestic	1,750	32	
Quantities discharged to:			
Wastewater treatment plants	180	3	500,000 m^3 d^{-1}
Yangtze River by Western Transmission Main	700	13	
Yangtze River by Southern Transmission Main	300	5	
Directly to Huangpu River and its branches	4,320	79	

According to the Strategic Study of Urban Waste Water Treatment in Shanghai (Shanghai EPB, 1985), the proposed control measures included point source treatment at the industrial sources, centralised treatment at industrial parks, joint treatment at several suburban towns and industrial centres, large combined sewerage collection systems for urban centres, and disposing wastewater to the Yangtze River and making use of its assimilative capacity (Table II.7).

Table II.7   The urban sewerage system of Shanghai

System name	Design capacity (10^3 m^3 d^{-1})	Domestic sewage (10^3 m^3 d^{-1})	Industrial waste (10^3 m^3 d^{-1})	Groundwater (10^3 m^3 d^{-1})	Comments
Shidongkou	700	573	127		Completed
Zhuyuan	1,700	543	916	241	Completed in December 1993
Bailonggang	4,934	2,340	2,216	378	Under planning, including 700,000 m^3 d^{-1} of Minghang Wujin system
Total quantity	7,334	3,456	3,259	819	

Shanghai combined wastewater treatment — Phase One Project

Shanghai Combined Waste Water Treatment Project adopted the scheme recommended in the Urban Waste Water Treatment Strategic Study of Shanghai, i.e. to intercept the urban sewage and to discharge (after screening treatment) deep in the estuary of the Yangtze River. The first phase gave priority to the interception of the sewage discharged to the Suzhou River, to the improvement of the water quality of the Suzhou River and to the environmental quality of the web of Suzhou River, so as to reduce the pollution of the Huangpu River (Figure II.6). The effluent disposal site in the estuary of the Yangtze River was located 10 km downstream of Wusongkou. The first phase serves 70.6 km^2, 2.55 million people and more than 1,000 industrial plants. The designed average dry season waste flow for the system was 1,400,000 m^3 s^{-1}, the designed peak dry season waste flow was 2,730,000 m^3 s^{-1} and, because it is a combined sewerage system, it also receives surface run-off.

The feasibility study suggested that the sewage from each discharge point should be collected by gravity transmission mains by the manifolds of the combined sewerage system, and then transmitted to the transfer pump station (Shanghai Environment Project Office, 1993). The wastewater should be lifted and passed through a siphon beneath the Huangpu River taking wastewater to the other side the Pudong Area for pre-treatment with screening. During pre-treatment, particles and suspended substances more than 5 mm in diameter are eliminated. Finally, the wastewater should be lifted and pumped to the Yangtze though an outfall diffuser system at Zyuyan (Anon, 1990).

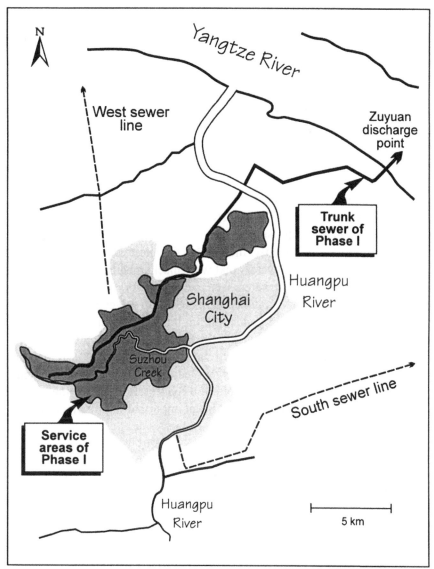

Figure II.6 Plan of the service area and trunk sewer line of Phase One of the Shanghai Sewerage Project

The construction of phase one began in August 1988. The main structures were completed and trial operation began on December 1993. The total cost of the project was 1.6×10^9 RMB yuan (about US$ 200 million). The project was partly financed by the World Bank.

Environmental benefit of the first phase

Before phase one of the sewerage project, urban sewage discharged to the Shuzhou River, including domestic and industrial wastewater and surface run-off, was carried into the Huangpu River at downtown Shanghai. According to statistics collected in the 1980s, the pollution load from Shuzhou River amounted to 46 per cent of the total pollution load received by the Huangpu River from the Shanghai urban area. Thus, intercepting the sewage discharged to Shuzhou River will improve the water quality of both the Shuzhou River and the Huangpu River.

Before the project, the water quality of the Shuzhou River was worse than the lowest water quality class (Class V) of the National Environmental Quality Standards of Surface Water. However, with the completion of the intercepting sewers along the Suzhou River in phase one, the water quality is expected to improve significantly. Included in the phase one components are collection of industrial wastewaters that were discharged to receiving water bodies and collection of wastewaters from several major river outfalls. With these sub-projects, the water quality of the Suzhou River will be further improved as a result of the reduction in total pollution loading. The unsanitary conditions that have existed for many years in the Suzhou River will finally cease and the ambient environment along the river will also be significantly improved. The Suzhou River feeds into the Huangpu River and, as a result of the reduction of the pollution load in the city section of the Suzhou River by 70 per cent, there will be an important improvement in the water quality of the Huangpu River.

Environmental impact around the outfall area in the Yangtze River

The deep water dispersion method was selected for wastewater discharge on the basis of modelling results. The dilution ratio at the mixing zone is 100 times the wastewater quantity and the water quality at the mixing zone can still achieve Class III water quality standard for most variables. According to physical and mathematical modelling of the wastewater dispersion, the key factors affecting the effectiveness of dispersion at the mixing area are flow rate and tidal condition. The combination of low flow rate in the dry season with low tide create the worst conditions for mixing. As a result the mixing area would have to be enlarged to as much as 4 km^2 in order to meet the required dilution ratio. Thus the outfall dispersion points must be situated sufficiently far from the bank to ensure that the mixing zone does not approach the near side of the river and create a "sewage belt". Avoiding the creation of the sewage belt is also important for fish migration within the channel.

II.4.3 Shanghai combined wastewater treatment — Phase Two Project

The scope of the second phase of the project includes wastewater collection from the additional areas of city centre that were not covered under phase one, including the new Pudong industrial centre and the many wastewater discharges to the inner canals in the suburban areas. It is hoped that with the completion of the second phase of the project the city will finally have an acceptable water environment.

The areas covered under the second phase include 21 km² of the Shuhus and Luwan Districts, 155 km² in the south of the new Pudong industrial area, and 92.1 km² of the upstream Huangpu River areas of Minghong and Wujin Districts. The total service area will be 269.6 km² with 4.68 million people. The second phase plan consists of a wastewater collection system and a pre-treatment system, with a discharge point to the Yangtze River at Bailon-gan (Shanghai Municipal Urban Construction Design Institute, 1993). The implementation of the second phase has been divided into four stages which correspond to the four collection trunk sewer lines. The total investment of the second phase is estimated to be 4.885×10^9 RMB yuan (US$ 58.6 million) and is expected to be completed by the end of 1998. The project feasibility study and the EIA are both underway.

II.4.4 The Zhonggang sewerage project

To protect the water quality from the upper stream of Huangpu River, the Zhonggang sewerage project has been proposed (Figure II.7). The service area will cover many rural industries, including mechanised animal farms, and the Xinhuo Industrial Area.

II.5 Other major measures used in cleaning the Huangpu River

The Huangpu River Pollution Control Project takes an integrated approach by including engineering and non-engineering measures. Besides the main engineering works mentioned above, some other activities include domestic wastewater treatment sub-projects for the protection of Dianshanhu Lake (the source water of Huangpu River), waste treatment for mechanised cattle, hog and poultry farms in the area, and the establishment of a clean belt along the river to protect the water supply intake.

The non-engineering measures are mainly related to institutional strengthening for organisation and regulatory measures. Some examples are:

- The establishment of a special office, the Office for the Protection of Shanghai Huangpu River Source, under the Shanghai EPB, with special responsibility for the management and enforcement of pollution control in the upper reaches of Huangpu River.

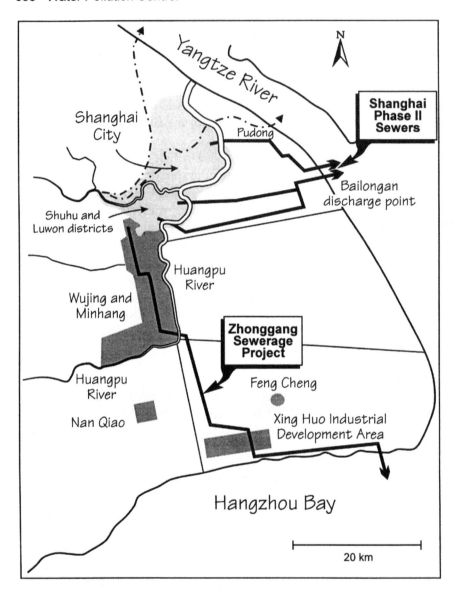

Figure II.7 Plan of the service area of Phase Two of the Shanghai Sewerage Project

- The publishing of the "Regulations for the Protection of the Water Source of the Upper Reaches of Huangpu River" and the corresponding rules for implementation, together with the authorisation of the Shanghai EPB as the responsible agency for organisation, implementation and enforcement of the regulations.

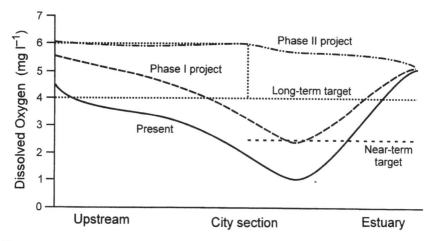

Figure II.8 Expected improvements in dissolved oxygen in the Huangpu River as a result of the Shanghai Sewerage Project

- The enforcement of the waste discharge permit system, based on control of waste loading, so as to limit the total amount of waste discharged to the natural system.
- The adoption of a pollution trading system that ensures there is always excess assimilative capacity in the river.
- The promotion of waste minimisation and the use of cleaner technology practices at pollution sources.

II.6   Conclusions

The main direct benefits of cleaning the Huangpu River Basin are social and environmental, although the economic benefit is also believed to be significant. A cleaner environment will be attractive to foreign investment which is critical to Shanghai's future economic development. Some of the main benefits are:

- Recovery of the ecological system of the Huangpu River and its tributaries due to increased concentrations of dissolved oxygen (Figure II.8) as the concentration of organic material decreases (Figure II.9).
- Improved drinking water quality, leading to a reduced rate of disease and improved hygienic conditions in the area.
- Elimination of the unsanitary and odorous conditions in the rivers, improving the aesthetic value of the river.
- The attraction of more outside investment to Shanghai as a result of its cleaner environment.
- An increase in the real estate price of the areas along the rivers and canals that are cleaner as a result of the project.

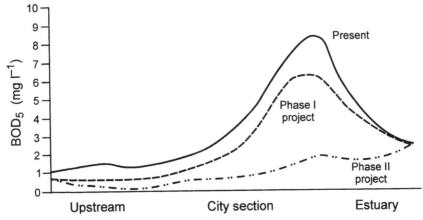

Figure II.9 Expected improvements in BOD5 in the Huangpu River as a result of the Shanghai Sewerage Project

II.7   References

Anon. 1990 Environmental impact of waste water discharge at Zuyuan. *Shanghai Environ. J.,* **19**(4).

Qu Geping 1991a *On Environmental Supervision and Management, Environmental Protection in China.* United Nations Environment Programme and China Environmental Science Press.

Qu Geping 1991b *The Evolution and Development of Environmental Protection Policy in China, Environmental Protection in China.* United Nations Environment Programme and China Environmental Science Press.

Shanghai Environment Project Office 1993 *Huangpu River Water Quality Protection Feasibility Study.* Shanghai Environment Project Office, Shanghai.

Shanghai EPB 1985a *Huangpu River Waste Water Integrated Prevention and Control Planning.* Shanghai Environmental Protection Bureau, Shanghai.

Shanghai EPB 1985b *Shanghai Municipal Waste Water Treatment Strategy.* Shanghai Environmental Protection Bureau, Shanghai.

Shanghai Municipal Urban Construction Design Institute 1992 *Feasibility Study of Extending the Shanghai Water Intake to the Up Stream of the Huangpu River.* Shanghai Municipal Urban Construction Design Institute, Shanghai.

Shanghai Municipal Urban Construction Design Institute 1993 *Feasibility Study of Waste Water Discharge at Bailongang.* Shanghai Municipal Urban Construction Design Institute, Shanghai.

Case Study III[*]

THE PASIG RIVER, PHILIPPINES

III.1   Country profile

The Philippines is a country of 65 million people, of whom around 8 million (equivalent to 13 per cent of the total population) reside in the National Capital Region (NCR), Metropolitan Manila (Figure III.1). The population has been growing at a rate of 2.3 per cent every year over the past 10 years and urbanisation has increased from almost 40 per cent in 1985 to 43 per cent in 1990.

Unemployment nationwide was 11.1 per cent in 1985 and declined to 8.6 per cent in 1989. However, the high influx of migrants from the provinces and the lack of employment opportunities in the metropolis brought the unemployment rate in the NCR to 26.1 per cent in 1985; this has since fallen to 17 per cent. The incidence of poverty has been decreasing, although it is recorded as high as 50 per cent for some provinces and at 32 per cent for the NCR.

When it assumed power in 1992, the Ramos administration embarked on an ambitious programme, called "Philippines 2000", to establish the country as a newly-industrialising economy by the turn of the century. Since then the Philippines have achieved a 5.1 per cent growth in 1994 from a low of 2.4 per cent in 1993. This has earned the country the respect, albeit prematurely, of the Asian business community and the *Far Eastern Economic Review* referred to the country as the "most improved economy" in its Year End Review of 1994.

Despite its laggard image, the Philippines remains the most politically stable country in Asia. Unlike its more economically stable neighbours which suffer from lack of succession laws, the Philippines has experienced a peaceful transfer of political power to a newly-restored democracy under Corazon Aquino, successor to the 20-year authoritarian regime of Ferdinand Marcos. This new democracy is marked by the reinstatement of democratic institutions, in particular a popularly mandated constitution, a legislature that has seen two terms since the dictatorship was overthrown, popularly elected local governments and at least three peaceful and credible elections (national and

---

*   *This case study was prepared by Renato T. Cruz*

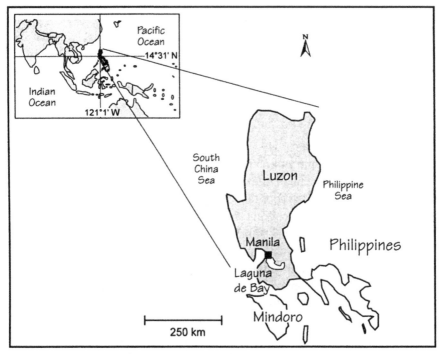

Figure III.1 Location map of the Philippines showing the national capital region of Manila

local) since 1986. By and large, the present government enjoys a relatively high level of support from the population. Its main dilemma has been sustaining the economic triumphs of the past two years while at the same time addressing the nation's advanced stage of poverty and environmental destruction.

III.2 Basin identification

The Pasig River system runs through five cities and four municipalities (Figure III.2) and connects two large, important bodies of water; Manila Bay in the west is the country's main port of maritime trade and travel and Laguna de Bay in the east is the largest freshwater lake in the country and connects 30 suburban towns to the metropolitan centre. Before the colonial period, the Pasig River was the main point of entry for international trade into what is now the City of Manila. Advancements in land transportation have changed the landscape considerably.

Traditionally, the municipalities upstream were fishing communities relying mostly on the Pasig River and Laguna de Bay, while the settlements downstream experienced rapid urbanisation with the influx of trade from other provinces and countries. Before pollution virtually extinguished

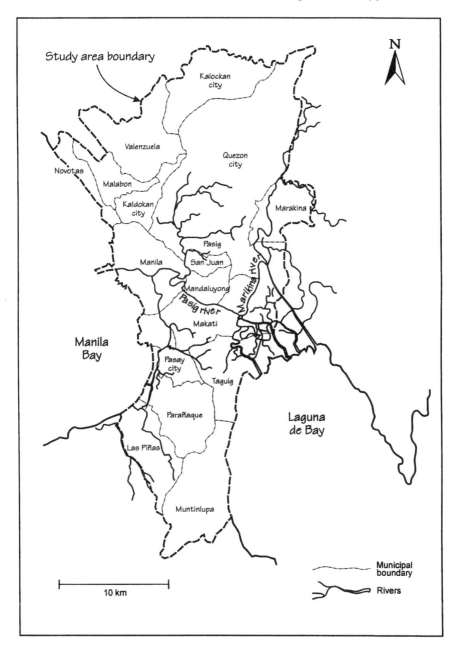

Figure III.2 Detailed map of the study area showing the Pasig River

aquatic life, the whole 25 km of the Pasig River between Laguna de Bay and Manila Bay served as a habitat for 25 varieties of fish and 13 different types of aquatic plant. Today, there are only six species of fish and two types of plants left that can tolerate the polluted water. The situation, however, is not irreversible. During the rainy months of June to December each year, fish from Laguna de Bay are carried by the floodwaters to the Pasig River. The flushing effect of the increased water levels in Laguna de Bay increase the dissolved oxygen content of the river to a level that increases its potential for some aquaculture activities. Unfortunately, during the dry summer months of March to May, the river is virtually dead because the water becomes stagnant with the much reduced flow.

The banks of the Pasig River are lined by squatter colonies consisting of approximately 12,000 households. About 2,000 families live in houses on stilts or under the bridges, in sub-human conditions, where they present a danger to themselves and to the vessels using the river. These settlements have no sanitary facilities and their liquid and solid wastes are discharged straight into the river.

The various subcultures existing in Metro Manila result in many problems that reflect the complex socio-economic characteristics of the city. With the continuous dumping of wastes, the river bed has become more and more silted with organic matter and non-biodegradable rubbish. This results in serious flooding along the river, affecting nearby communities and carrying polluted water to the households living close to the river.

III.3   Pre-intervention situation

A feasibility study conducted in 1991 by the Department of Environment and Natural Resources (DENR) with funding from the Danish International Development Agency (Danida) and technical assistance from the Danish consultancy company, Carl Bro International, established the levels of pollution and the overall condition of the Pasig River. The study was conducted between 1989 and 1990 and has provided the main reference point for the rehabilitation programme.

III.3.1 Pollution sources

Industrial pollution accounts for 45 per cent of the total pollution in the Pasig River. About 315 of the 2,000 or more factories situated in the river basin have been determined as principal polluters of the river, dumping an average of 145 t of biochemical oxygen demand (BOD) per day. This was established by determining the suspended solids in their treated and untreated wastewaters. According to records, the textile and food manufacturing industries are the greatest water polluters among those considered in the study. The pollution rate is expected to decrease by 2 per cent a year due to the limited commercial land available along the river and the increased requirements for container transport.

Domestic liquid waste contributes another 45 per cent of the pollution load in the Pasig River. There were approximately 4.4 million people living in the Pasig River catchment area during the study period and only 0.6 million, or 12 per cent, were serviced by the sewerage system which treats domestic wastewaters before discharging them into Manila Bay. Untreated waste-waters from the remaining 88 per cent of the population flow through canals and esteros into viaducts leading into the Pasig River. It is estimated that 148 t d^{-1} of BOD is added to the Pasig River purely from the sewage outlets scattered along its banks. The Metropolitan Waterworks and Sewerage System (MWSS) (the government agency responsible for domestic liquid waste) has been hampered in its task by a lack of funds. As it is also respons-ible for water supply in the metropolis, it has had to give water supply a higher priority than sewage management.

Solid waste contributes only 10 per cent of the pollution in the Pasig River. Although very visible, rubbish contributes only 30 t of BOD per day. How-ever, the solid waste deposited on the surface of the water blocks the penetration of sunlight to underwater plant life and the solid waste that sinks to the river bed suffocates the existing aquatic life.

Rubbish collection by the Metro Manila Authority (MMA) in the residen-tial areas of the 367 *barangays* (villages) in the study varied between 70 and 100 per cent per barangay depending on the accessibility of the area to land-based collection. Inaccessible areas occur mostly along the banks of the river and hence the rubbish from these was thrown into the water. The estimated 34 t of rubbish accumulated in these riverside areas in 1990 is expected to increase to 55 t by the year 2005.

III.3.2 Increasing urban migration and economic difficulties
From 1988 to 1990, the rate of migration into the squatter colonies along the riverbank was estimated at 73 per cent. A steady influx of migration into the metropolis has resulted in congestion and the exploitation of land and, ulti-mately, the Pasig River. Increasing poverty in the rural areas has driven rural people to migrate to Metro Manila to seek better income opportunities. The river banks are the most logical areas for new settlements because many of the other squatter colonies in the metropolis are already overpopulated.

The economic problems experienced by the government have prevented it from providing better housing facilities for the poor. Similarly they have been unable to address the deficient infrastructure or to introduce anti-pollution measures and this has resulted in the present state of the river and its environment.

III.3.3 Lack of a strategic programme for river rehabilitation

The feasibility study concluded that the pollution problems in the river have been deteriorating since the 1970s, or over the past 20 years. Previous administrations have embarked on river rehabilitation schemes but all of these, however, were short-lived because they failed to address the root of the problem.

The feasibility study also concluded that sufficient laws and regulatory responsibilities have invested in existing government agencies. Unfortunately, these agencies have not been able to exercise their regulatory functions effectively due to legal processes and circuitous bureaucracy. Among other reasons, it was discovered that there were government agencies with over-lapping responsibilities but without any single agency tasked with overall co-ordination. Worst of all, local government units have been negligent in enforcing a land-use and zoning ordinance established by the Metropolitan Manila Commission (predecessor of the MMA) in the early 1970s. Hence, a comprehensive development plan would have to be formulated and imple-mented to effect sustained progress in the improvement of the Pasig River, where much of the city's wastes end up.

III.3.4 Flooding

Flooding was also identified as a problem. The combination of old drains and rubbish result in blockages in the system. In a flood in 1986, the whole of Metro Manila was submerged in water reaching a depth of 7 feet (approxi-mately 2.1 m) in some areas. Investigations revealed that this was due to inadequate drainage and to serious clogging of the drainage system in areas it was supposed to serve. Since then, the Department of Public Works and Highways (DPWH) has engaged in declogging programmes, has constructed drains in low-lying areas and has renovated drains and river walls. For flood control activities alone, the government has spent an average of P100 million for each of the past five years.

III.3.5 Diminished use of the river

The Pasig River has been historically known for its recreational and transport functions. With its gradual degeneration, this aspect has been reduced to use for rowing by some enthusiasts only. The river was classified as Class D and, therefore, secondary water-contact sports were discouraged. When it is upgraded to Class C, sports such as rowing and sculling can be encouraged.

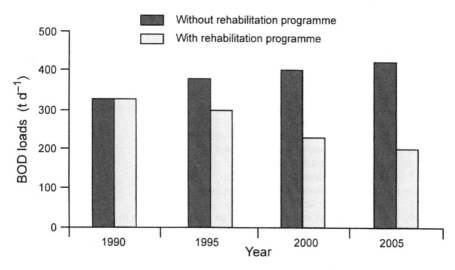

Figure III.3 Projected loadings of BOD for the Pasig River system with and without the rehabilitation programme

A river transport system was established in the early 1990s partly as an effort to provide alternative transport routes in the metropolis. The private company managing the ferries had to terminate their services after a few years due to heavy financial losses. Apparently, the foul odour and the unsightly floating debris made travel on the ferries very uncomfortable for the passengers.

On the whole, most aquatic life in the river has declined as the levels of pollution have increased. The feasibility study concluded that the river is presently at Class D. Mathematical model simulations indicated the BOD loading should be reduced from its 327 t d^{-1} in 1990 to 200 t d^{-1} in order to restore the river's ecology (Figure III.3).

III.3.6 Previous programmes on the Pasig River

Efforts to revive the Pasig River have been attempted before. These have generally failed because the programmes did not recognise the importance of involving the communities and the private sector.

One such effort was that of the former First Lady and Acting Governor of Metro Manila, Mrs Imelda R. Marcos. Her plan was a grandiose spectacle to attract tourists with floating casinos and restaurants, like Hong Kong's Aberdeen, and gondolas fashioned after those of Venice and others. The river walls were painted and trees were planted to initiate the improvement programme. Very quickly, however, these plans fizzled outdue to lack of support.

Box III.1   A summary of proposed projects under the Pasig River
Rehabilitation Program

1.   River Rehabilitation Secretariat
2.   Flushing of the Pasig River
3.   Industrial Waste to Energy
4.   Secondary industry for resource recovery
5.   Hazardous hospital waste treatment
6.   Collection of solid waste in rivers
7.   Upgrading of squatter settlements
8.   Upgrading of water quality laboratories
9.   Absorption capacity of Manila Bay
10.   Construction of sanitation sewerage system (Metross II)
11.   Septic tank maintenance programme
12.   Local treatment of sewage from high-income residential areas
and complexes
13.   Diversion of San Juan River
14.   Collection of solid waste in inaccessible barangays
15.   Integrated solids waste management programme
16.   East-West Mangahan
17.   Pasig-Marikina floodway
18.   Development for the National Capital Region
19.   Removal of sunken wrecks
20.   Riverside parks
21.   Urban renewal of Escolta district

III.4   The intervention scenario

III.4.1 Objectives, strategies and targets

The feasibility study proposed a Pasig River Rehabilitation Program (PRRP) which aims:

- To improve the quality of the water in the Pasig River.
- To improve the environmental conditions in and along the river.

The programme has two strategies:

- Physically clean up the river in the short-term.
- Stop pollution at source in the long-term.

The plan of operation suggested by the study recommends 21 different projects (see Box III.1) over a period of 10–15 years at a cost of a little over US$ 420 million. This is a comprehensive programme that would address the main sources of pollution in the river as well as the attendant problems that have contributed to the deterioration of the surrounding environment. The

PRRP is a multi-agency programme with the Department of Environment and Natural Resources acting as the lead agency. The programme has the following targets:

- Completely eliminate the offensive odour in the dry season through the reduction of the level of pollutants discharged into the river.
- Reduce the BOD load of the Pasig River from the estimated 330 t d^{-1} to 200 t d^{-1}.
- Reduce the amount of solid waste dumped into the rivers and creeks of the Pasig River System with regular waste collection activities.
- Increase and control the flow of the water through the Pasig River system especially during the dry season.
- Reduce the frequency of flooding along the Pasig River and its main tributaries.
- Strengthen the content, and improve the enforcement, of the Zoning Ordinance of 1981 for the National Capital Region.
- Remove the sunken vessels from the bed of the river.
- Develop parks along the Pasig River.
- Relocate the squatters living along the Pasig River and its main tributaries.

III.4.2 Activities and strategies
The following activities are being carried out to achieve the targets listed above.

Establishment of the River Rehabilitation Secretariat
Recognising the need for a distinct body to co-ordinate the efforts to rehabilitate the Pasig River, the PRRP required the establishment of the River Rehabilitation Secretariat (RRS) as a project office under the DENR. The RRS is the instrument responsible for establishing the co-ordination system, providing technical support to programme management and paving the way for the transfer of such responsibility to an existing government agency. As the official secretariat of the PRRP, the RRS is responsible for:

- The review of plans, programmes and targets and the implementation of the programme.
- Monitoring and co-ordination of activities between and among partners.
- Evaluation and assessment of the effectiveness of the programmes under the PRRP to ensure that these follow the precepts and mandate of the programme and their respective organisations.
- Screening and endorsement of the technical and financial viability of projects proposed for the programme.
- Identifying deficiencies in resources, issues and concerns affecting the programme.

- Reviewing and recommending improvements in policies, laws and rules affecting the programme.

The RRS has a pivotal role in the co-ordination of efforts in the programme and is, therefore, also required to set-up the partnership mechanisms with all those involved. Its structure involves training that is open to the staff of partners and other activities that will make the working relations conducive to co-operation by all concerned.

Industrial pollution abatement

Two projects are being implemented in the Plan of Operations to address industrial pollution: "Waste to Energy" and "Secondary Industry from Waste Recovery". The RRS and the Metropolitan Environment Improvement Program (MEIP) took an alternative approach for dealing with polluting companies. The two projects engaged the 25 top industrial polluters of the Pasig River in a Clean River Pact. Under this agreement, the participating companies pledged to support the PRRP and were committed to comply with the DENR standards of effluent. For its part, the PRRP provided the companies with technical assistance. The pact ensures that the industries will either treat their wastewaters or minimise their waste discharges to ensure that they can be absorbed by the Pasig River. In this way, the co-operation of the "partner companies" was encouraged and the RRS was mandated to take the lead in encouraging the industries. Under this arrangement, the RRS, the National Capital Regional Office of the DENR and the Laguna Lake Development Authority conduct regular monitoring of the riverside establishments.

Liquid waste management

The existing sewerage system must also be upgraded, entailing huge amounts of public funds. The MWSS has a long-term sewerage improvement programme to increase the coverage of its present treatment facilities. This was included as one of the major projects under the Plan of Operation. Initially, the MWSS addressed this problem with its Septic Tank Management Program (STAMP). Through STAMP, domestic and commercial septic tanks in selected areas are desludged to prevent outflows into the main drainage system of Metro Manila. Financial constraints have hampered the implementation of this project but efforts can be improved as more public funds are generated.

Solid waste management

To address the problem of solid waste, the programme linked up with a parallel government programme on solid waste management. The Waterways

Sanitation Service of the MMA leads the physical clean-up of the river, assisted by local government units and other agencies such as the Philippine Coast Guard, which erected boom traps in strategic sites along the river to help trap floating debris for eventual collection. Intensive awareness-raising campaigns are being carried out within the riverside communities to motivate them to organise waste management and waste recycling. The young people of the area have also been mobilised to help in the dissemination of information on the efforts to improve the Pasig River and on the help that everybody can give in this effort.

Infrastructure development
To increase the flow of the river, shallow areas are being dredged and the 22 identified sunken vessels have been resurfaced. Dredging is limited to the areas at the mouth of the Pasig River but river walls are being renovated at several sites. The Rehabilitation Program also supports the development of riverside parks to help to discourage the settlement of new squatters and to encourage an appreciation of the river. Recognising the hazard to the communities encroaching the river, and their direct contribution to the pollution of the river, squatters are being relocated to sites outside of Metro Manila.

The water quality laboratory of the Environmental Management Bureau will also be upgraded. The existing capabilities and facilities of the laboratory will be developed to encourage its use as a National Reference Laboratory for water quality analyses.

Information, education and communication
Knowing the previous attempts to rehabilitate the Pasig River have failed because of the lack of support from the private sector, the RRS established a Public Information and Activation Unit. The purpose of this unit is to raise sufficient support from the private sector (e.g. communities, business sector, schools) to promote the programme and its projects. A comprehensive communication plan has been prepared to manage these activities.

Communication materials are prepared and disseminated. These include a television commercial, documentaries, posters, stickers, leaflets, brochures, pamphlets, primers and regular newsletters. Audio-visual presentations are also being developed for use in briefing seminars on the PRRP and its programmes, and in training. The media are also provided regularly with updated information regarding the programme.

Along with these materials, the RRS supports the organisation of private sector groups for activities that help the programme. The Department of Education, Culture and Sports is a critical link between the RRS and students in public schools, especially for raising awareness and for the implementation of the PRRP Schools Program in which exhibits and competitions are held within campuses, culminating in inter-school competitions. The inclusion of ecology and environmental conservation in school timetables is also being encouraged. Communities are being organised to implement waste management programmes with the help of the Sagip Pasig Movement (Save the Pasig Movement) whose honorary Chairperson is First Lady Amelita M. Ramos. The RRS also supports the Linis Ganda Movement in organising "junkshop co-operatives" engaged in waste recycling. Clean-up campaigns have also been launched in communities, especially those along the Pasig River and its main tributaries.

Personnel development
Realising the success of the PRRP hinges on the efficiency of human resources, a Manpower Development Unit was set up in the RRS with two basic strategies: placement of qualified personnel only within the organisational network and the enhancement of the capabilities of the current personnel. Specific areas of training focus on the development of skills in co-ordination, project management, resource management, environmental education, communication and specialised technical skills. This ensures that partners will participate fully in the programme at the capacity in which they were trained.

Water quality monitoring
To gauge the success of the programme, the waters of the Pasig River are being analysed twice a month and the pollution levels are being determined. The programme uses 10 sampling stations along the Pasig River system (including San Juan River, Marikina River, Manila Bay and Laguna de Bay) to gauge the degree of pollution based on BOD, dissolved oxygen (DO), coliform bacteria counts, salinity, phosphates, nitrates and others (see examples in Figure III.4).

The water quality experts of the programme are assisted by the Mike 11 System model. Measurements of the physical attributes of the river and the pollution levels taken at regular intervals are fed into the system. The data gathered are processed by the model which simulates the river and its flow based on mathematical equations. With this, the experts may be able to predict high water levels in the river or to simulate the flow of a large volume of water from one end of the river to the other, together with the levels of pollution in the river under the simulated conditions.

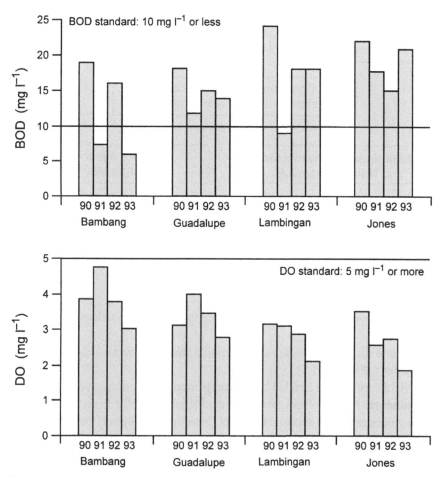

Figure III.4 Annual average BOD and dissolved oxygen (DO) concentrations in the main Pasig River, 1990–93

III.4.4 Present structure

President Fidel V. Ramos has included the PRRP as one of his priority agenda items during his administration. Therefore, it comes under the government's administrative network, with the President of the Republic and the Congress at the very top. Moving the programme towards its goals is the Presidential Task Force for Pasig River Rehabilitation which was created by President Ramos in July 1993. It is composed of leading government agencies directly concerned with the efforts of the PRRP and is chaired by the Secretary of the DENR. The task force is the main body to which general programme

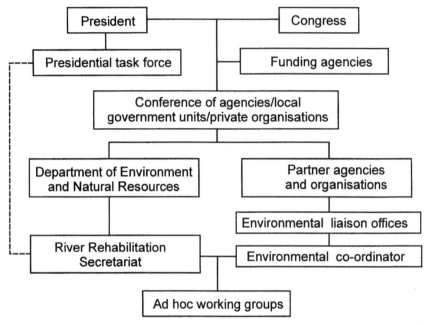

Figure III.5 The organisational structure of the Pasig River Rehabilitation Program (PRRP)

concerns and directional issues are addressed and it is directly responsible to the President (Figure III.5).

The RRS structure is headed by a management team composed of a Chief Environmental Adviser from Danida and a Project Director from the DENR. They are assisted by the Assistant Project Directors for Support Programs, for Administration and for the Working Groups (Figure III.6). The whole programme is supported by administrative staff composed of the accounting, computer operation, secretarial, transport and utility staff. There are also Action Officers serving as co-ordinators for the nine working groups of the PRRP. They hold the vital link between the RRS, as the co-ordinating office, and the different offices of the partners as represented by their Environmental Co-ordinators, Pollution Control Officers and liaison staff involved in the programme.

Supporting these working groups are the units for Planning and Monitoring, Manpower Development, and Public Information and Activation. These support units provide the personnel and logistics for overall co-ordination, training support for the personnel involved in the PRRP, public awareness campaigns and the effective implementation of non-priority projects through the mobilisation of the general public and the communities.

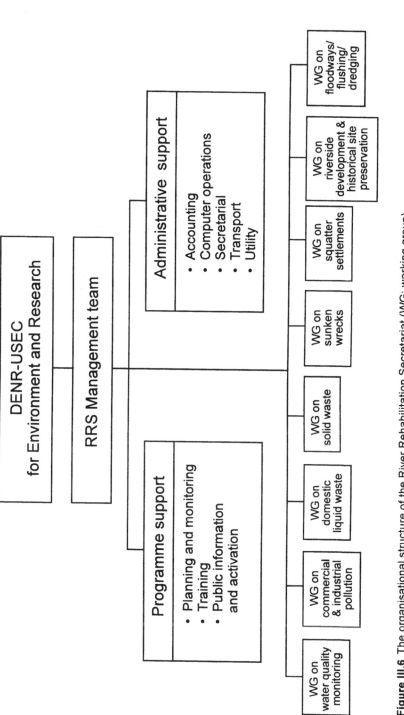

Figure III.6 The organisational structure of the River Rehabilitation Secretariat (WG: working group)

III.4.5 Major accomplishments in the first year

Within the first 18 months of the programme, a network comprising more than 100 government and non-government groups was involved. Thirty-five government agencies signed a Memorandum of Agreement clearly stating their acceptance of, and support for, the Plan of Operation of the Pasig River Rehabilitation Program. The Memorandum also identifies key responsibilities to which the signatories have committed themselves.

The Memorandum of Agreement was the springboard for establishing a planning and monitoring system for the inputs of each of the agencies involved in the plan of operation. The RRS facilitated the formulation of the work programme by gathering the various agencies into eight technical working groups. The committees meet regularly to discuss implementation of plans as well as policy recommendations to the Presidential Task Force. All the agencies meet twice a year to exchange information and discuss bottlenecks and policies that affect all the participating agencies.

Political support was generated through the creation of the Presidential Task Force for the Pasig River Rehabilitation. As a result, most of the participating agencies were compelled to live up to their commitments, although several have been constrained by finances and by varying priorities. This coordination system has also attracted the participation of some private sector groups that have taken an interest in the Pasig River.

Of the 25 partner companies that signed the Clean River Pact in September 1993, 10 companies have already complied with the DENR standards of effluent by October 1994, five had been issued Cease and Desist Orders and the rest were improving or constructing their waste treatment facilities. Since that time, the MEIP has assisted industries to comply with the pact. The experience gained in this respect will eventually be applied to other companies along the river. In preparation for this, the RRS, Laguna Lake Development Authority and National Capital Regional Office of the DENR have embarked on an integrated Industrial Data Base Project detailing the companies and their operations along the Pasig River and Laguna Lake.

Floating debris and the dumping of solid waste in the river system has been partially controlled through multi-agency efforts. Local Government Units in the concerned areas intensified their law enforcement activities and their awareness campaigns. The Philippine Coast Guard set up boom traps to prevent the debris from spreading out into the river. The Waterways Sanitation Service of the MMA revived their rubbish collection activities on boats with designated pick-up stations along the river.

In the meantime, the Save the Pasig Movement and the RRS have been setting-up support programmes to address the problem of solid wastes. The Save the Pasig Movement has been training communities along the Pasig River in waste management and has created a multi-sectoral network in at least two communities to support their community-based waste management programmes. The RRS has also supported the Linis Ganda Movement, which collects recyclable materials from communities and re-sells them to companies with recycling systems. Both organisations have also helped in the organization of "junkshop co-operatives" in some of the municipalities and cities involved in the programme.

Twenty-five sunken wrecks have been successfully removed from the riverbed through the efforts of the Philippine Coast Guard. Some newly sunk vessels have been identified and efforts are already being made to have these refloated as soon as possible.

The city administration of Manila converted a former waste disposal site into a park, and 2 km of a 20 km stretch of the river have been developed by the municipal government of Marikina into a park with benches, jogging lanes and park facilities.

River walls and other structures along the river have also been repaired and constructed to maintain the banks. Shallow portions of the river have been dredged with the help of the Department of Public Works and Highways.

More than 1,000 squatter families along the banks of the Pasig River have been relocated to various sites in Cavite and Marikina. This was achieved through the collective efforts of the appropriate Local Government Units, National Housing Authority, DPWH and the Office of the President.

There has been an increased awareness amongst the general public of the programme aided by the media taking a considerable interest in the Pasig River. Comments, suggestions and "letters to the editor" have been printed in various daily newspapers and public expressions of interest and concern have been conveyed to the offices of the First Lady, the DENR and the RRS.

III.5 Lessons learned, constraints and opportunities

The initial phase of the programme was quite instructive. For its first year alone, much of the difficulties were centred on the availability of technology, bureaucratic procedures and a general lack of funds for the implementation of projects.

It was found that the polluting industries had adequate waste treatment facilities but still could not comply with the DENR standards due to the inefficiency of their operations. In response, the RRS has conducted training for waste treatment plant operators for these companies. Combined with continuous monitoring, this has helped to boost efforts to curb industrial pollution. Over-all the co-operative approach seems to be working well in dealing with industrial pollution.

The presence of laws and regulations against littering and dumping have helped the programme in its drive to reduce floating wastes on the river. The main constraint remains the enforcement of such laws. Logistical requirements are barely met and bureaucratic procedures have hampered the implementation of the projects. There was a need for more collection boats to help revive the river waste collection programme because the rental contract for the fleet of 12 boats expired in December 1994. Successful waste collection depends on the dissemination of information about waste reduction and the education of the riverside communities in waste management. Current efforts are minimal when compared with the gravity of the problem. Training for waste management in the communities needs more personnel.

As exemplified by the municipal government of Marikina, the development of the riverbanks depends mainly on the local government. The political will to evict people from illegally built establishments and structures and to maintain the developed areas along the river has driven local governments to lengthy debates with concerned groups. In addition, funding for the construction and maintenance of parks along the river is scarce. This is aggravated by the fact that the zoning ordinance that stipulates that waterways must have a 10 m clearance on both sides, is hardly put to effect.

Although the MWSS carried out regular desludging of septic tanks, there has been a shortage of sludge disposal sites complying with the environmental standards. There is also a need to secure funds for the second phase of the Metro Manila Sewerage System which is still in the planning stage. Despite what has already been spent, the expenditure allocated for dredging is a minuscule amount compared with the overall amount required to create an impact on the flow of the river. This measure is, therefore, only palliative.

Like all relocation efforts, the PRRP squatter relocation programme faces the problems of funding, logistics and the constant struggle with the community organisations of the squatter groups. As with the solid waste programme, this also needs education of the riverside communities, especially the squatters eligible for relocation. The administration and the programme both agree that the squatter families who encroach the river and build structures over the waterways present a danger to themselves and to those travelling along the river. The problem is further aggravated by the growing influx of migrants from other parts of the country and the metropolis, coupled with the ever-decreasing space available for them.

III.6   Conclusions and recommendations

The first 18 months of the programme concentrated on building consensus among the organisations concerned on the master plan for rehabilitation and on setting-up the implementation system to meet the objectives of the long-term programme. This in itself has been a most gruelling but equally rewarding experience. Once all the agencies, public and private, agreed on the objectives and strategies of the Rehabilitation Program, getting them to align their respective programmes and projects into an overall system was less difficult. Unfortunately, the PRRP has to grapple with the attendant problems of co-ordinating a multi-agency, long-term programme which will cross the term of three Presidents. The Philippine Government has a habit of changing priorities with every administration.

The long-term success of the programme also hinges on the capability of its managers to obtain the resources required to meet its objectives. Logically, the consistent implementation of the master plan would build a formidable credibility for the programme which, in turn, could attract support from donors. Unless, however, the Rehabilitation Program can be rationalised to be financially beneficial, it will be dependent on grants and soft loans and will not be able to attract profit-orientated private sector investment. Unfortunately, this is a circular argument. The huge financial gap in the programme will continue to plague its successful implementation.

Social pressure will be an important element in the future of the programme. The continuous, direct participation of private sector organisations will compel the Government to pursue the long-term objectives of the programme. Public opinion and the vigilance of the media will certainly escalate this pressure. This will be a function of a consistent and aggressive information, education and communication campaign and of the transparency of the programme. So far, the information, education and communication efforts of the PRRP have roused public awareness but have not brought it to a level of concern that can mount pressure on the Government to pursue the programme.

The following will be critical areas requiring careful attention in the next phase of the programme:

- Increased co-ordination between the agencies and organisations involved in the programme through closer review of the plans for implementation, common efforts at capacity building, and critical support to key projects.

Another important element is the institutionalisation of the co-ordination system that has been established, either through the establishment of a new agency with a limited tenure or the strengthening and incorporation of this function in an existing government agency. A strong law needs to be passed by the legislature in the immediate future to realise this.

- Constant review and upgrading of the plan of operation. The programme should be flexible in order to respond to rapid changes in the economic and political environment. If there is an effective system of co-ordination among all the agencies involved, it should not be difficult to amend plans, to rectify errors and to take advantage of new opportunities.
- An aggressive campaign to raise resources to ensure the implementation of the key projects in the Plan of Operation. It will be impossible to secure funding support for all the projects in a short period of time and, therefore, resource generation should be prioritised. If the co-ordination system has been put in place, major efforts should be made to obtain the necessary funds to ensure programme implementation. Lack of funds should not be used as an excuse for delays in project implementation. Instead, creativity should be exercised in revising plans or breaking up the projects into more implementable sizes to prevent delays. The worst thing that could happen to the programme is for it to lose its momentum and, in the process, to lose public and political interest.
- Strengthening public participation in the programme. Private organisations are usually more capable than a government of sustaining initiatives because they are less affected by political considerations. The active participation of more private organisations, especially those that can provide special technical expertise (for example, in the form of community mobilisation and research) not normally inherent in government, will ensure continuity of the programme.

NIGERIA

IV.1 Introduction

Nigeria is located approximately between latitude 4° and 14° North of the Equator, and between longitudes 2° 2' and 14° 30' East of the Greenwich meridian (Figure IV.1). It is bordered to the north by the Republics of Niger and Chad, to the south by the Atlantic Ocean, to the east by the Republic of Cameroon and to the west by the Republic of Benin. The population is more than 100 million, spread unevenly over a national territory of 923,770 km^2. Nigeria has the eighth largest national population in the world and about a quarter of the total population of all the countries in Sub-Sahara Africa.

The climate, which affects the quality and quantity of the country's water resources, results from the influence of two main wind systems: the moist, relatively cool, monsoon wind which blows from the south-west across the Atlantic Ocean towards the country and brings rainfall, and the hot, dry, dust-laden Harmattan wind which blows from the north-east across the Sahara desert with its accompanying dry weather and dust-laden air. The mean temperature is generally between 25 and 30 °C (77 and 86 °F), although because of the moderating influence of the sea the mean daily and annual maximum temperatures increase from the coast towards the interior. In the dry season the temperatures are more extreme, ranging between 20 and 30 °C (68 and 86 °F).

IV.1.1 Water resources

Nigeria has abundant water resources although they are unevenly distributed over the country. The highest annual precipitation of about 3,000 mm occurs in the Niger Delta and mangrove swamp areas of the south-east, where rain falls for more than eight months a year. There is a progressive reduction in precipitation northwards with the most arid north-eastern Sahelian region receiving as little as 500 mm a^{-1} precipitation from about 3–4 months of

---

*   *This case study was prepared by Lawrence Chidi Anukam*

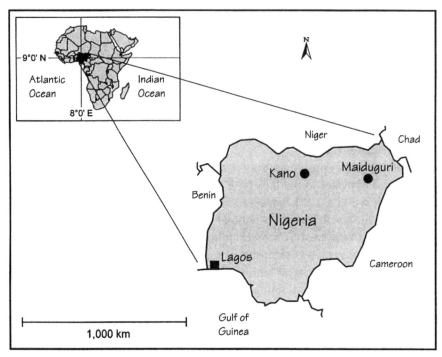

Figure IV.1 Location map of Nigeria

rainfall. Widespread flooding occurs in the southern parts of the country, while the northern parts experience chronic water shortages during the dry season when rainfed springs, streams and boreholes dry up.

There are four major drainage systems in the country (Figure IV.2):

- The Niger River Basin Drainage System with its major tributaries of Benue, Sokoto-Rima, Kaduna, Gongola, Katsina-Ala, Donga, Tarabe, Hawal and Anambara Rivers.
- The Lake Chad Inland Drainage System comprising the Kano, Hadejia, Jama'are Misau, Komadougou-Yobe, Yedoseram and Ebeji Rivers.
- The Atlantic Drainage System (east of the Niger) comprising the Cross, Imo, Qua Iboe and Kwa Rivers.
- The Atlantic Drainage System (west of the Niger) made up of the Ogun, Oshun, Owena and Benin Rivers.

Apart from the Lake Chad Inland Drainage System, the remaining three drainage systems terminate in the Atlantic Ocean with an extensive network of delta channels (Figure IV.2).

Groundwater resources are limited by the geological structure of the country (Figure IV.3), more than half of which is underlain by the Pre-Cambrian

Figure IV.2 Map of Nigeria showing major rivers and hydrological basins: 1 Niger North, 2 Niger Central, 3 Upper Benue, 4 Lower Benue, 5 Niger South, 6 Western Littoral, 7 Eastern Littoral, 8 Lake Chad

Basement Complex, composed mainly of metamorphic and igneous rocks. However, there are fairly extensive areas of fractured schists, quartzites and metamorphosed derivatives of ancient sediments from which water is often available at great depth. The sedimentary formations such as the Tertiary deposits of the Chad-Sokoto basins, the Cretaceous deposits of the Niger and Benue troughs, and the sedimentary formation of the Niger Delta, yield groundwater in varying quantities.

IV.1.2 Water pollution
Water pollution in Nigeria occurs in both rural and urban areas. In rural areas, drinking water from natural sources such as rivers and streams is usually

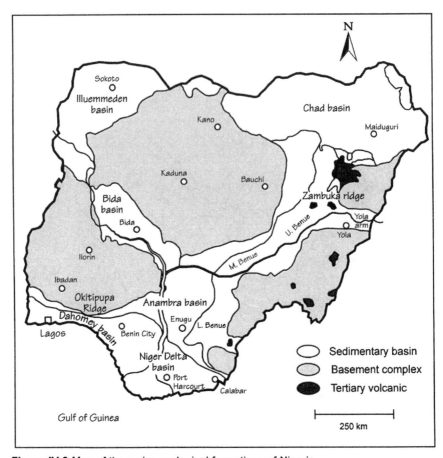

Figure IV.3 Map of the major geological formations of Nigeria

polluted by organic substances from upstream users who use water for agricultural activities. The most common form of stream pollution associated with forestry activities is increased concentrations of soil particles washed into the stream by land disturbance. The large particles sink to the bottom and increase the bed load while, depending on the stream velocity, smaller particles remain in suspension. In the river Niger, for example, studies have shown that the suspended matter can obstruct the penetration of light and limit the photosynthetic zone to less than 1 m depth. Suspended sediments in watercourses have become a serious concern for the water supply authorities because they lead to increased water treatment costs.

Many factories in Nigeria are located on river banks and use the rivers as open sewers for their effluents. The major industries responsible for water pollution in Nigeria include petroleum, mining (for gold, tin and coal) wood and pulp, pharmaceuticals, textiles, plastics, iron and steel, brewing, distillery fermentation, paint and food. Of all these, the petroleum industry presents the greatest threat to water quality. From time to time accidental oil spillages occur which endanger local sources of water supply and freshwater living resources, especially in the rural areas.

The problems associated with the lack of adequate water resources in the country threaten to place the health of about 40 million people at risk. Recent World Bank studies (World Bank, 1990) suggest that it would cost in excess of US\$ 10^9 a year to correct such problems if ground and surface water contamination goes unchecked. The people most affected tend to be the urban and landless poor. In the long-term, the present level of environmental degradation could create health problems from water-borne diseases for most of this population. Many people are already affected by having to consume unsafe drinking water. Water contamination also places other resources at risk; fisheries and land resources, for example, have already been affected significantly. Most of the environmental pollution problems arise from anthropogenic sources, mainly from domestic and industrial activities.

IV.2 National environmental policy

IV.2.1 The Federal Environmental Protection Agency
The Federal Military Government has placed great importance on the environment and established the Federal Environmental Protection Agency (FEPA) by Decree 58 of 30 December 1988 (FGN, 1988a). The FEPA has statutory responsibility for overall protection of the environment and its initial functions and priorities included:
- Co-ordinating all environmental activities and programmes within the country.
- Serving as the national environmental focal point and the co-ordinating body for all bilateral and multilateral activities on the environment with other countries and international organisations.
- Setting and enforcing ambient and emission standards for air, water and noise pollution.
- Controlling substances which may affect the stratosphere, especially the ozone layer.
- Preventing and controlling discharges to air, water or soil of harmful and hazardous substances.

IV.2.2 The National Policy on Environment
The National Policy on the Environment was launched by the President in Abuja on 27 November 1989 (FEPA, 1989). The goal of that policy was to achieve sustainable development in Nigeria and, in particular to:

- Secure for all Nigerians a quality environment adequate for their health and well-being.
- Conserve and use the environment and natural resources for the benefit of present and future generations.
- Restore, maintain and enhance ecosystems and ecological processes essential for the functioning of the biosphere and for the preservation of biological diversity and to adopt the principle of optimum sustainable yield in the use of living natural resources and ecosystems.
- Raise public awareness and promote understanding of essential linkages between environment and development and to encourage individual and community participation in environmental improvement efforts.
- Co-operate in good faith with other countries, international organisations and agencies to achieve optimal use of transboundary natural resources and effective prevention or abatement of transboundary environmental pollution.

The introduction of guidelines and standards was part of the implementation of the policy and the environmental pollution abatement strategy contained therein. The guidelines and standards relate to six areas of environmental pollution control:

- Effluent limitations.
- Water quality for industrial water uses at point of intake.
- Industrial emission limitations.
- Noise exposure limitations.
- Management of solid and hazardous wastes.
- Pollution abatement in industries.

Environmental protection measures are only meaningful if the environment to be protected is adequately understood. Neither over-protection nor under-protection of the environment is desirable. Ideally, standards should be set based on nationally generated, environmental baseline data. Such data are scarce in Nigeria in the present circumstances. An alternative approach is to adapt standards and guidelines adopted by the World Health Organization (WHO) and the developed nations of Europe and America. The water quality component of the guidelines are based on the WHO guidelines. However, in transposing data between countries, socio-economic and climatic differences must be taken into account.

IV.2.3 Establishment of environmental monitoring programmes

With the establishment of the guidelines and standards, the FEPA is initiating a monitoring programme to ensure that the set standards are met. The objectives of the programme include:

- Establishment of an environmental baseline.

Figure IV.4 Map of Nigeria showing the major administrative divisions and their populations densities and the location of proposed zonal laboratories

- Detection and evaluation of environmental trends.
- Provision of advance warning of approaching critical conditions.
- Detection of accidental critical events which may exceed the rate of recovery of the environment.
- Prevention of potential threats to the human environment.
- Provision of a means of data storage and retrieval.

Efforts are being made to build zonal laboratories in various parts of the country to provide adequate monitoring coverage for domestic, recreational and industrial causes and effects of environmental degradation. Six zonal laboratories were proposed at the following cities (Figure IV.4): Lagos, Abuja, Benin, Kano, Jos and Port Harcourt. The Lagos laboratory has already been commissioned and is also serving as the national reference laboratory.

IV.2.4 The national environmental reference laboratory

The FEPA's Lagos Office and Zonal Laboratory Complex were commissioned in October 1990. The Lagos Complex is acting as a national environmental reference laboratory and is serving the environmental monitoring activities of the States and the Federal Capital Territory. The Lagos Laboratory Complex is made up of six units:

- Water and wastewater laboratory.
- Analytical instrument laboratory.
- Toxic chemical laboratory.
- Microbiology laboratory.
- General purpose laboratory, including bioassay techniques.

Once adequately equipped, the laboratory complex will provide the FEPA with the capability to generate reliable data for determining compliance with the National Interim Guidelines and Standards which were set up by the government to monitor and control industrial domestic and industrial pollution.

IV.3   Water resources management

The turning point for water resources development and management in Nigeria occurred after the severe drought of the 1960s. The Government's response to the catastrophe was the initiation of strategies for co-ordinated and effective water resources development, culminating in the mid-1970s in the creation of the Federal Ministry of Water Resources and the River Basin Development Authorities. The activities of these institutions were further strengthened in 1981 by the establishment of the National Committee on Water Resources, and by the Water Boards at the state level. These bodies were charged with taking an inventory, and ensuring rational and systematic planned management and conservation, of the country's water resources.

In the 1970s and early 1980s, water resources management in Nigeria was faced with a lot of problems which slowed down the development of the resource. Some of these problems included:

- The deficiency of the resource itself.
- Unnecessary duplication and overlap in organisations, structures and functions of the relevant bodies.
- The ill-defined and uncoordinated roles of the Federal, State and Local Government agencies responsible for water resources development.
- Failure to recognise the inter-relationship between surface and ground waters, and between water resources and land use.
- Lack of effective water and environmental protection laws, and the means to enforce the already existing laws.

In the late 1980s, Nigeria began to make serious efforts to address these problems: a national body was created to co-ordinate all environmental protection activities in the country (see section IV.2.1); a comprehensive national environmental policy was formulated which, among other things, addressed the issue of water resources (see section IV.2.2); and the Hazardous Waste Decree was promulgated with the intention of discouraging reckless and illegal dumping of hazardous and harmful wastes on land and into water courses (FGN, 1988b).

IV.3.1 Strategies under the National Policy on Environment

Implementation of the Nigerian National Policy on Environment depends on specific actions directed towards major sectors and towards problem areas of the environment (FEPA, 1989). The management approach adopted in the policy is based on an integrated, holistic and systematic view of environmental issues. The programme activities of this policy are expected to establish and strengthen legal, institutional, regulatory, research, monitoring, evaluation, public information, and other relevant mechanisms for ensuring the attainment of the specific goals and targets of the policy. They will also encourage environmental assessment of proposed activities which may affect the environment or the use of natural resources prior to their commencement. The strategies put forward for effective water resources management in Nigeria include:

- Promulgation of a national water resources law to co-ordinate water resources development.
- Formulation of a water resources master plan.
- Improvement of water use efficiency for sustainable development.
- Implementation of water conservation measures including inter-basin water transfer.
- Establishment and enforcement of national water quality and emission standards to protect human health and aquatic ecosystems and species.
- Establishment of environmental monitoring stations or networks to locate and monitor sources of environmental pollutants and to determine their actual or potential danger to human health and the environment.
- Continuous data collection for resource monitoring and management.
- Introduction of economic incentives.

The on-going programmes to assess the available water resources of the country are being strengthened to provide, among other things, data on:

- Hydrological features affecting surface water resources.
- The location of groundwater resources and their characteristics in terms of depths, yields, permeabilities, storage and recharge.

- Per capita water use and requirements.
- Changes in hydrological regimes resulting from human activities, such as water use or extraction, pollution and the effects of mining and lumbering.
- The management of small and large dams.
- Irrigation problems with regard to crop water requirements, salinity, drainage and pollution from fertilisers, pesticides and cultivation activities.
- Existing freshwater living resources.

As part of the strategies for the implementation of the National Policy on Environment in the water sector, a comprehensive national water resources master plan has now been drawn up with support from the Government of Japan, through the Japan International Cooperation Agency (JICA). For the first time, a decree on water resources protection and management has been promulgated (FGN, 1993), with the purpose of:

- Promoting the optimum planning, development and use of the Nigeria's water resources.
- Ensuring the co-ordination of such activities as are likely to influence the quality, quantity, distribution, use and management of water.
- Ensuring the application of appropriate standards an techniques for the investigation, use, control, protection, management and administration of water resources.
- Facilitating technical assistance and rehabilitation for water supplies.

IV.4 Industrial water pollution control programme

Industrialisation is considered vital to the nation's socio-economic development as well as to its political standing in the international community. Industry provides employment opportunities for a large proportion of the population in medium to highly developed economies. The characteristics and complexity of wastes discharged by industries vary according to the process technology, the size of the industry and the nature of the products.

Ideally, the siting of industries should achieve a balance between socio-economic and environmental considerations. Relevant factors are availability and access to raw materials, the proximity of water sources, a market for the products, the cost of effective transportation, and the location of major settlements, labour and infrastructural amenities. In developing countries such as Nigeria, the siting of industries is determined by various criteria, some of which are environmentally unacceptable and pose serious threats to public health. The establishment of industrial estates beside residential areas in most state capitals and large urban centres in Nigeria is significant in this respect.

Surface water and groundwater contamination, air pollution, solid waste dumps and general environmental degradation, including the loss of land and aquatic resources, are major environmental problems caused by industrialisation in Nigeria. Improper disposal of untreated industrial wastes has resulted in coloured, murky, odorous and unwholesome surface waters, fish kills and a loss of recreational amenities. A significant proportion of the population still rely on surface waters for drinking, washing, fishing and swimming. Industry also needs water of acceptable quality for processing.

Economic development can be compatible with environmental conservation and the present problems of environmental resource degradation need not arise within the framework of sustainable development. Failure to halt further deterioration of environmental quality might jeopardise the health of a large proportion of the population, resulting in serious political and socioeconomic implications.

IV.4.1 Industrial effluent standards

The latest issue of the *Directory of Industries in Nigeria* published by the Federal Ministry of Industry indicates that over 3,000 industrial establishments exist in the country. These industries vary in process technology, size, nature of products, characteristics of the wastes discharged and the receiving environment. Presently, there are 10 major industrial categories readily discernible in Nigeri: metals and mining; food, beverages and tobacco; breweries, distilleries and blending of spirits; textiles; tanneries; leather products; wood processing and manufacture, including furniture and fixtures; pulp, paper and paper products; chemical and allied industries; and others.

Ideally, each effluent should be detoxified with the installation of pollution abatement equipment based on the best practical technology (BPT) or best available technology (BAT) approach. The high cost of imported BPT and BAT, and the lack of locally available environmental pollution technology, normally requires that Uniform Effluent Standards (UES) are based on the pollution potential of the effluent or the effectiveness of current treatment technology. This approach is easy to administer, but it can result in overprotection in some areas and under-protection in others. To overcome this problem, uniform effluent limits based on the assimilative capacity of the receiving water have been drawn up for all categories of industrial effluents in Nigeria (Table IV.1). Additional effluent limits have been provided for individual industries with certain peculiarities (FEPA, 1991a).

Specific regulations to protect groundwater from pollution have also been issued by the FEPA (FEPA, 1991b,c). Industrial sites have to meet concentration limits for their effluents, as given in Table IV.2. These are specified in facility permits issued to the industries and enforcement takes place by compliance monitoring.

Table IV.1 Guidelines for interim uniform effluent limits in Nigeria for all categories of industries (mg l^{-1} unless otherwise stated)

Variables	Discharge to surface water	Land application
Temperature	< 40 °C within 15 m of outfall	< 40 °C
Colour (Lovibond Units)	7	–
pH units	6–9	6–9
BOD$_5$ at 20 °C	50	500
Total suspended solids	30	–
Total dissolved solids	2,000	2,000
Chloride (as Cl$^-$)	600	600
Sulphate (as SO$_4^{2-}$)	500	1,000
Sulphide (as S^{2-})	0.2	–
Cyanide (as CN$^-$)	0.1	–
Detergents (linear alkylate sulphonate as methylene blue active substances)	15	15
Oil and grease	10	30
Nitrate (as NO$_3^-$) NO$_3$	20	–
Phosphate (as PO$_4^{3-}$)	5	10
Arsenic (as As)	0.1	–
Barium (as Ba)	5	5
Tin (as Sn)	10	10
Iron (as Fe)	20	–
Manganese (as Mn)	5	–
Phenolic compounds (as phenol)	0.2	–
Chlorine (free)	1.0	–
Cadmium, Cd	< 1	–
Chromium (trivalent and hexavalent)	< 1	–
Copper	< 1	–
Lead	< 1	–
Mercury	0.05	–
Nickel	< 1	–
Selenium	< 1	–
Silver	0.1	–
Zinc	< 1	–
Total metals	3	–
Calcium (as Ca^{2+})	200	–
Magnesium (as Mg^{2+})	200	–
Boron (as B)	5	5
Alkyl mercury compounds	Not detectable	Not detectable
Polychlorinated biphenyls (PCBs)	0.003	0.003
Pesticides (Total)	< 0.01	< 0.01
Alpha emitters (µC ml^{-1})	10^{-7}	–
Beta emitters (µC ml^{-1})	10^{-6}	–
Coliforms (daily average MPN/100 ml)	400	500
Suspended fibre	–	–

– Not applicable or none set   Source: FEPA, 1991a

Table IV.2   Maximum permitted concentrations of toxic substances in industrial
effluents in Nigeria for the protection of groundwater

Variable	Maximum concentration (mg l^{-1})
Arsenic	0.05
Barium	1.0
Cadmium	0.01
Chromium	0.05
Lead	0.05
Mercury	0.002
Selenium	0.01
Silver	0.05
Endrin	0.0002
Lindane	0.004
Methoxychlor	0.1
Toxaphene	0.005
2,4-D	0.1
2,4,5-TP Silvex (tree killer)	0.01

Source: FEPA, 1991a

The Nigerian guidelines require industries to monitor their effluents
in-house while the FEPA cross-checks the effluents characteristics to ascertain the degree of compliance with the set standards. Analytical methods
commonly used for the determination of significant variables in waters and
wastewaters are prescribed by the FEPA for all parties involved in the monitoring exercises. Well-tested, standard methods for water and wastewater
analysis used by United States Environmental Protection Agency (EPA), the
UK Department of Environment (DOE), the American Public Health Association (APHA) or the American Society for Testing and Materials (ASTM)
were adopted for monitoring purposes. For reporting purposes, the analytical
method(s) used have to be specified.

IV.5   Conclusions
Towards the end of the 1980s, Nigeria began to place a high priority on environmental matters, particularly water-related issues. This is reflected in recent
environmental policy, legislation, action plans and programmes introduced by
the Government. In all these programmes, environmental monitoring activities,
especially water quality aspects, are given strong consideration.

With the creation of the FEPA as the central co-ordinating body for all environmental matters within the country, Nigeria has evolved a mechanism that will monitor adequately and will keep records of all relevant environmental variables. The new integrated water resources management concept adopted by the Government will, without doubt, improve all aspects of water use and conservation within the country if the political will and financial resources for the implementation are sustained.

IV.6   References

FEPA 1989 *Our National Environmental Goals*. Special Publication No. 3. Federal Environmental Protection Agency, Lagos.

FEPA 1991a *Guidelines and Standards for Environmental Pollution Control in Nigeria*. Federal Environmental Protection Agency, Lagos.

FEPA 1991b *S.I.8 National Environmental Protection (Effluent Limitation) Regulations of 1991*. Federal Environmental Protection Agency, Lagos.

FEPA 1991c *S.I.9 National Environmental Protection (Pollution Abatement in Industries and Facilities Generating Wastes) Regulation of 1991*. Federal Environmental Protection Agency, Lagos.

FGN 1988a *Federal Environmental Protection Agency Decree No. 58 December 30, 1988*. Federal Government of Nigeria, Government Press, Lagos.

FGN 1988b *Harmful Wastes Decree No. 42 of November 30, 1988*. Federal Government of Nigeria, Government Press, Lagos.

FGN 1993 *Water Resources Decree No. 101 of August 1993*. Federal Government of Nigeria, Government Press, Lagos.

World Bank 1990 *Towards the Development of an Environmental Action Plan for Nigeria*. Report number 9002-UNI. World Bank, Washington, D.C.

Case Study V[*]

THE WITBANK DAM CATCHMENT

V.1   Introduction

The Witbank Dam catchment is located in the upper portion of the Olifants River basin. The Olifants River is one of South Africa's major water resources. The water quality in the Witbank Dam catchment is rapidly deteriorating, mainly due to coal mining. If this trend were to continue, the water in the Witbank Dam would be unfit for use, for most of the recognised users, by the end of this century. Poor water quality in the dam has meant that power generation, which is the largest industrial activity in the catchment, already has to rely largely on other sources of water from outside the catchment.

This case study describes the water quality management approach of the Department of Water Affairs and Forestry (DWAF) to ensure that the surface water quality in the Witbank Dam remains fit for use and that the resource is secured adequately for the future. Many aspects of this approach are currently still being implemented. Nevertheless, further deterioration in water quality has been arrested and has been evident since October 1993 (see Figure V.3). Indications are that the implementation of this approach will result in water fit for use in the Witbank Dam catchment for at least the next 10 years. Other strategies will have to be employed to address water quality in the longer term.

V.2   Background information

South Africa is a country of great diversity. Its society comprises under-developed, developing and developed components. The annual disposable income per capita for all population groups is approximately US$ 2,000, but varies between US$ 1,020 and 7,750 for different population groups. Furthermore, the country is characterised by great disparities concerning access to adequate water supplies. Water-related issues are, therefore, a central aspect in the country's political arena; so much so that water supply and sanitation for all the inhabitants of South Africa is a key element of the

---

[*]   *This case study was prepared by S.A.P. Brown*

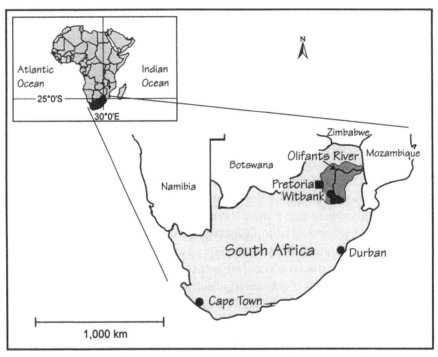

Figure V.1 Location map of South Africa showing the Olifants River basin and the Witbank Dam catchment

Reconstruction and Development Programme (RDP). This programme was initiated by the Government of National Unity to restore the social and economic imbalances in the country.

South Africa is, however, a semi-arid country with limited water resources. Water is geographically unevenly distributed throughout the country and is not consistently available throughout the year. Generally, more water is available in the eastern portion of the country and availability gradually declines westwards. The Olifants River, one of South Africa's major water resources, is situated towards the east. The availability of water is further compounded by the fact that the demography of the South African population is changing rapidly. Vast numbers of people are moving to cities where they live in areas of poor water supply and sanitation services, or none at all.

The DWAF is the authority, in South Africa, responsible for overall water resource management and it has to ensure the supply of adequate quantities of water of acceptable quality to recognised water users. However, in practice, part of this responsibility is delegated to other levels of government, other agencies, water users and to those who have an impact on the water resource.

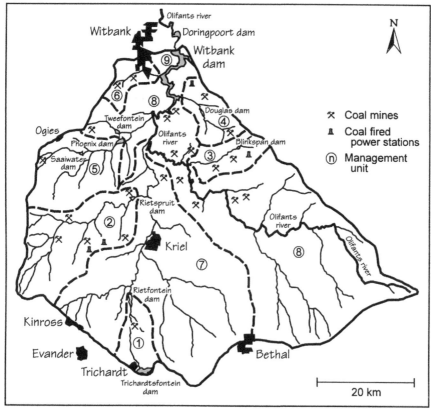

Figure V.2 Detailed map of the Witbank Dam catchment and its nine management units, showing urban development and major industrial and mining activities (After Wates, Meiring and Barnard, 1993)

As a result of the dynamic political situation in South Africa, roles and responsibilities are being redefined and reallocated. In carrying out its mandate, DWAF often has to reconcile, integrate and co-ordinate conflicting and diverse interests within the framework of sustainable and equitable use of South Africa's water resources.

V.3   The Witbank Dam catchment
The Witbank Dam catchment is located at the headwaters of the Olifants River (Figure V.1). A more detailed representation of the catchment is provided in Figure V.2 which indicates the location of Witbank Dam, urban development and major industrial and mining activities, as well as the nine management units (see section V.5.1). The Witbank Dam catchment covers an area of 3,256 km^2 and has a mean annual run-off of 125×10^6 m^3 a^{-1}.

Land-use practices in the catchment are varied and include the following:
- Agriculture, of which maize is of strategic importance to South Africa's national staple food supply. Dry-land cultivation of maize is practised on 24 per cent of the catchment area.
- Power generation, which is the largest industrial activity in the catchment and includes four of the country's major coal-fired power stations.
- Coal mining. A total of 29 major collieries and a number of smaller operations are active in the catchment, producing approximately 47 per cent of the country's coal production.
- Urban development, which is limited to a number of smaller towns.

V.4   Pre-intervention situation

V.4.1   The strategy

The general approach to pollution control and environmental management in South Africa entails a management strategy based on a single environmental medium which is either air, water or land. The regulatory authorities responsible for the management of the environment are organised as follows:
- Air: Department of National Health.
- Water: Department of Water Affairs and Forestry.
- Land: Department of Agriculture together with various other Departments. For example control over mining activities is exercised by the Department of Mineral and Energy Affairs (DMEA).

The three main regulatory authorities have developed different organisational structures to suit their regulatory approach. Furthermore, the present arrangement fails to recognise the transfer of pollution across environmental boundaries. It also does not provide for a regulatory mechanism to ensure that environmental management is effective and efficient. The result is the absence of clear responsibilities, the overlapping of institutional boundaries, the exclusion of areas which require attention and a duplication of effort.

Prior to 1991, the water quality management strategy of DWAF was based on the Uniform Effluent Standards (UES) approach. In applying this strategy, the focus was mainly on point source effluents. Diffuse sources of water pollution and the receiving water body were not given the necessary attention. With respect to control over mining activities, these shortcomings were further compounded by several factors:
- The DWAF addressed water quality management concerning mining in isolation, i.e. control of water quality in relation to mining activities was not integrated with other activities. This approach even included separate offices dealing with mining-related matters within the same management area.

- There was a lack of co-ordination between DWAF and the DMEA which has the primary responsibility regarding the influence of mining activities on land use.
- The mining community was not aware of the detrimental effect of coal mining on the water environment.

V.4.2 Water quality issues

Coal mining is a major potential source of diffuse water pollution. Sulphate is a good indicator of salinity arising from this form of pollution. Approximately 70–80 per cent of the sulphate load in the Witbank Dam catchment emanates from diffuse sources and can be attributed to coal mining. This increase in diffuse pollution has resulted in a gradual decline in water quality in the Witbank Dam catchment. Water quality in the dam itself has declined from 50 mg l^{-1} sulphate and 100 mg l^{-1} total dissolved solids (TDS) to over 150 mg l^{-1} sulphate and 400 mg l^{-1} TDS. The concentrations of these two variables over a 16-year period are given in Figure V.3. In some reaches of rivers and streams in the catchment this deterioration has been more pronounced and, in some cases, water quality has deteriorated from a natural baseline level of approximately 50 mg l^{-1} to over 1,500 mg l^{-1} sulphate.

The other major water quality issues are:

- Eutrophication. Phosphorus is the limiting nutrient in the dam and the concentration of total phosphorus has not changed significantly over the past decade. However, the transparency of the water in the dam has increased by an order of magnitude over the same period of time. This is mainly due to the increase in TDS which has enhanced the flocculation of clay particulates resulting in an increase in light penetration.
- Elevated levels of compounds toxic to the natural aquatic environment occur in some reaches of streams in the catchment. These compounds are predominantly metals and ammonia. Acid mine drainage associated with coal mining mobilises metals. Aluminium, iron and manganese are the main metals of concern. Ammonia originates from sewage effluents.

V.5   Intervention with a new approach

During 1991, DWAF adopted a new water quality management strategy. This strategy focuses on the receiving water body and considers all sources of water pollution. Catchment water quality management plans and sectoral specific management strategies are central to the new strategy. As part of this strategy, the water quality management approach in the Witbank Dam catchment was reviewed. The aim of the new approach is to arrest deterioration of water quality, to ensure fitness for use by the recognised water users and to secure the water resource. The main thrust of this approach consists of:

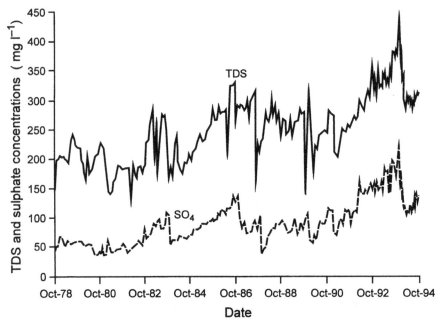

Figure V.3 Total dissolved solids and sulphate concentrations in Witbank Dam, 1987–94 (After Wates, Meiring and Barnard, 1993)

- Development and implementation of a catchment water quality management plan.
- Prevention and minimisation of pollution arising from mining activities wherever possible.

These two activities are inter-related and are implemented concurrently. The catchment water quality management plan has to provide, amongst other things, compliance requirements for each activity based on the level of pollution that can be accommodated by the water without impairing its suitability for use. Pollution prevention is a key issue in the catchment strategies embodied in the management plan to ensure that set water quality management objectives are met.

V.5.1 Catchment water quality management plan

A catchment plan will provide a framework to manage water quality coherently and consistently; to influence present and future land use, particularly those uses over which DWAF does not have direct control; and to integrate other resource management efforts and environmental media issues with water quality. In order to provide this framework, the following steps were taken:

- Water quality objectives were set at strategic locations in each catchment. The water quality objective at a particular location is a quantitative statement of the water quality that must be maintained at that particular point to ensure suitability for use. In the Witbank Dam catchment, the recognised water uses are domestic, power generation, mining, recreation, the natural aquatic environment and irrigation.
- Flexible catchment strategies were formulated to ensure that water quality objectives can be attained.
- Compliance requirements were set for those activities that could adversely affect water quality. Sites giving rise to both point and diffuse sources of pollution were considered "single" sources and the compliance requirements were determined and stipulated accordingly.
- The collective powers and influence of other authorities, agencies and the public were co-ordinated in a co-operative manner to implement catchment strategies. This applies particularly to control over future land use and adjustment to existing land-use practices to reduce diffuse sources of water pollution.
- Monitoring and auditing systems were provided to ensure the implementation of catchment strategies. The effectiveness of the catchment water quality management plan and of water quality management efforts undertaken by other role-players was also monitored.

The catchment water quality management plan is being developed to such a level of detail that it specifies what must be achieved and implemented, where and when it will be implemented, how it will be administered and managed and who will be responsible for the specific activities. However, the catchment cannot be managed as a single management unit. Sub-catchments upstream of the dam have different water-use requirements. For this reason, the catchment was subdivided into nine management units on the basis of the sub-catchments (see Figure V.2). At the lower end of each management unit, in-stream water quality management objectives are set.

The management strategies and water quality objectives embodied in the plan focus on:

- Salinity, with sulphate as the selected indicator of salinity.
- Eutrophication, with phosphorus as the limiting nutrient.
- Toxic constituents, particularly heavy metals and ammonia.

Salinity

The sulphate management objective for the dam itself was set at 155 mg l^{-1} (95 percentile value). This is approximately 23 per cent lower than the user requirement of 200 mg l^{-1} . This margin allows for:

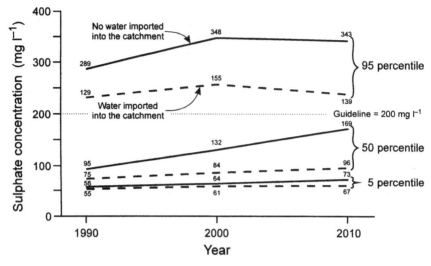

Figure V.4 Present and predicted sulphate concentrations in Witbank Dam assuming power stations operating at zero discharge facilities and with a 45 per cent reduction in non-point source colliery pollution (After Wates, Meiring and Barnard, 1993)

- Further mining, industrial and agricultural development.
- Potential malfunctioning of water pollution control systems.
- Current knowledge gaps on the future potential impact of atmospheric deposits, high- extraction coal mining and open-cast mining technique.
- Lack of adequate information to establish accurately water user requirements, particularly that pertaining to the natural aquatic environment.

Attainment of the sulphate management objective depends on zero discharge from power stations, 45 per cent reduction in diffuse pollution from collieries and additional water imported into the catchment. The projected improvements resulting from applying these strategies are indicated in Figure V.4. The sulphate management objectives for the nine management units are outlined in Table V.1.

Eutrophication
Eutrophication control is aimed at limiting the available phosphorus compounds in order to reduce algal growth in the Witbank Dam. Natural anthropogenic weathering and agriculture are the largest diffuse sources of phosphorus in the catchment. Substantial losses of phosphorus applied to agricultural land take place in the catchment; 32 per cent of this reaches the Witbank Dam. It was not considered practical to control these diffuse sources of phosphorus. Municipal sewage treatment plant effluents provide 38 per cent of the recorded catchment sources and 44 per cent of the recorded

Table V.1 In-stream sulphate management objectives

Management unit[1]	Most sensitive user requirement (mg l^{-1} SO$_4$)	Management objective	
		50 percentile	95 percentile
1	30	24	34
2	200	70	120
3	*	620	1,200
4	*	830	1,450
5	200	220	390
6	*	260	380
7	200	160	260
8	200	190	460
9 (Witbank Dam)	200	84	155

[1] Management units are based on sub-catchments of the Witbank Dam catchment

* None identified
Source: Wates, Meiring and Barnard, 1993

phosphorus discharge to the dam. Phosphorus control is principally aimed at these sources by means of imposing a special phosphate standard of 1 mg l^{-1} PO$_4$-P on all sewage plant effluents.

Metals and ammonia
Control of metals and ammonia will be effected by restricting the maximum allowable free and saline ammonia and metal concentrations in discharges from mining and industrial complexes. These maximum allowable concentrations are given in Tables V.2 and V.3.

V.5.2 Prevention and minimisation of pollution
The essence of the approach to pollution prevention and minimisation is the use of regulatory instruments which facilitate direct intervention to prevent pollution at source. The Water Act, which is the statutory component of the regulatory instruments applied by DWAF directly, has limited power to exert influence on land use affected by mining. Thus direct intervention to prevent diffuse pollution, in particular, is not always possible. In order to address this, as well as other shortcomings mentioned earlier in relation to mining activities, co-ordination between the regulatory systems of DWAF and DMEA was effected. Co-ordination was accomplished by participation within an integrated environmental management system for prospecting and mining activities.

Table V.2 Maximum allowable concentrations of free and saline ammonia in discharges from mining and industry

pH of discharge	Free and saline ammonia concentration (mg l^{-1} NH_3-N)		
	15 °C[1]	20 °C[1]	25 °C[1]
6.5	10.0	10.0	10.0
7.0	3.4	4.4	4.8
7.5	1.1	1.7	1.3
8.0	0.62	1.1	0.76
8.5	0.36	0.37	0.27
9.0	0.13	0.14	0.12

[1] Temperature values apply to the effluent or discharge. Ideally, they should have been applied to the receiving water body. However, for prolonged periods the base flow in receiving streams can be very low to negligible and therefore the temperature requirement has been applied to discharges.

Source: Wates, Meiring and Barnard, 1993

Table V.3 Maximum allowable concentrations of heavy metals in discharges from mining and industry

Heavy metal	Maximum allowable concentration (mg l^{-1})
Aluminum	150
Cadmium	2
Chromium	200
Copper	20
Iron	1,000
Lead	10
Manganese	500
Mercury	0.1
Nickel	100
Selenium	30
Zinc	200

Source: Wates, Meiring and Barnard, 1993

The integrated system plays a key role in the regulatory systems of both departments in the following ways:

- Placing the departments in a position to address anticipated effects on the water environment before mining proceeds.
- Placing the departments in a position to ensure that environmental objectives are met constantly.
- Ensuring that mining proponents have understood the magnitude and nature of the effect which their activities will have on the environment, and have committed themselves to a practical means of dealing with these effects before commencing a mining venture.
- Providing the authorities with an opportunity to satisfy themselves that the proponents have the means to ensure that the management measures proposed to control the environmental effect of their activities will be implemented.

In order to fulfil the requirements of the integrated management system, each mine in the catchment has to carry out the following:

- Implement an approved Environmental Management Programme, pertaining to a particular mine, that explicitly prescribes measures necessary to prevent and minimise pollution.
- Implement adequate measures before closure to prevent pollution and to provide for sustainable use of the water resource.
- Only discharge polluted water from point or diffuse sources in accordance with the conditions prescribed by DWAF.
- Make adequate financial provisions to ensure that the impact management measures planned can be implemented.

V.6   Shortcomings of the approach

Various shortcomings were identified as a result of the experience gained during the development and implementation of the water quality management approach in the Witbank Dam catchment. The following issues require attention:

- The development of a catchment plant that focuses on water quality alone serves a limited purpose. Water supply and demand issues must be included in the development of a catchment water management plan to ensure effective water resource management.
- Commitment to water-related issues are currently voluntary, particularly those outside the direct influence of the Act administered by DWAF. In most cases, these pertain to influence on land use which is crucial to the success of a plan. Mechanisms have to be established to ensure that commitments are fulfilled. Some of these will be addressed by amendments to the Water Act envisaged in the near future.

- The institutional capacity of the DMEA particularly needs to be improved. This department, the lead agency in the implementation of the integrated management system for prospecting and mining, has shown shortcomings with respect to its awareness of the needs of the water environment, and with respect to managing a complex system, such as the integration system for prospecting and mining.
- Participation of stakeholders must be ensured right from the initiation of the development of a catchment water quality management plan. In the case of the Witbank Dam plan, the technical aspects were developed with limited participation. It became clear recently that, for this reason, difficulties could be experienced with the implementation of the plan. Stakeholders do not feel they have "ownership" of the plan, and this results in a lack of desire to be part of its implementation and execution or to contribute to its success. This shortcoming is currently being addressed and participation is encouraged at all levels of development and implementation of the water quality management approach in the Witbank Dam catchment.

V.7 Conclusions

The key aspects of the water quality management approach in the Witbank Dam catchment are currently being implemented. Notwithstanding the shortcomings already identified, the indications are that the implementation of this approach will result in water fit for use and that the water resource will be secured for at least the next 10 years. This prognosis is based on the result achieved with a similar approach in the Klipspruit, a small catchment adjacent to the Witbank Dam catchment. In the Klipspruit catchment, two levels of water quality management objectives were set. The second level of objectives were only expected to be attained after the implementation of various long-term strategies. However, implementation of the approach has resulted in immediate profound improvements in water quality. In fact, the level of water quality in the Klipspruit has improved to such an extent that the second level of objectives have mostly been reached before implementation of the required long-term strategies.

V.8 References

Wates, Meiring and Barnard 1993 Technical support document for Witbank Dam Water Quality Management Plan. Prepared for the Department of Water Affairs and Forestry, South Africa.

Case Study VI[*]

THE UPPER TIETÊ BASIN, BRAZIL

VI.1   Introduction

The São Paulo Metropolitan area, located in the Upper Tietê River basin, comprises 38 cities in addition to the city of São Paulo. The spectacular growth that has occurred in this area has been accompanied by an enormous increase in population and associated serious environmental problems related to water pollution. The water supply system provides about $60 \text{ m}^3 \text{ s}^{-1}$ for this area, about 80 per cent of which is returned untreated to the main water courses. Water quality problems are compounded by the fact that the rivers form part of a system designed exclusively for electric power generation. This system requires the flow to be reversed and, consequently, a mixture of untreated wastewater and the natural river flow remain permanently within the boundaries of the metropolitan area.

Public outcry has forced the State of São Paulo government to take action towards improving the environmental quality of its waters. The Tietê Project was launched in 1991 with the ambitious goal of treating 50 per cent of the total wastewater by 1996. This goal would be accomplished with three new wastewater treatment plants, with the expansion of an existing plant and with the implementation of several others accessory works, such as sewer collection networks and interceptors. As part of this project, industries are also required to comply with emission standards set in 1976 and which have never before been enforced.

VI.2   The metropolitan region of São Paulo

The metropolitan region of São Paulo (Figure VI.1), which includes the city of São Paulo and 38 adjacent cities, occupies 8,000 km^2 of which 900 km^2 is urbanised. The whole area is situated about 700 m above sea level and is mostly part of the Upper Tietê basin. The Tietê River is the largest river in the State.

*   This case study was prepared by Roberto Max Hermann and Benedito
       Pinto Ferreira Braga Jr

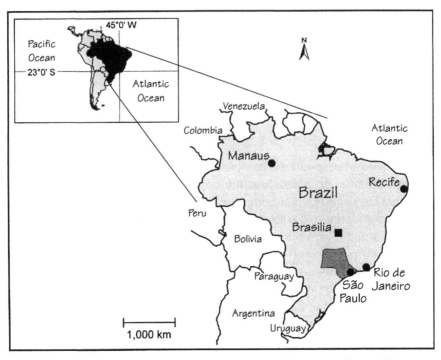

Figure VI.1 Location map of Brazil showing the Upper Tietê basin, State of São Paulo, Brazil

The present population of the area is about 16 million and is estimated to reach about 19 million by the year 2000. In 1880 the population was about 4,000, increasing to 200,000 in 1930, 1 million in 1940 and 6.5 million in 1970. This growth is also reflected in the urbanised area which, in 1880, was 2 km^2, growing to 130 km^2 in 1940 and 420 km^2 in 1954. In addition, the demand for municipal water supply is growing exponentially, from 5 m^3 s^{-1} in 1940 to a projected 65 m^3 s^{-1} by the year 2000 (Figure VI.2).

This region has the largest urban concentration in the whole of South America and the largest industrial complex in Latin America. The industrial output is 27 per cent of the national total and 62 per cent of the State total. The motivation for this rapid development arose during the 1940s in an effort to substitute imported goods with indigenous products. The consequences of this level of production and the concurrent population increase are a high population density (0.1 per cent of the total country area is occupied by 12 per cent of the total population), a high energy demand of 7,000 MW (25 per cent of the total Brazilian demand) and, especially, through several problems of conflicts over water use. The fast industrial development has resulted in rapid

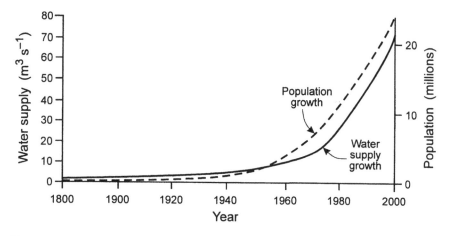

Figure VI.2 Past and projected growth in population and water supply in the São Paulo region

urbanisation, demanding electric power generation, water supply and flood control. A lack of capital resources has induced serious environmental problems. Only 10 per cent of the total sewage is treated at secondary level and, as a consequence, the urban rivers are highly polluted with a variety of industrial and municipal wastes.

VI.3   Pre-intervention situation

Geography and history have been influential in shaping early water resources development in the São Paulo metropolitan area. The first large hydraulic project was conceived purely for the purpose of generating hydroelectricity and stemmed from the need to supply cheap energy for industry. The system was designed to take advantage of a hydraulic head of about 700 m and was completed in the late 1950s (Figure VI.3). It includes several dams, two pumping stations that reverse the flow direction of the Pinheiros river, and two power plants located at the foothills of Serra do Mar, at sea level.

The implementation of this system resulted in very low velocities in the Pinheiros and Tietê rivers, which between them receive almost all the sewage generated in the region. Only about 10 per cent of this sewage is treated at the secondary level and, therefore, severe environmental problems were experienced. The complexity of the system grew as the need to increase municipal water supplies resulted in some of the reservoirs (originally planned for hydroelectric generation, e.g. the Guarapiranga reservoir) being used to

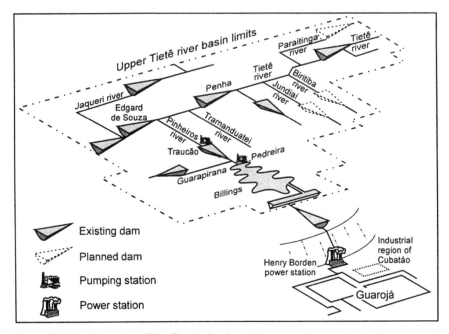

Figure VI.3 Configuration of the first hydroelectricity generation system in the Tietê basin

supply water. The untreated sewage, flowing through the main channels, was then used for power generation.

Several attempts have been made in the past to control water pollution in this river basin. As early as 1953 a plan involving the construction of six wastewater treatment plants at secondary level was proposed by the city of São Paulo. Many other plans have been proposed, but in the late 1980s and early 1990s construction started on the gigantic SANEGRAN Project. This project included, among other features, a wastewater plant with a final treatment capacity of 63 m^3 s^{-1}.

At present, the main rivers in the region receive a daily discharge of about 1,200 t organic load and 5 t inorganic load. Of these, about 370 t d^{-1} organic load and the entire 5 t d^{-1} inorganic load are believed to be generated by industry.

VI.4 The Tietê Project
Public outcry against the problems caused by the very poor environmental quality of the water bodies in the area reached a climax during the late 1980s. The media played a very important role in organising several objections

against the degradation; a petition to the State Government demanding action had over one million signatures.

In September 1991, the State Government launched the Tietê Project, to clean up the rivers and reservoirs of the São Paulo area. Two publicly owned companies are involved in this process:

- Companhia de Saneamento Basico de São Paulo (SABESP) which is a utility company responsible for planning, building and operating the water supply and sewage systems in the state of São Paulo, including the São Paulo metropolitan area.
- CETESB (Companhia de Tecnologia de Saneamento Ambiental) which is in charge of environmental control at the state level.

To manage the Tietê Project, the government of the State of São Paulo created, by a special decree, a task force with selected professionals from both of these companies and six other State departments. The directive committee is chaired by the State Governor himself.

The Tietê Project began with a master plan for sewage collection and disposal which had been prepared during the period 1983–87. Under this plan, five wastewater treatment plants were considered with a total capacity of 53.2 $m^3 s^{-1}$. This first plan was reviewed and updated to increase the treatment capacity. Three new treatment plants were also designed and one of the existing plants was considerably enlarged. All five plants use the activated sludge treatment method. Sewer collection networks and interceptors were also enlarged. Figure VI.4 shows the Tietê Pollution Abatement plan as it is being implemented.

The Tietê Project is funded by a loan from the InterAmerican Development Bank (IADB) of about US\$ 450 million and matching funds provided by the State of São Paulo of US\$ 600 million over the three years from 1994 to 1996. During this period, the industries, which are being enforced through a special programme co-ordinated by CETESB, have invested about US\$ 200 million in the implementation of treatment systems. The operating costs are also being met by the industries.

VI.5   Industrial wastewater management

Brazilian law requires industries to discharge their wastewater into the public sewer network whenever feasible. With the expansion of the collection system, a large number of new industries connected their sewage outlets to the public network and, as a result, overloaded the treatment plants. To avoid such problems and to preserve the treatment process, all industries are required to comply with permits issued against strict standards. As the agency

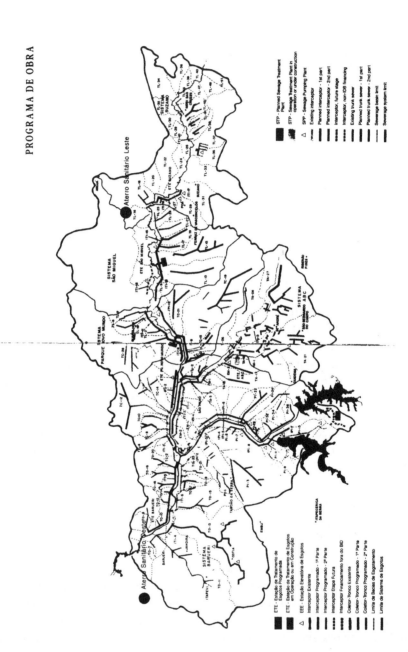

Figure VI.4 The Tietê Pollution Abatement Project

in charge of environmental control at the state level, CETESB is responsible for enforcing state laws requiring permits for industrial effluent discharges into the collection system. Although this State Law, number 997, was passed in 1976 it was not strictly enforced until 1991. Starting in 1991, CETESB began a major programme to assess the industrial effluents of every industry located in the São Paulo metropolitan area. There are about 40,000 licensed industrial plants in the area but only 1,250 are believed to be responsible for about 90 per cent of the organic and inorganic loads. Based on information provided during the licensing procedure these 1,250 industries were selected for closer investigation. Intense negotiations were undertaken with all the 1,250 main polluters. They were asked to submit plans, and a schedule for implementing treatment plants, that would enable them to comply with the emission standards and permits required by the State Law 997.

After collecting all the information, CETESB defined a system called STAR (Sistema de Tratamento de Águas Residuárias, or Wastewater Treatment System) which is an information protocol establishing the treatment processes for each industry together with the schedule for the implementation of the treatment and the permit system, under the agreement signed by CETESB and the industries. The information gathered under STAR was stored in a data bank at the CETESB headquarters.

Industries were supported through loans, which were provided if needed, and were drawn from funds from two different sources: CETESB itself, which was in charge of managing a special line of credit directly from the World Bank (PROCOP), and BNDES, a Brazilian federal agency conceived to help industries to improve their performance. During the implementation of the treatment systems, CETESB monitored the effluent discharges closely using mobile equipment and also the receiving water bodies at fixed points. Industries are gradually introducing self-monitoring and CETESB is establishing a compliance monitoring system, in order to check the results reported from the self-monitoring as an aid to the implementation of enforcement actions. This procedure is followed continuously thereby assuring long-term compliance with legal standards. The process is illustrated schematically in Figure VI.5.

Figure VI.6 shows the number of industries with effluent control at different stages by the end of September 1994 and Figure VI.7 shows the gradual increase in the number of industries which had achieved the effluent treatment targets set by CETESB. Of the 1,250 targeted industries, 1,007 had their treatment systems working satisfactorily. The resulting decrease in pollution loads between 1991 and 1994 are shown in Figure VI.8. The

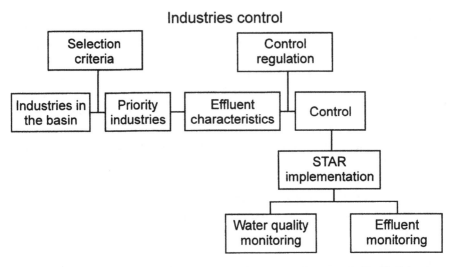

Figure VI.5 Schematic representation of the pollution control process in the Tietê basin

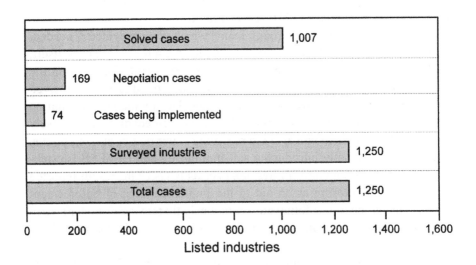

Figure VI.6 Number of industries with effluent control at different stages by September 1994

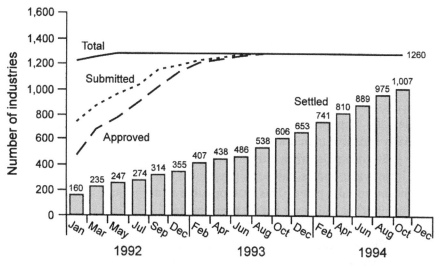

Figure VI.7 Progress in the implementation of industrial effluent control, 1992–94

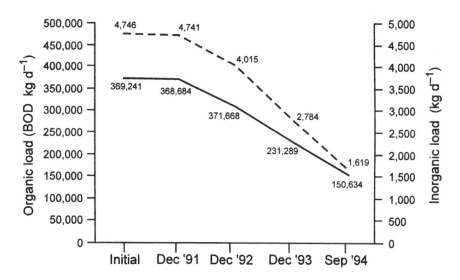

Figure VI.8 Decreases in industrial pollution loads resulting from the installation of effluent control, 1991–94

organic load of 370,000 kg BOD d^{-1} at the outset of the process was reduced to 150,000 kg BOD d^{-1} in September of 1994. The inorganic load decreased from 4,700 to 1,600 kg d^{-1}.

VI.6   Conclusions

This successful case study of industrial wastewater management and control illustrates the importance of public participation. Elected officials are particularly sensitive to public opinion, in order to satisfy their voters. From the late 1980s onwards, when citizens began protesting against the degradation of local water bodies, the permit system was enforced and compliance action began to take place. This was possible under a State law that was passed in 1976 but had never before been enacted.

In addition, many industries decided to support the programme, by adopting efficient treatment methods to promote pollution control, and also to win a better public image as a result of public pressure. Finally, it should be noted that credit was available, where and when necessary, which made the investment decisions much easier.

VI.7   References

CETESB 1994 *Projeto Tietê, Despoluição Industrial*. Relatório de Acompanhamento, Setembro 1994. Companhia de Tecnologia de Saneamento Ambiental, São Paulo.

Alonso, L.R. and Serpa, E.L. 1994 *O Controle da Poluição Industrial no Projeto Tietê, 1994*. Companhia de Tecnologia de Saneamento Ambiental (CETESB), São Paulo.

Case Study VII[*]

THE MEZQUITAL VALLEY, MEXICO

VII.1 Introduction

Mexico is a federal republic composed of 31 states and a federal district. The country has a surface area of nearly 2×10^6 km^2 and an annual rainfall of 777 mm, which is equivalent to $1,522 \times 10^9$ m^3 a^{-1} of water. This volume of water should be sufficient for all the needs of its population but the poor geographical and temporal distribution of the water resources result in a shortage of water for 75 per cent of the country. These areas are classified as arid or semi-arid (SEMARNAP, 1996).

The national population of 89 million has an annual growth rate of 1.9 per cent and 70 per cent of the population reside in urban areas. The metropolitan area of Mexico City, with 18 million inhabitants, contrasts greatly with the highly dispersed rural population of 20 million living in 149,000 communities of less than 1,000 inhabitants. The general level of education is low and there are more than 56 ethnic groups speaking indigenous languages. Life expectancy is 69–72 years. The prevalence of infectious disease and parasitism is superimposed on that of chronic degenerative illnesses. The infant mortality rate continues to be high.

The gross national product was US\$ 3,750 per capita in 1993 (INEGI, 1994) and Mexico is currently facing a severe economic crisis. This aggravates the poverty experienced by 50 per cent of the population living in rural areas and the marginal zones of large cities. This situation is generally considered to be only temporary and it is expected that economic development will begin again in Mexico as it has done in the past. In 1994, Mexico signed the North American Free Trade Agreement with the USA and Canada. Mexico is also a member of the Organisation for Economic Co-operation and Development (OECD).

*   *This case study was prepared by Humberto Romero-Alvarez*

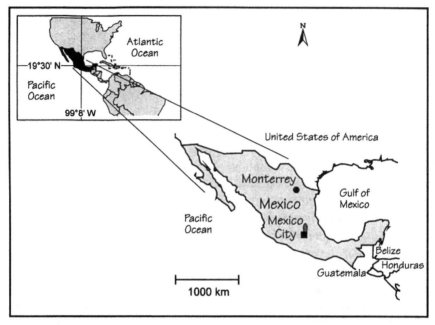

Figure VII.1   Location map of Mexico showing Mexico City and the Mezquital Valley to the north of the City

VII.2   The Mezquital Valley

The Mezquital Valley is within the bounds of the state of Hidalgo. It is situated in the Mexican high plateau, 60 km north of Mexico City (Figure VII.1), with an altitude between 1,700 m and 2,100 m above sea level. The 495,000 inhabitants of the valley are principally involved in agricultural activities, complemented by livestock breeding. Their standard of living is higher than that of the population without access to wastewater for use in irrigation (Romero, 1994).

Irrigation districts 03-Tula and 100-Alfajayucan use raw wastewater from the metropolitan area of Mexico City (Figure VII.2). This wastewater has received no conventional treatment. Due to the immense size of the cultivated area (83,000 ha in 1993–94) and its antiquity (91 years in continual operation), the region represents a unique example of wastewater irrigation (Table VII.1). The wastewater, whether raw, partially treated or mixed with rainfall, is highly valued by the farmers because of its ability to improve soil quality and because of its nutrient load that allows increased productivity (Table VII.2) (SARH, 1994; CNA, 1995). In 1990, the maize–alfalfa crop covered a surface 10-times bigger than the vegetable crop, but the productivity was six-times lower.

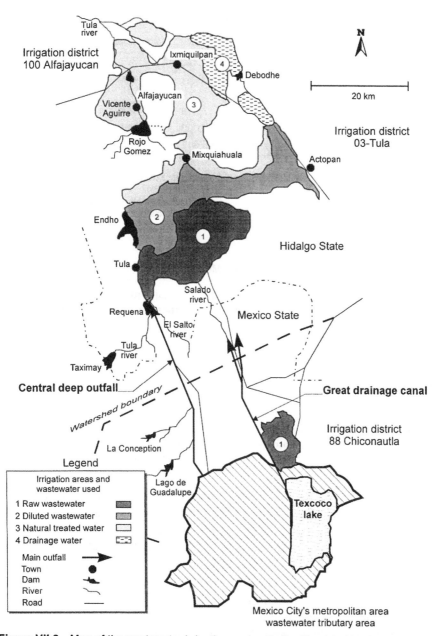

Figure VII.2 Map of the wastewater irrigation system in the Mezquital Valley, Mexico

Table VII.1 Irrigation data for the Mezquital Valley, 1993–94

Irrigation systems	Area (ha) covered[1]	Cultivated[2]	No. of users	Water volume (10^6 m^3 a^{-1})	Production value (10^6 N$)[3]
District 03 (Tula)	45,214	55,258	27,894	1,148	255
District 100 (Alfajayucan)	32,118	22,380	17,018	651	85
Private units	5,375	5,450	4,000	96	0
TOTAL	82,707	83,088	48,912	1,895	340

[1]   Covered area refers to irrigable land with irrigation infrastructure
[2]   Cultivated area includes some areas with more than one crop per year
[3]   Average exchange rate for that period was N$ 3.5 per US$ 1
Source: National Water Commission (CNA), Irrigation Districts Headquarters, Mixquiahuala, Hidalgo, México, 1995

Table VII.2 Agricultural productivity in the Mezquital Valley, 1990–92 (t ha^{-1} a^{-1})

Crops	National mean	Mezquital mean	Hidalgo State irrigation area	Rainfed area
Sweet corn	3.7	5.1	3.6	1.1
Kidney bean	1.4	1.8	1.3	0.49
Oat	4.7	3.7	3.6	1.7
Barley (fodder)	10.8	22.0	15.5	13.5
Lucerne	66.3	95.5	78.8	0.0

Sources: Agricultural and Hydraulic Resources Secretary (SARH), México 1994 (National values)   National Water Commission (CNA), Irrigation Districts Headquarters, Mixquiahuala, Hgo., México 1995 (Mezquital Valley data)

The wastewater is contaminated with pathogenic organisms and toxic chemicals that constitute a health risk for both farmers and consumers of agricultural products. The principal crops grown are alfalfa, maize, wheat, oats, beans, tomatoes, chillies and beetroot. There is a small but valuable production of restricted crops in the lower section of the Valley (District 100), including lettuce, cabbage, coriander, radish, carrot, spinach and parsley. This crop restriction is part of the management policy for reuse of wastewater with adequate health safeguards.

During its use in the Mezquital Valley irrigation districts, the Mexico City wastewater (a mixture of domestic and industrial waste) receives natural "land" treatment which is equivalent or superior to conventional secondary wastewater treatment. The environmental effects that could be experienced due to the water pollution that would result if this irrigation scheme was not available are:

- The raw wastewater would lead to gross environmental pollution estimated at $1,150 \text{ t d}^{-1}$ organic matter, expressed in terms of their biochemical oxygen demand (BOD), which would affect the land and water resources downstream in the Panuco River basin, including several coastal lagoons and the Gulf of Mexico.
- Municipal and rural water supplies, hydroelectric plants, fishery developments, aquatic ecosystems and a rich biodiversity would be affected.
- Nutrient rich wastewater flowing downstream to the river basin would cause excess aquatic weed and vector infestation as a result of eutrophication.
- The aesthetic value of the natural environment and the landscape would be affected by foaming and other effects, such as odour.
- Without this huge, natural land treatment process it would be almost impossible to accomplish and to integrate sustainable development of land and water resources in a very important region of Mexico.

VII.3 Pre-intervention situation

At present, there are legal and institutional guidelines that ensure sustainable agricultural development in the Mezquital Valley. The National Water Law, in force since 1993, has one section dedicated specifically to the prevention and control of water contamination. In addition, Ecological Technical Standards 32 and 33 (now Official Mexican Standards) set down the requirements for wastewater use in agricultural irrigation (Diario Oficial de la Federacion, 1993). The National Water Commission (Comisión Nacional del Agua; CNA) was officially created in 1989 as a federal government entity responsible for promoting construction of the hydro-agricultural infrastructure, as well as for its operation, and for ensuring that the laws and standards relating to efficient use of water and control of its quality are upheld.

The Federal Government, specifically CNA, has been in charge of the irrigation districts since 1949. Each district is under the administration of a chief engineer appointed by CNA, and being under the control of a single authority greatly facilitates management of the irrigation scheme. There is also a management board composed of representatives of central and state governments, water users associations and local credit banks. Some farmers work in co-operatives managed by themselves, although most are individual workers who own very small parcels of land (an average of 1.5 ha per user).

Farmers lodge their water demands with the local District Office, specifying where and when the water is required. The District Manager then prepares a first draft of the irrigation schedule, analysing the different factors involved, such as the amount of water available, water demand timetables, the crop preferences of the farmers, agricultural authority policies, crop restrictions and resources available. The resultant irrigation programme (plan de riego) is implemented following discussion with, and approval by, the farmers who will take part in it.

A fee is charged to the users (farmers) by CNA to recover some of the operational costs, although government subsidies remain high. Efforts are being made to eliminate these subsidies. The real operational and maintenance costs are around N$ 4.42 (4.42 new pesos) per thousand cubic meters and the farmers are paying only N$ 1.46 (33 per cent), plus N$ 0.75 (17 per cent) estimated as labour costs for small maintenance works (the average exchange rate for the 1993–94 agricultural cycle was N$ 3.5 per US$ 1). Therefore only 50 per cent of the operational costs are covered by the farmers using the wastewater. Every year since the beginning of this century, the government has provided funding for continuous extension of the irrigation infrastructure. It is rather difficult to estimate these construction costs as a component of the wastewater economic value because insufficient information is available. However, the farmers profits are often about 60 per cent from marketed crops and some salad vegetables can be more profitable (70 per cent and even 80 per cent).

In the last four years, due to the spread of cholera, CNA has enforced restriction on crops irrigated with wastewater and whose products are consumed uncooked, such as salad crops. This decision, taken as a preventative measure, caused social conflict with farmers who saw their income severely reduced by the restriction of their cash crops without other viable alternatives being proposed.

The volume of wastewater generated has increased over time. It is distributed in the Mezquital Valley by a complex system of tunnels, reservoirs and canals, which themselves have a purifying effect on the wastewater. The result is that different areas are irrigated with water of different quality. For example, at the entrance to the Valley, the wastewater has a maximum of 6×10^8 faecal coliforms per 100 ml, whereas at the outflow from the Vicente Aguirre reservoir the count is reduced to a minimum of 2×10^1 (Table VII.3). The same reduction occurs with helminths; the concentration of *Ascaris* eggs is reduced from 135 per litre at the Valley entrance to less than one per litre at the outflow of the lowest reservoir (Cortés, 1989; Cifuentes *et al.*, 1994). This situation has stimulated the interest of academic institutions, which carry out epidemiological studies in the Mezquital Valley. Their first results (Figure

Table VII.3   Faecal coliform concentrations in the Mezquital Valley reservoirs (MPN[1] per 100 ml)

Reservoir	Geographic mean[2]	Maximum[3]	Minimum[3]
Endho			
Inflow	2.6×10^7	6×10^8	3×10^4
Effluent	6.1×10^4	3×10^6	4×10^4
Rojo Gomez			
Inflow	5.3×10^5	3×10^4	5×10^3
Effluent	1.4×10^4	2×10^5	1×10^1
V. Aguirre			
Inflow	5.9×10^3	1×10^4	1×10^2
Effluent	3.3×10^2	3×10^4	2×10^1

[1]   Most probable number
[2]   Source: Cortés, 1989

[3]   Source: Cifuentes *et al.*, 1995

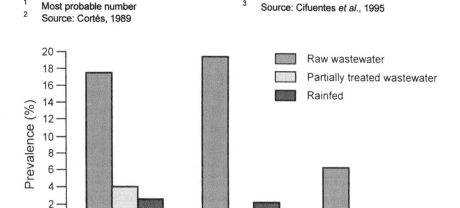

Figure VII.3   Percentage *Ascaris lumbricoides* infection in different age groups of children according to the method of irrigation used for agriculture (After Cifuentes *et al.*, 1994)

VII.3) demonstrated that there is a higher risk of *Ascaris lumbricoides* infection in the infants of farm workers using raw wastewater than for those using partially treated wastewater from storage reservoirs, and that the risks for both groups were considerably higher than for those in the rain-fed control area. By contrast, the risk to children and adults in the reservoirs group was similar to that observed in the controls (rain-fed area). As expected, the age group 5–14 years, especially males, had the highest intensity of *Ascaris* infections when exposed to raw wastewater (Cifuentes *et al.*,

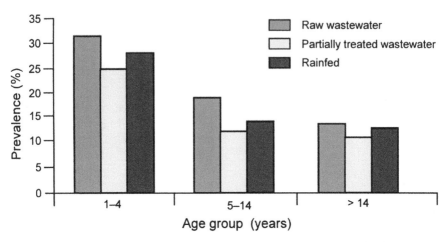

Figure VII.4 Percentage of diarrhoeal disease in different age groups of children according to the method of irrigation used for agriculture (After Cifuentes *et al.*, 1994)

1995; Blumenthal *et al.*, 1996). In addition, these studies suggested an association between the prevalence of diarrhoeal disease and the exposure of the farmers' children to wastewater of different quality; children from households exposed to raw wastewater had a small but significantly increased risk. The higher rates of diarrhoeal diseases found in infants (1–4 years old), who mostly depend on their mothers, could be explained by crowded households, deficient hygiene practices and unsanitary conditions in the farmers' domestic environment (Figure VII.4) (Ordóñez, 1995). These results support the view that parasite infection is more effective as an indicator of the effects of wastewater use on the health of an exposed population.

VII.4   Intervention scenario

In 1993, Mexico hosted a regional workshop to analyse the issues surrounding agricultural wastewater use and to propose appropriate interventions to ensure public and occupational health and safety. The workshop was organised by the Mexican Institute for Water Technology (IMTA), with the assistance of the World Health Organization (WHO), the Pan-American Health Organization (PAHO), the Food and Agriculture Organization of the United Nations (FAO), the United Nations Environment Programme (UNEP) and the United Nations Centre for Human Settlements (UNCHS/HABITAT). Representatives from 12 countries in Latin America and the Caribbean participated. The workshop recommended the creation of a study and reference centre in the Mezquital Valley with the aim of promoting, co-ordinating and integrating investigative studies carried out in the favourable conditions found in that area.

With regard to wastewater treatment as a measure for the protection of health and the environment, CNA is conducting detailed engineering studies in relation to the possible construction of conventional treatment plants in the Great Drainage Canal, in the metropolitan area of Mexico City, and in the discharge point from the Central Deep Outfall (Emisor Central) in the Mezquital Valley. In this respect, CNA has existing experience with treatment plants, both large and small, currently operating in the metropolitan area and whose effluents are used to irrigate green areas and to fill recreational lakes in the urban area.

On a smaller scale, it may be possible to convince farmers to invest in treatment plants at the plot level to ensure safe production of salad vegetables and other high risk crops. At present, CNA is concentrating on assisting the farmers who use wastewater to build their own stabilisation ponds, to adapt the quality of the wastewater to the requirements for cropping restrictions and to demonstrate that the practices being used are safe. To ensure that these safe practices are used correctly, a strict wastewater quality certification programme is needed.

Two events in the politics and administration of the country have facilitated more direct intervention in the future to improve the conditions under which wastewater is used in the Mezquital Valley. First, recent changes in the organisation of federal public administration, have placed the overall management of water (i.e. through CNA) under the newly created Ministry of Environment, Natural Resources and Fisheries. This will allow more emphasis to be given to environmental problems, which are precisely the central issue in the Mezquital Valley and which could affect downstream water resources in the Panuco River basin (as mentioned above). The second important event was the proposal to create the regional study centre in the Mezquital Valley. The specific objective of this centre is to enhance technical and scientific understanding in order to enable rational and safe use of wastewater and thereby to assist the development of sustainable agriculture. In order to assist the many and varied investigations in the Mezquital Valley, the reference centre should provide two basic facilities:

- An information system including data generated by the field studies and environmental monitoring network.
- Various demonstration units of an experimental and educational nature, to facilitate training and technology transfer.

VII.5 Lessons learned, constraints and opportunities

The project to create a study centre in the Mezquital Valley faces obstacles commonly found in developing countries. These are:

- High levels of poverty and unemployment which are aggravated by excessive demographic growth, and a currency (the peso) weighed down by external debt and a shortage of financial resources.
- Persistent conditions of environmental deterioration. Above all, the need for basic domestic sanitation in rural areas demands attention and competes for scarce funds.
- Strong market pressure to adopt developed country solutions which are inappropriate (technically, economically and financially) for developing countries. The treatment of wastewater is a good example of this.
- The process of administrative decentralisation. In its initial phase this results in serious difficulties with co-ordination, usually because there are few well-prepared professional and technical personnel available at the local level.

Nevertheless, there are factors that favour the implementation of the project, such as:

- Many institutions and researchers, both national and international, are interested in carrying out appropriate studies.
- There is political will to halt environmental deterioration and to revert present trends in order to ensure sustainable development.
- The basic institutional infrastructure exists to implement interventions for improving agricultural production and water sanitation in the irrigation districts.
- Some international co-operation agencies are interested in giving technical and financial assistance to the proposed study centre, because of its regional relevance for countries in Latin America and the Caribbean. The InterAmerican Development Bank, for example, has indicated its interest in the project. The Bank, together with the Japanese government has approved a US$ 800 million credit for large-scale wastewater treatment plants in the metropolitan area of Mexico City, as well as for the necessary hydraulic infrastructure.

VII.6 Conclusions and recommendations

- The rational use of wastewater for irrigation in agriculture and forestry, or in aquaculture, is a highly useful and productive practice that contributes to sustainable development which is the central objective of Agenda 21 as approved at the United Nations Conference on Environment and Development (UNCED) in Rio de Janeiro in 1992.

- The interventions necessary to improve the efficiency of wastewater use, in order to protect health and safeguard the environment, require a full understanding of local socio-cultural and economic conditions. Such understanding must result in action, which should be translated into guidelines and applied promptly.
- In the Mezquital Valley, irrigation conditions are ideal for carrying out field research. The results of this research could be used at the national level and eventually in other developing countries.

Taking the above points into account, it is proposed:

- To support the creation of a Regional Study and Reference Centre for the rational and safe use of wastewater in the Mezquital Valley.
- To enforce crop restrictions and other wastewater use regulations, based on recent epidemiological findings.
- To introduce simultaneously a pilot intervention programme of basic housing sanitation in the irrigation area.

VII.7 References

Blumenthal, U.J., Mara, D.D., Ayres, R.M., Cifuentes, E., Peasey, A., Scott, R., Lee, D.F. and Ruiz Palacios, G. 1996 Evaluation of the WHO nematode egg guideline for restricted and unrestricted irrigation. *Wat. Sci. Tech.* **33**(10–11), 277–83.

Cifuentes, E., Blumenthal, U.J., Ruiz-Palacios, G., Bennett, S. and Peasey, A. 1994 Escenario epidemiológico del uso agricola del agua residual: el Valle del Mezquital, México. *Salud Públ. Méx.*, **36**(1), 3–9.

Cifuentes, E., Blumenthal, U.J., Ruiz-Palacios, G. 1995 *Riego Agrícola con Aguas Residuales y sus Efectos sobre la Salud en México, del libro Agua, Salud y Derechos Humanos.* Iván Restrepo. México.

CNA 1995 *Información proporcionada por la Jefatura de los Distritos de Riego del Valle del Mezquital.* Comisión Nacional del Agua, Mixquiahuala, Hidalgo, México.

Cortes, J. 1989 *Caracterización Microbiológica de las Aguas Residuales con Fines Agrícolas. Informe del estudio realizado en el Valle del Mezquital.* Mexican Institute of Water Technology (IMTA), Jiutepec, México.

Diario Oficial de la Federacion 1993 NOM-CCA-032-ECOL/1993 and NOM-CCA-033-ECOL/1993. México, 18 Octubre 1993.

INEGI 1994 *Sistemas de Cuentas Nacionales de México.* Instituto Nacional de Estadística, Geografía e Informática (INEGA), Mexico.

Ordoñez, B.R. 1995 Personal communication, Mexico.

Romero, A. H. 1994 Estudio de Caso (Valle del Mezquital). In: Proceedings *Taller Regional para las Américas sobre Aspectos de Salud, Agricultura y*

Ambiente, Vinculados al Uso de Aguas Residuales. Mexican Institute of Water Technology (IMTA), Jiutepec, México,

SARH 1994 *Anuario de la Producción Agrícola.* Ministry of Agricultural and Hydraulic Resources, México, D.F., Mexico.

SEMARNAP 1996 *Programa Hidráulico 1995–2000.* Secretaria de Medio Ambiente, Recursos Naturales y Pesca (SEMARNAP), Mexico.

Case Study VIII[*]

LERMA-CHAPALA BASIN, MEXICO

VIII.1 Introduction

In many of its regions, Mexico currently faces an imbalance between water demand and availability, primarily due to natural water scarcity as well as uneven water quality distribution. Rapid urban and industrial growth, among other economic and social factors, have made this worse. Water needs have grown, water users are fiercely competing with each other and conflicts are emerging as a result. Water quality has also deteriorated as urban and industrial effluents are often discharged with no previous treatment. Furthermore, Mexico is slowly overcoming a severe economic and financial crisis which has limited hydraulic infrastructure development and impoverished large population sectors.

Mexico covers 1.97 million km^2 of the North American continent (Figure VIII.1), with a population of 91.12 million growing at 1.8 per cent a year. Politically, Mexico is divided into 31 autonomous states (each one with its own elected government) and a federal district, which includes Mexico City. A complex system of mountain ranges create 310 hydrological basins which experience different degrees of hydraulic development and water pollution. Of all the Mexican basins, Lerma-Chapala is the most important. Consequently, it receives priority attention at all three government levels, federal, state and municipal, and especially from the National Water Commission (Comisión Nacional del Agua; CNA) which is the sole federal authority entrusted with overall national water resources administration. Public awareness on water issues in Lerma-Chapala has led to the active participation of water users, non-governmental organisations (NGOs) and social institutions with a plethora of interests directly or indirectly linked with the water sector.

---

[*]   *This case study was prepared by José Eduardo Mestre Rodríguez*

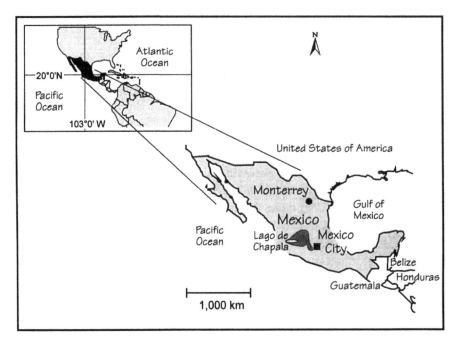

Figure VIII.1   Location map showing the positon of Mexico and the Lerma-Chapala basin

VIII.2 The Lerma-Chapala basin

The River Lerma with a length of 750 km originates in Mexico's central high plateau at an altitude above 3,000 meters above sea level (masl). The river ends in Lake Chapala (1,510 masl) which is the largest tropical lake in Mexico (Figure VIII.2), 77 km long and 23 km wide. The maximum storage capacity of the lake is 8.13 km³ and the surface area is about 110,000 ha. The lake is also rather shallow; its average depth is 7.2 m, with a maximum of just 16 m. The Lerma River basin, is a tropical region with an average temperature of 21 °C, an area of 54,400 km² (less than 3 per cent of Mexico's entire territory) and an average rainfall of 735 mm a⁻¹, mainly concentrated in the summer, from which a mean run-off of 5.19 km³ is derived. The River Santiago arises from Lake Chapala and flows westwards finally reaching the Pacific Ocean. The Santiago River basin is less developed in terms of population and economic activity, except for Guadalajara, the second largest city in Mexico, and with a metropolitan area with more than 3.5 million inhabitants.

Some 26,000 deep water wells operate within the Lerma-Chapala basin, with very low efficiency rates, due to their high electricity consumption and rather low water yields. Almost 70 per cent of all 38 aquifers in the region are overexploited (Figure VIII.3).

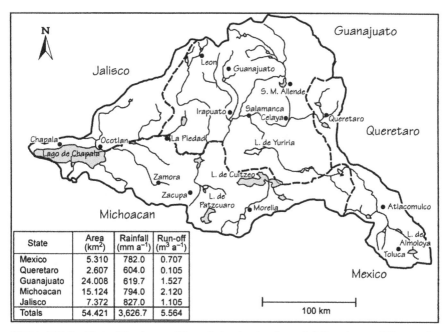

State	Area (km^2)	Rainfall (mm a^{-1})	Run-off (m^3 a^{-1})
Mexico	5.310	782.0	0.707
Queretaro	2.607	604.0	0.105
Guanajuato	24.008	619.7	1.527
Michoacan	15.124	794.0	2.120
Jalisco	7.372	827.0	1.105
Totals	54.421	3,626.7	5.564

Figure VIII.2   Map of the Lerma-Chapala basin showing rainfall and run-off figures for each state included in the basin

The current basin population is 9.35 million with an annual growth rate slightly less than the national average. The population is distributed between 6,224 localities, 18 of which have a population greater than 50,000 inhabitants; the rural population is currently 32 per cent. Regional socio-economic development has been triggered by water availability and industrial and agricultural production per capita have surpassed national levels. This region boasts 6,400 industries which generate one third of the GNP and 20 per cent of all national commerce occurs within this basin. Furthermore, it currently comprises one eighth of all the irrigated land in Mexico. The agriculture in this area is of such importance that national farm produce exports rely heavily on the performance of this tiny region. With the three economic sectors highly developed and with a superior transportation network, partially financed by private investors, this area is, undoubtedly, one of the richest regions in Latin America.

The Lerma-Chapala basin includes fractions of the central states of Guanajuato, Jalisco, Mexico, Michoacan and Queretaro (Figure VIII.2). Conflicts derived from surface run-off uses (mainly for irrigation and potable water supplies), combined with the general discharge of untreated effluents,

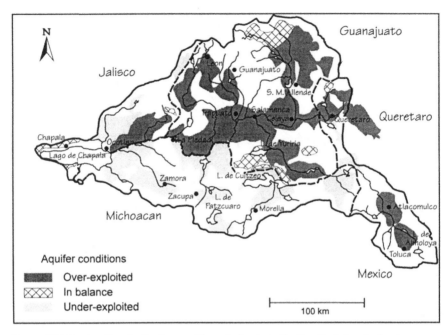

Figure VIII.3   Aquifers in the Lerma-Chapala basin indicating their level of water abstraction

have given rise to serious regional, and local, pollution problems. Frequent conflicts over water quality occur in Chapala Lake which plays a key role as the main water source for Guadalajara.

VIII.3 Pre-intervention situation

Before 1989, the regulatory and legal framework provided clear procedures for surface run-off measurement and the related information systems and analysis tools; but there were serious deficiencies in water quality monitoring and recording. In addition, institutional structures, mostly centralised at the federal level, were unable to slow down water quality deterioration throughout the basin. Eventually, this situation became acute, dramatically reducing water availability for many uses. There was, nevertheless, public and official awareness of the key issues relating to water quality and sustainable development. Hence, in 1970, under the Secretaria de Recursos Hidráulicos (Ministry of Hydraulic Resources), the first technical and administrative unit was created to prevent and control water pollution from different sources. The Lerma-Chapala basin was a natural choice for the pilot area to carry out the first water quality assessment and to lay the foundation for future intervention.

From an economic and financial perspective, the hydraulic services in the Lerma-Chapala basin did not differ from the general scheme prevalent in the rest of the country. Funding was insufficient to meet demands. Water pricing and actual payments made by users were below real water costs, restricting capital investment and management expenditures. This, in turn, limited the possibility of providing a reasonable water service for irrigation, for industry and for households. Furthermore, such a situation fostered the limited participation of water users and generated a negative attitude towards water resources management and supply. Even today, when changes are currently being implemented, many users (at all levels and sectors) are still reluctant to pay for water.

Potable water supply had reached acceptable levels of coverage in urban areas but not in rural areas. In townships with a population above 50,000 inhabitants, service coverage was usually close to 85 per cent or more and large cities usually boasted coverage of around 95 per cent. Chlorination of the water was rather uncommon, except in large cities. Water quality control was also extremely limited, notwithstanding the efforts of the water and health sectors. The Limnological Studies Center, established in Chapala in 1975, and the regional laboratory for public health, set up in Leon, Guanajuato in 1981, backed up efforts to promote water quality control.

Urban sewage systems had lower coverage levels than the potable water systems. Untreated effluents were discharged directly into rivers and reservoirs. Furthermore, when treatment facilities did exist, like in the city of Querétaro, their operation was usually inefficient, as a result of faulty design and mismanagement related to financial aspects. Few social sectors were willing to pay for effluent treatment.

The Mexican economy grew considerably after the Second World War. National and international investments promoted industrial growth and this was further aided by a domestic market unable to purchase imported goods. Simultaneously, irrigated agriculture grew steadily in terms of surface area, economic importance and water demand. National and regional economic development policies did not allow for a long-term water conservation strategy and as a result irrigated agriculture is responsible for 81 per cent of all water abstractions in the Lerma-Chapala basin.

This region includes 16 large reservoirs which help to regulate erratic run-off from year to year. They have also helped considerably to reduce flooding risks. However, as a result of an excess of nutrients derived from untreated effluents, the reservoirs were seriously affected by massive infestations of water hyacinths.

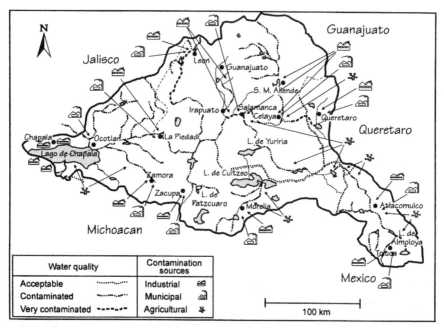

Figure VIII.4   Map of the Lerma-Chapala basin indicating the water quality classifications for the main river stretches and the associated sources of water contamination

Industries as well as most towns, located in the basin are mainly supplied by groundwater sources (90 per cent). The most important industries concentrate their activities on meat, dairy and other agricultural produce, beverages, pulp and paper, leather goods, petrochemical and chemical products, all with little or no emphasis on wastewater treatment and recycling.

Development in the Lerma-Chapala basin is largely sustained by intense water use. Industries in the basin generate around 0.608×10^3 m^3 a^{-1} wastewater with 130,500 t a^{-1} biochemical oxygen demand (BOD) coming from urban waste and 424,260 t a^{-1} chemical oxygen demand (COD) coming from industrial discharges. These organic and inorganic pollutant loads and a scarcity of wastewater treatment capacity have intensified water quality problems and severely reduced water availability (Figure VIII.4). Diffuse pollution caused by drainage containing fertiliser and insecticide residues from irrigated areas, together with solid waste washed away by rain from rural households lacking domestic waste disposal systems for excreta and rubbish, have also contributed to the water quality problems.

Lake Chapala is the most important water distribution centre in the region and was seriously threatened by growing biological and chemical water pollution. This generated a public outcry in the state of Jalisco and eventually became a matter of national concern.

VIII.4 Intervention scenario

By the end of 1988 it had become apparent to society and government institutions that a complex and serious situation existed in Lerma-Chapala basin. Water demands were higher than natural availability and to such extent that even all the effluents were also already committed for use. Water allocation was a chaotic process because most water rights were granted with no clear strategy to protect water users downstream or to cope with regional water scarcity caused by frequent droughts. Users were competing with each other, usually industry and cities were exerting heavy pressure on irrigated farmland. Conflicts were not uncommon at all levels including disputes for water among neighbouring states. In general, water quality had fallen to a new, unacceptably low level. In specific locations, water quality had deteriorated so badly that life itself, in all its forms and manifestations, was challenged. River basin protection was almost non-existent. Erosion had increased in former forest areas and grasslands were disappearing at an astounding rate as a result of irrational livestock practices. Silt sedimentation eventually reduced the hydraulic capacities of streams, rivers and reservoirs and dramatically reduced the lifespan of several dams.

Society began demanding swift and effective executive action to remedy the situation in the basin. In April 1989, the Federal Government and the governments of the five states which share the basin formally, agreed to co-ordinate their efforts to carry out a "Program for Water Allocation among Users" under a new set of rules and simultaneously to undertake a "Large-Scale Sewage Treatment Program in the Lerma-Chapala Basin" (Programa de Ordenamiento de los Aprovechamientos Hidráulicos y el Saneamiento de la Cuenca Lerma-Chapala). The four main objectives derived from this dual programme were:

- To reduce water pollution.
- To establish a new system in water allocation.
- To give a thorough impetus to all activities that may help raise water efficiencies.
- To establish some sound basic rules for soil and water management, to enable and encourage biological canopy protection and recuperation, practical (and profitable) approaches for rational soil management and other preventative action.

These four objectives were accepted and adopted by society which, in turn, has played a key role in reviewing the results, evaluating the actions and even by arguing for the introduction of changes proposed by social sectors.

Government agencies installed a "Consulting Council for Evaluation and Follow-up" of all sub-programmes and activities derived from the basin programme. The Council was integrated by Federal Government ministers, state governors and chairmen from decentralised public enterprises (mainly petroleum refining and electricity). This Council was, in fact, a predecessor of the present River Basin Councils.

The Consulting Council resulted in continuous social pressure and gave rise to a paramount change in government policy on prevention and control of water pollution because the administrative decentralisation process was accelerated. As its functions and responsibilities grew with time, the Consulting Council eventually became a River Basin Council. A Work Group was created as a flexible instrument to review conflicts and all actions in detail, and to raise proposals to the Council. It had representatives of each Council member; these representatives were empowered to vote and to establish commitments on behalf of the institution he or she represented. A chairman was elected who was always a public servant from the National Water Commission. The Work Group met every two months, whereas the Council had a solemn public session every year or so, usually with the President of the Republic present. The Council work agenda for every session had been discussed previously and had been approved by the Work Group. All key issues, such as financing or law enforcement, which were voted on by the Council had already been approved either in the Work Group itself or by means of bilateral lobbying. Hence, all key issues were always approved by consensus. This mechanism itself has proved invaluable. Many potential, bitter confrontations and outdated standpoints were avoided.

The Consulting Council created an appropriate atmosphere that eventually attracted water users. Hence, within the Consulting Council, a Water Users' Assembly was created as a powerful body that could listen to a plethora of water demands, as well as provide a swift vehicle for raising to the Council level the needs, hopes and means of water users for contributing to the improvement of the hydraulic situation in the Lerma-Chapala Basin. Eventually, water users' representatives became Council members with identical rights to speech and vote as Government members.

Three years later the new National Water Act (December, 1992), inspired by the Consulting Council process, enforced the creation of basin councils throughout the country to improve institutional co-ordination and to enhance all forms of fruitful relationships amongst users and water institutions. The water act assigned CNA a key role in regional water management within the federal government. Furthermore, it encouraged greater participation by state and municipal authorities (Article 13). Hydrological basins (defined either by surface or groundwater borders) were finally, and legally, recognised as the ideal geographical unit for rational water management. The National Water Act could perhaps have gone further with its definition of Basin Councils because, for all practical purposes, the Consejo de Cuenca Lerma-Chapala was already further advanced than was required by law.

For the first time in Mexican history, the Water Act included a single chapter on water pollution prevention and control. This section clearly holds CNA responsible for promoting and, when necessary, operating federal infra-structure and services essential to preserve, conserve and improve water quality in hydrological basins and aquifers (Article 86). All purveyors of water supply and effluent treatment have a direct responsibility to comply with the law. In effect, a large-scale decentralisation process has been under way in the water sector for the past two decades. If unpredictable events occur, and for the sake of public interest direct intervention by CNA is required, then (and only then) the Federal Government will provide water services until such extreme events cease or are brought under control.

As direct result of a Master Water Plan (an achievement in itself, derived from public hearings and intense discussion amongst council representatives to the Work Group) and in close co-ordination, CNA and the Lerma-Chapala Basin Council have implemented an ambitious "Large-Scale Sewage Treat-ment Program" to clean up the region. This is the first large-scale water treatment programme in Mexico, undertaken as a result of widespread participation and not only as a federal programme. The programme deals with freshwater supply disinfection and building treatment facilities able to cope with urban-industrial effluents. The projects were mostly generated by State and Municipal authorities and funding was raised by federal water rights (a payment similar to tax), subsidies (both federal and state originated), domestic and foreign credits, private sector investments and water supply savings derived from water pricing strategies. All construction activities were usually run by local authorities via contractors and by private sector investors.

In the case of treatment facilities the decision-making process was clearly defined; several key townships were identified by the Council as those most directly responsible for domestic pollution levels either on a general or local basis. These city authorities were invited to consider joining the Sewage Treatment Program and those that agreed (and a large proportion did agree) had technical, financial and institutional support provided when required. The details of this scheme were rather complex given that, for example, sewage systems were incomplete in several cases and billing procedures were underdeveloped in some other sites.

Before the Clean Water Program was enacted in April 1991, potable water was mainly disinfected using chlorine. On a regional basis 5,763 l s^{-1} were disinfected water, equivalent to 31 per cent of the total water supply, to service 2.2 million inhabitants at 10 sites. By the end of 1994, chlorine disinfection had increased to 18,000 l s^{-1}, which represented 85 per cent of the total water supplied to 5.7 million inhabitants in 594 localities.

A permanent monitoring system is run by CNA based on residual chlorine determination. Regular maintenance is also provided to chlorinators exclusively when required, without interfering with local water supply policies and responsibilities. Other organisations are responsible for operating the systems. In order to preserve standards in its drinking water sources, CNA has updated its source inventory. Presently, 498 sources are protected, i.e. 20 per cent of all registered water sources.

The wastewater treatment programme was planned in three stages. The first stage, which ended in December 1994, was aimed at reducing the organic pollution impact on the Lerma River basin by 50 per cent and by 65 per cent in Lake Chapala. The goal was to build and operate 48 plants for municipal wastewater treatment, with an overall capacity of 3,700 l s^{-1}. Global capital investments have been close to 367 million pesos (approximately US$ 80 million).

By 1997, 45 plants with a treatment capacity of 5.72 m^3 s^{-1} were operating on a regular basis with an average running efficiency of around 70 per cent. Furthermore, 40 per cent of the operating plants have to improve their efficiencies whereas the remainder are discharging within legal BOD limits. Six further treatment facilities were under construction to raise the regional capacity to 9.56 m^3 s^{-1} (on a regional level the present domestic effluents are close to 17 m^3 s^{-1}). On the shores of Lake Chapala, 17 municipal plants have been completed (treating a total of 643 l s^{-1} at 90 per cent efficiency in BOD removal). In this particular zone, to ensure the operation of the facilities, given that most plants are quite small, a special technical administrative unit was created entirely run and funded by the local state government. This scheme has now evolved to a point where most expenditure is provided by municipal authorities and funded through integral water tariffs. On average in May 1989, almost 90 per cent of all water in Lake Chapala had been reported as poor quality (Figure VIII.5). By contrast, 85 per cent is now considered of good quality and 15 per cent of adequate quality (Figure VIII.6). These

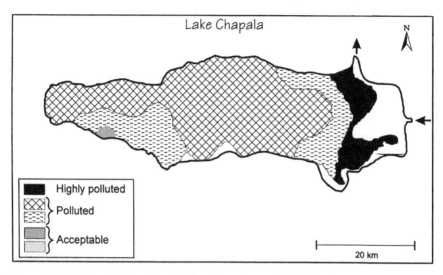

Figure VIII.5   Map of Lake Chapala showing water quality distribution determined by a water quality index in 1989, prior to the sewage treatment programme

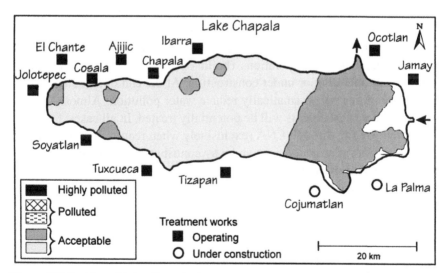

Figure VIII.6   Map of Lake Chapala showing the improvement in lake water quality in May 1996 (determined by a water quality index) as a result of the introduction of the Sewage Treatment Program, First Stage, together with the location of effluent treatment plants. Compare with Figure VIII.5

results clearly indicate actual achievements in reversing Lake Chapala's former severe environmental deterioration.

Sewer systems have expanded as a consequence of newly-constructed treatment plants. Furthermore, as a general rule, primary treatment systems and stabilisation lagoons in particular, are now the "preferred" method of wastewater treatment, providing clear-cut technical and financial advantages over other conventional methods. Since early 1997, the decision-making process has been directly affected by newly enacted Federal Official Regulations (Normas Oficiales Mexicanas), promoting realistic discharge standards according to present economic and financial parameters throughout the country. In most situations, raw domestic sewage effluents may meet the new standards after primary treatment.

Federal Government-owned electrical and petroleum industries in the basin have also built large-scale treatment plants to purify and reuse their wastewater; their overall capacity is 415 l s^{-1}.

The second stage of the Lerma-Chapala clean-up programme, which is already under way, aims to increase treatment capacity to 10,670 l s^{-1} of municipal and industrial wastewater by means of constructing and operating 52 new plants and expanding five existing facilities, with a total investment of 1,200 million pesos (US\$ 150 million). Funding is provided by federal, state and private investment as well as by credits and water supply enterprise savings. Several turnkey operations (build–operate–transfer schemes) are either already operating or under construction. At the end of this stage, 100 treatment facilities will dramatically reduce water pollution. Almost 85 per cent of all domestic effluents will be potentially treated. In all cases, Federal Government, acting through CNA (exclusively when required) may provide technical support in project design and may contribute to supplement investment funding. Almost half of all funding will have been furnished by private investors and their participation in design, construction and operation activities will be of paramount importance.

A third stage of the Large-Scale Sewage Treatment Program includes building 50 additional facilities orientated to meet the needs of small townships and rural communities. These plants will boast a total treatment capacity of 1,833 l s^{-1}.

In order to control and monitor water quality in the basin, CNA keeps a regional water agency with headquarters in Guadalajara. This agency regularly inspects and maintains a network of 50 monitoring stations, 22 of which are located in the Lerma River and 28 in Lake Chapala. It also runs two specialised water quality laboratories in the region. All information is systematically processed and analysed with digital model tools, some of which were developed through joint ventures with the International Institute of Applied Systems Analysis (IIASA) at Laxenburg, Austria, Thames Water International in Reading, England, and Canada's Centre for Inland Waters near Toronto, Canada. By means of such models, a detailed Lerma River classification that complied with the Water Act has been produced and officially published for the various river stretches. The models allow forecasting based on alternative scenarios derived from constructing new facilities and modifying water quality policies.

Information systems are kept by CNA and by the Lerma-Chapala Basin Council. They can be accessed and queried via the Internet and are periodically being overhauled to improve information and to offer user-friendly systems. Regional water sector statistics are now being offered either in a printed form, following a similar pattern to the French Water Information Network (Réseau National des Données sur l'Eau) managed by the International Office for Water (Office International de l'Eau) in Limoges, France, or on CD-ROM, through proprietary procedures provided by the Mexican Institute for Water Technology (Instituto Mexicano de Tecnología del Agua).

Efforts are being made by CNA and the Lerma-Chapala Basin to improve water use efficiency in the basin, mainly in agricultural and urban use systems. Water pricing policy that keeps in touch with reality and adjusts billing and collecting systems to increase payments has proved a successful strategy. Irrigation service payments, for example, have increased by 500 per cent since 1990. These actions are aimed at increasing treated water reuse, at constitutionally strengthening operating agencies and at controlling physical water loss, amongst other things.

The transfer of irrigation districts to users' control has also contributed to improved efficiency. To date, 214,000 ha have been transferred, i.e. 74 per cent of all the irrigated area in the basin. Prior to this, irrigation districts were rehabilitated and modernised, with an overall investment of 445 million pesos (US$ 55.6 million). Users are particularly encouraged to participate in decision-making and planning processes, as well as in water management. Furthermore, they are also invited to develop and to conserve infrastructure and to provide services directly. This experience, linked to the Lerma-Chapala Basin Council, has been of paramount importance. Irrigation farmers have acquired a mature approach to water issues; they now successfully run their irrigation districts, most former federal employees are no longer needed, water distribution has improved and money collection has

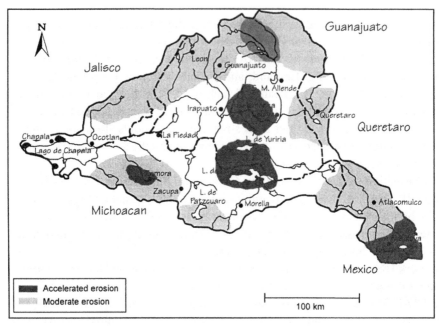

Figure VIII.7 Map of the Lerma-Chapala basin showing areas affected by different levels of soil erosion

increased. Problems do arise from time to time but most of them are solved locally with little or no government intervention.

Pilot programmes have also been implemented in several micro-regions to tackle and prevent soil erosion and hence to reduce accelerated sedimentation in water bodies (Figure VIII.7). Aquatic weed infestations, which currently cover over 11 per cent of all water surface, are another problem that has been successfully addressed by CNA, specially in Lake Chapala where less than 4 per cent of the water surface is presently covered.

There is a permanent campaign, through the media and the Internet, focused on widespread knowledge and understanding of the objectives and activities of the Basin Council. These activities are also helpful in promoting different independent user organisations, with a view to integrating an even more powerful Water Users' Assembly, whose representatives would continue to support and participate jointly in the Council activities.

VIII.5 Conclusions and lessons for the future

The Lerma River water quality, and especially the present condition of Lake Chapala, has shown a considerable improvement in the last seven years as a result of integrated action within the hydrological basin (conceived as a management unit) (Figures VIII.5 and VIII.6). The most important lessons learned will refocus the attention of CNA, the Basin Council and society itself towards:

- The need to increase political willingness towards resource allocation, administrative decentralisation, co-ordination of efforts and undertaking commitments.
- Completion of a new institutional and legal framework in which CNA is the regulatory agency at the national level, and the Basin Council at the regional level, and with the Federal Water Authority resting exclusively in CNA.
- Strengthening Basin Council's role in the water sector, establishing clear regulations for their individual participation and their joint collaboration with municipal, state and federal government institutions.
- Pursuing and completing an integrated water information system, that is now available to authorities, and pushing forward the expansion of the number and versatility of measuring equipment and sites, and the power and flexibility of analysis and decision-making tools.
- Improving planning and evaluation tasks, encouraging joint and effective water users' participation and fostering a permanent commitment by society on regional water issues.
- A new water culture within society; individuals and communities most become aware of water scarcity, pollution and erosion; they should also be willing to accept that they have to pay the price for a better future in terms of water availability and quality; and sustainable development should become a matter of general knowledge, for politicians, scientists, technicians, lawyers and lay persons alike.
- Expanding, and improving, the Lerma-Chapala experience (both institutional and non-governmental, with all its complex technical, political, financial, legal, social and human features) to other hydrological basins throughout Mexico.

VIII.6 Final reflections

The continuity of Lerma Chapala's sewage treatment programme must be ensured because water quality goals can only be achieved through time and with effort. Treatment plants not only need to be constructed, but they need to be operated efficiently and permanently. As in many other places in the world, the key issue is financial. People must be willing to pay for water treatment, and water companies (whether official or private) must evolve to reduce water losses, to raise efficiencies and to improve substantially metering, billing and collecting procedures.

The results achieved so far must be consolidated by complementary action guaranteeing the operation of treatment plants through widespread and permanent training and certification of operators, through an effective system of discharge permits (both to sewers and natural water bodies) and other preventative measures to restrain industrial pollution by encouraging in-house pre-treatment, and through greater emphasis on widespread non-point source pollution generated by irrigated farmland effluents and inadequate sanitary conditions (i.e. excreta disposal) in rural dwellings. In conclusion, water quality improvement will be triggered whenever an effective approach to law enforcement is seriously adopted.

The master plan, its activities and results, must be systematically evaluated so that positive results can be incorporated into other basins in Mexico.

Water quality goals established by users must be consistent with their willingness to pay the cost to fulfil such objectives. Concern over water quality deterioration must be raised, stimulating public awareness of current pollution problems.

Finally, there is still a long way to go to achieve success in this, or in any other, region in Mexico. However, steps are being taken in the right direction and the momentum is gradually increasing.

Case Study IX[*]

THE DANUBE BASIN

IX.1　Introduction

The Danube river basin is the heartland of central and eastern Europe (Figure IX.1). The main river is among the longest (ranked 21) in the world and the second longest in Europe. It has a total length of 2,857 km from its source at a height of 1,078 m in the Black Forest, Germany, to its delta on the Black Sea, Romania. The watershed of the Danube covers 817,000 km^2 and drains all or significant parts of Germany, Austria, the Czech Republic, the Slovak Republic, Hungary, Croatia, Slovenia, Bulgaria, Romania, Moldova, Ukraine and parts of the Federal Republics of Yugoslavia, Bosnia and Herzegovina. The watershed represents 8 per cent of the area of Europe (Figure IX.2).

Between the source and the delta, the main Danube river falls a total height of 678 m and its character varies, therefore, from a mountain stream to a lowland river. Upstream of the Danube delta the mean flow of the river is about 6,550 m^3 s^{-1} with maximum and minimum discharges of 15,540 m^3 s^{-1} and 1610 m^3 s^{-1} respectively. About 120 rivers flow into the Danube, such as the Tisza and Sava which have their own significant flow. The contribution from the main tributaries is given in Figure IX.3.

The mean altitude of the river basin is only 475 m, but the maximum difference in height between the lowland and alpine peaks is over 3,000 m. However the basin can be conveniently divided into an upper, middle and lower region (according to its geological structure and geography), and the Danube delta. The range of mean monthly temperature increases in an easterly direction from 21 °C in Vienna to 23 °C in Budapest and to 26 °C in Bucharest. The average annual precipitation in the Danube river basin varies from 3,000 mm in the high mountains to 400 mm in the delta region. The mean annual evaporation varies between 450 mm and 650 mm in lower regions.

Approximately 80 million people are living in the basin (Table IX.1). The economic conditions vary from the highly developed countries of Germany

---

[*]　*This case study was prepared by Ilya Natchkov*

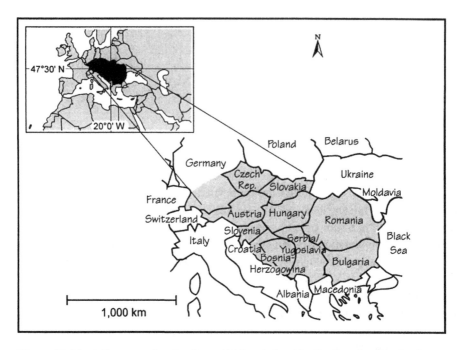

Figure IX.1 Location map showing the countries drained by the Danube river basin

and Austria, to countries with modest economical and technological possi-bilities. Most of the countries in the region are in transition after recent political changes and are suffering severe economic and financial constraints.

IX.2 Economic activities in the basin

Throughout the basin, the tributary rivers and the main Danube river provide a vital resource for water supply, sustaining biodiversity, agriculture, industry, fishing, recreation, tourism, power generation and navigation. In addition, the river is an aquatic ecosystem with high economic, social and environ-mental value. A very large number of dams and reservoirs, dikes, navigation locks and other hydraulic structures have been constructed in the basin to facilitate important water uses; these include over 40 major structures on the main stream of the Danube river. These hydraulic structures have resulted in significant economic benefits but they have also caused, in some cases, significant negative impacts downstream. These impacts include, for example, increased erosion and reduced assimilative capacity where river diversions have resulted in reductions in flow below the minimum required for desired water uses, such as fisheries and maintenance of aquatic ecosystems.

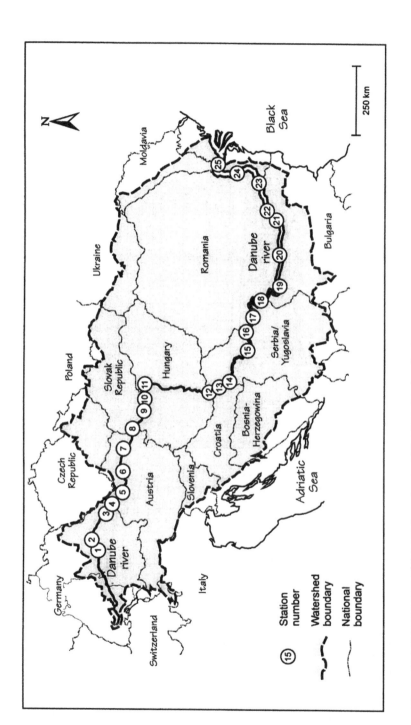

Figure IX.2 Detailed map of the catchment area of the Danube river basin

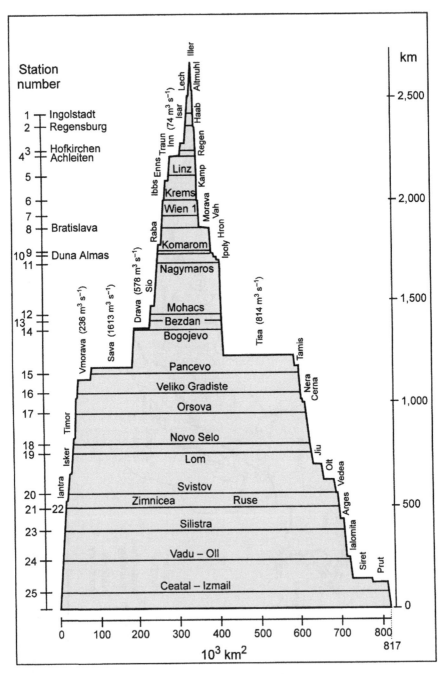

Figure IX.3 The contribution made to the total flow of the Danube river by the various tributaries along its length

Table IX.1 Area and population data for the countries included in Environmental Programme for Danube river basin[1]

Country	Total area (10^3 km^2)	Area within Danube basin (10^3 km^2)	Area within Danube basin (% of the total)	Population (10^6)	Population in Danube basin (10^6)	Population in Danube basin (% of the total)	Urban population (%)
Germany	356.9	59.60	16.7	80.0	9.00	11.25	
Austria	83.9	80.70	96.2	7.42	7.42	100.00	
Czech Republic	78.9	22.49	28.5	10.36	2.74	26.45	78
Slovak Republic	49.0	48.68	99.3	5.31	5.10	96.00	78
Hungary	93.0	93.0	100.0	10.60	10.60	100.00	61
Croatia	56.5	33.75	59.7	4.76	3.25	68.30	51
Slovenia	20.25	15.20	75.0	2.00	1.76	87.75	56
Bulgaria	111.0	48.20	43.4	8.80	4.07	46.25	68
Romania	238.0	233.20	98.0	22.76	22.00	96.70	54
Moldova	33.7	8.79	26.1	4.34	0.77	17.60	47
Ukraine	604.0	36.31	6.0	51.70	3.00	5.80	67
Total in the area of EPDRB		679.92			69.7		
FR Yugoslavia, Bosnia and Herzegovina		134.25			10.5		
Others		2.83			80.2		
Total		817.00					

[1] Some data are estimates because statistics were not available specifically for the Danube river basin

The main water uses in the basin are domestic drinking water supply, industry and irrigation. Many water works along the Danube and its tributaries use bank-filtered water. The Nussdorf water works provides about 15 per cent (150,000 m^3 d^{-1}) of Vienna's water demand from bank wells. The water supply of Bratislava relies on bank-filtered water (160,000 m^3 d^{-1}). In Hungary, most drinking water (90 per cent) actually comes from sub-surface water resources. The total pumped amount is approximately 6,000 $\times$ 10^6 m^3 d^{-1} of which 70 per cent provides public water supplies and 30 per cent is used for irrigation and industrial purposes. In Bulgaria, the total water abstraction from the Danube is 1,142 $\times$ 10^6 m^3 a^{-1} (surface and bank-filtered) of which 70 per cent goes to irrigation, 20 per cent serves industry and 10 per cent provides public water supplies. In the downstream countries the main user is agriculture, which accounts for 85 per cent of total use in Moldova. In upstream countries, such as Slovakia, the main water user is industry (accounting for up to 71 per cent of total surface water withdrawals).

Economic activities and land use in this large river basin are very diverse, including numerous large urban centres and a wide range of industrial, agriculture, forestry and mining activities. There are also numerous important natural areas, such as wetlands and flood plain forests. The water resources and the environmental quality of the basin are under great pressure from these activities. Microbiological contamination is evident throughout the river system and is generally due to the discharge of urban wastewater and storm water. Urban and industrial discharges from inadequate waste treatment and disposal facilities also contribute significant quantities of oxygen depleting substances (measured in terms of their biochemical oxygen demand (BOD)). Nutrients from domestic and industrial sources, chemical fertilisers used in agriculture, and manure from intensive and large-scale livestock operations, have leached into the groundwater and into the surface waters and their sediments. The resultant increases in nutrient levels have stimulated eutrophication and degraded the aquatic ecosystem. Water quality for the eight countries of the Danube basin is summarised in Table IX.2.

The countries of the middle and lower Danube basin are undergoing a major restructuring and transformation of their political, social, administrative and economic systems. From an environmental perspective, some of the most important changes will be in the industrial sector, where the nearly exclusive emphasis on production in the past resulted in significant pollution and waste of resources. Some institutional changes, such as the decentralisation of management and financial responsibility for water supply and

Table IX.2   Proportion of river network conforming to different water quality classes in eight countries of the Danube basin (according to national classification systems)

Country	Water Quality Class				
	I	II	III	IV	V
Austria[1]	23	71	6	0	
Bulgaria[2]	37	22	24	16	1
Czech Republic[3]					
Oxygen regime	0	22	19	36	23
Basic physical and chemical indicators	0	0	0	1	99
Biological and microbiological parameters	4	26	66	4	0
Germany[4]					
Baden Württemberg	17	75	7		
Bavaria	8	87	4		
Hungary[5]	31	54	15		
Romania[6]	42	24	24	12	22
Slovakia					
Oxygen regime	0	22	33	16	29
Basic physical and chemical indicators	0	0	17	27	56
Chemical components	16	26	11	26	21
Biological and microbiological parameters	0	0	13	18	69
Slovenia[7]	0	50	32	12	6

Unless otherwise noted the water quality classification is based on five classes.

[1]   I & I-II, II & II-III, III & III-IV, IV system for 1992; Source: IUCN, 1994

[2]   Source: IUCN, 1994

[3]   Source: Haskoning, 1994

[4]   I & I-II, II & II-III, III & III-IV, IV system; Source: IUCN, 1994

[5]   I, II, III system, 1991 figures; Source: IUCN, 1994

[6]   Source: IUCN, 1994

[7]   I & I-II, II-III, III-IV, IV system for Drava basin only; Source: Haskoning, 1994

wastewater management to local authorities, are creating opportunities for substantial improvements in water services and in environmental benefits.

IX.3   The Environmental Programme for the Danube river basin

Recognising the growing regional and transboundary character of water resources management and the related environmental problems, the Danube countries (together with the interested members of the international community) met in Sofia in September 1991 to consider a new regional initiative to support and to enhance national activities for the management of the Danube basin. The countries agreed to develop and to implement a programme of priority actions and studies in preparation for the eventual agreement of a new convention that would provide an effective mechanism for regional co-operation. The countries also agreed to form a Task Force to oversee this programme, and the Commission of the European Communities (CEC) agreed to provide support and co-ordination for the Task Force.

The international community agreed to assist the participating countries to develop a three-year programme of pre-investment activities, data collection, studies and fact finding to support the development of a strategic action plan. The Environmental Programme for the Danube River Basin (EPDRB) includes national reviews, basin-wide studies of point and non-point sources of pollution and biological resources, institutional strengthening and capacity building activities, and pre-investment studies in selected tributary river basins. Many activities are ongoing, such as the development of international systems for monitoring, data collection and assessment and emergency response systems. International funding for these activities is provided by the European Bank for Reconstruction and Development (EBRD), CEC-PHARE, the Global Environmental Facility (GEF) partners (including the United Nations Development Programme (UNDP) and the United Nations Environment Programme (UNEP)), the World Bank, several bilateral donors (including the Austrian, Netherlands and USA governments), and the private Barbara Gauntlett Foundation.

Furthermore, to secure the legal basis for protecting the water resources, the Danube river basin countries and the European Union (EU) signed the *Convention on Cooperation for the Protection and Sustainable Use of the River Danube* (the Danube River Protection Convention) of 29 June 1994, in Sofia. The Convention is aimed at achieving sustainable and equitable water management. In parallel, the development of the strategic action plan has been a major task of the environmental programme for the Danube river basin. The action plan makes a significant contribution to efforts to improve water and environmental management in the Danube basin as defined in the Convention, and contributes to the implementation of the Environmental Action Programme for Central and Eastern Europe.

IX.4   The strategic action plan

The action plan provides direction and a framework for achieving the goals of regional integrated water management and riverine environmental management for the period 1995–2005. It also aims to provide a framework in support of the transition from central management to a decentralised and balanced strategy of regulation and market-based incentives. The action plan lays out strategies for overcoming the environment problems related to water in the Danube river basin. It sets short-, medium- and long-term targets and defines a series of actions to meet these targets.

Despite the diversity of problems, interests and priorities across the Danube river basin, the countries share certain important values and have agreed on principles that underlie the goals and actions of the plan. They include the precautionary principle, the use of best available techniques (BAT) and best environmental practice (BEP) for the control of pollution, the control of pollution at source, the polluter-pays-principle; and a commitment to regional co-operation and shared information among the partners implementing the action plan.

The action plan has four equally important goals:

- Reduce the negative impacts of activities in the Danube river basin on riverine ecosystems and the Black Sea.
- Maintain and improve the availability and quality of water in the Danube river basin.
- Establish control of hazards from accidental spills.
- Develop regional water management co-operation.

The approaches to be taken are set out in a series of strategic directions covering key sectors and policies, including phased expansion of sewerage and municipal wastewater treatment capacity; reduction of discharges from industry; reduction of emissions from agriculture; conservation, restoration and management of the wetland and flood plain areas of the tributaries and main stream of the Danube river basin; integrated water management; environmentally sound sectoral policies; control of risks from accidents; and investments.

IX.5 Problems and priorities

Five priority problems that affect water quality, water use and ecosystems were identified in the basin. These were:

- Microbiological contamination.
- Contamination with substances that enhance the growth of heterotrophic organisms and with oxygen-depleting substances.
- High nutrient loads and eutrophication.
- Contamination with hazardous substances including oil.
- Competition for available water.

Table IX.3 indicates the relationship between these five water management problems in the Danube river basin and the primary water uses of drinking water, fisheries, industry, irrigation and recreation.

Microbiological contamination is probably the most important health-related water quality problem in the region. The generally agreed conclusion, based on available data, is that the Danube and its tributaries are heavily polluted with faecal bacteria and viruses in most river reaches. The overall

Table IX.3 Relations between key water management problems and the primary water uses in the Danube river basin

Problem	Drinking water supply	Fisheries	Industry	Irrigation	Recreation
Nutrient load and eutrophication; Factor: nitrogen and phosphorus; Sources: municipal wastewater, industry, agriculture	Increased cost of treatment; consumer acceptance problems; nitrate contamination of groundwater	Loss of sensitive species	Increased cost of treatment and reduction in some uses, e.g. cooling		Degradation of environmental quality and loss of opportunities and benefits
Hazardous substances, including oils; Sources: industry, agriculture, transport	The presence of these pollutants in significant concentrations would seriously affect drinking water, fisheries and the riverine ecosystems. However, present data and monitoring systems are inadequate to establish current levels in most areas of the basin and to determine the overall priority for dealing with these pollutants. At the local level, serious problems may already exist in some tributary river basins. Metals and some micro-pollutants that are readily absorbed onto fine particles may be stored in the sediments trapped by the numerous hydraulic structures in the Danube basin				
Microbiological pollution; Factor: bacteria, viruses, etc.; Sources: municipal wastewater, livestock, lack of adequate sanitation	Renders surface waters and groundwater unfit for water supply or increases the cost of treatment		Increases cost of treatment in some types of processes, particularly food processing	Water unfit for certain crops	Loss of opportunity, including elimination of some uses such as bathing and other contact activities
Growth of heterotrophic organisms and oxygen depletion; Factor: Organic matter, ammonia; Sources: municipal wastewater, industry, livestock	Surface water unfit for water supply; reduced groundwater infiltration and lower quality water	Severe loss of habitat when O_2 conc. drop below minimum required; fish loss due to toxic conc. of ammonia	Increased cost of water treatment	Modern irrigation equipment may clog	Loss of opportunity and economic benefits
Competition for available water; Factor: water planning, allocation, and operation; Sources: sectorial authorities	Reduced or intermittent supplies	Loss of habitats; disrupted migration and spawning patterns	Reduced or intermittent water supply	Reduced water supply during the critical crop growth period	Loss of opportunity and economic benefits

Source: Strategic Action Plan for the Danube River Basin, 1994

situation is that the Danube should not be used as a drinking water source without treatment, such as extensive sand filtration, and that bathing in the river should be discouraged. Current health statistics are believed to record only a limited number of the actual incidents of water-born diseases. Some information suggests that there are a number of epidemics each year and that thousands of people in the basin suffer each year from water-born diseases including dysentery, hepatitis A, rotavirus and cholera.

Microbiological contamination is normally a local problem, because most pathogens have a limited survival time in water. However, there are reported situations where regional or transboundary impacts occur such as in the Koros river flowing between Romania and Hungary.

Hazardous and toxic substances are of particular concern, particularly pesticides, other organic micropollutants such as PCBs and polyaromatic hydrocarbons (PAH), and heavy metals. There are serious concerns about pollutants accumulated in sediments in reservoirs and in river reaches down-stream of industrial areas. A survey of 55 sites in 1991, along the Danube River, revealed that 23 of these sites should be treated as hazardous waste. The main sources of such pollution are industry and mining.

Transport activities appear to be important sources of oil pollution, and the main source of lead, to the Danube and its tributaries. The transport of oil in pipelines has also created continuos and accidental spills into the rivers of the basin. The most recent accident occurred when an oil pipeline in the Ukraine led to contamination of the River Tisza and threatened water supplies in Ukraine and Hungary.

Diffuse discharges from agriculture are important sources of micro-pollutants. About 300–500 different active agents of pesticides have been used in the basin.

Serious health concerns also exist due to the high levels of nitrogen found in drinking water and that can lead to methaemoglobinaemia. High levels of nitrate have been reported in groundwaters from aquifers in several parts of the basin, particularly in the intensively cultivated areas of Hungary, Romania and Slovakia. The nitrate level in the Danube has increased four to five times in recent years. If this is allowed to continue, the region will face a serious health problem.

Organic materials discharged into a water body enhance the growth of heterotrophic organisms which consume the available dissolved oxygen. This can lead to changes in natural biodiversity as has been observed in some Danube tributaries; for example, the Vit River in Bulgaria is unable to support fish downstream of the city of Pleven, primarily due to the discharges from a sugar factory.

Competition for available water is a serious problem in some regions of the Danube river basin, particularly in Hungary and the tributaries in Romania and Bulgaria. The numerous diversions of water, combined with a large seasonal variation in flow, often result in a water supply shortage. A number of reservoirs have been constructed on the tributaries but the allocation of the available water resources among the users causes many conflicts and problems for reservoir operation. The challenges and problems of multipurpose water allocation have been growing in recent years because of a 10-year drought experienced in the lower Danube region. The city of Sofia is now supplied from the bottom of an almost empty reservoir and suffers from severe shortage in water supply. In addition, the water quality does not meet current standards, but no alternative is available.

Practices and policies in different sectors can be a cause of environmental problems or a constraint to effective action. Some of the sources of the pollution problems, and therefore the water quantity problems, result from the activities of cities, rural towns and villages, industry, energy production and transport, and agriculture.

In all sectors, the key actions required to bring about change must come from the public authorities, public and private enterprises, NGOs and the general public (as both citizens and as consumers). The relationships between these "actors" and the principal sectors and sources of pollution in the Danube river basin are outlined in Table IX.4.

IX.6   Strategic directions

The action plan provides long-term strategies and direction for developing detailed measures and programmes in each sector, and for the necessary management infrastructure and institutions that will be needed. The impact of the plan will be incremental and its success will be measured in step-by-step improvements.

Achieving the goals of the plan will occur through sustained and integrated action in the long-term. Although the countries in transition have seen declines in industrial production and changes in the agriculture sector that have resulted in reduced emissions and nutrient run-off, the resulting improvements in water and environmental quality may be only short-lived once economic activity in the countries picks up again. Unless there is a concerted effort to promote modernisation and restructuring in the industrial sector, based on cleaner technologies and production processes, and a policy shift in the direction of a more sustainable agriculture, the recent

Table IX.4 Management actions required by the three groups involved in use and control of water resources according to the main sectors and sources of pollution in the Danube river basin

Actors	Actions required in			
	Cities	Rural towns and villages	Industry	Agriculture and livestock
Public authorities	Invest in infrastructure. Establish standards of drinking water service. Insure adequate tariffs. Optimise water allocation and distribution	Manage sanitation and drinking water protection programmes. Optimise water allocation and distribution	Regulate hazardous waste. Regulate waste water discharges. Administer effective water and pollution fees. Optimise water allocation and distribution	Administer training and extension programmes. Administer effective water fees. Optimise water allocation and distribution
Public and private enterprises	Operate wastewater treatment facilities. Pre-treat industrial waste	Control seeping from solid waste disposal into groundwater. Dispose of hazardous waste safely	Pre-treat industrial waste. Reduce and treat industrial waste	Adopt imported practices for use of fertilisers and agrochemicals. Manage livestock manure
General public and NGOs	Pay for service. Conserve water. Adopt environmental consumption standards. Manage household hazardous wastes. Support effective regulations	Pay to protect drinking water sources. Adopt environmental consumption standards. Manage household and farm wastes. Support effective regulations	Support water quality objectives. Support effective regulations	Support water quality objectives. Manage livestock manure. Promote organic farming. Support effective regulations

Source: Strategic Action Plan for the Danube River Basin, 1994

improvements in the middle and lower basin of the Danube will be short-lived. Progress in such areas as municipal wastewater treatment and control of industrial emissions, has apparently been much greater in Austria and Germany. However, the quality of the Danube and several of its tributaries in the upper basin suggests that far more must be done to achieve reasonable ambient water quality objectives.

An effective water management system requires an efficient monitoring strategy. Nearly all the countries in the basin need to improve their existing monitoring systems. In the meantime the countries have agreed to harmonise monitoring and assessment methods, to develop joint monitoring systems, to implement joint programmes, and to elaborate an interconnected data base management system. The international monitoring system that was being developed and initiated in 1993, funded by the EPDRB, consists of 224 stations for meteorological, hydrological, water and sediment quality monitoring.

The lessons learned during the implementation of the EPDRB and the development of the strategic action plan show that the institutional and the policy issues are fundamental to its success. There are three important participants in that process: public authorities, public and private enterprises, and the general public and NGOs. The public authorities have to play the critical role as regulators and facilitators. The greatest contribution to its success may come from a sound institutional and policy framework, including modern laws, water management practices and administrative arrangements. The policy framework varies considerably throughout the basin. The following five key areas indicate where institutional and policy reform could have broad beneficial impacts on water management in the basin:

- Realistic and achievable emission limits and water quality standards. In central and eastern European areas of the basin, the water quality standards on which discharge limits are based are, in some cases, too stringent to be measured and, in many other cases, they are too stringent to be met in. They are, nevertheless, arguably correct from a scientific point of view. The result is that these standards have often been ignored because of the technical and financial difficulties in achieving them. The development of a coherent system of water classification, of water quality objectives tailored to meet local needs, i.e. the water uses to be supported in a particular river reach, and of water quality standards, would provide a better basis for water management and regulation.

- Implementation and enforcement. The choice of approach for implementing water quality objectives and standards is often represented as the choice between the "command and control" approach and the "market based" approach. The former has been relied upon heavily in the past in western countries in the basin. A combined approach of water quality standards, discharge limits for individual facilities and financial instruments can be most effective in bringing about improvement in an industrial enterprise's environmental performance.

- Incentives and disincentives. Nearly all countries in the basin have some form of discharge fee and penalty system in place. Water-use and pollution charges can be used as incentives for large or medium-sized industrial and municipal wastewater treatment plants to improve their performance. In the past, these charges or fines have usually been too low to cause any change in behaviour by enterprises.
- Monitoring and information systems. Information is needed to develop integrated water management plans, to assess ambient water quality, to monitor wastewater discharges, to implement and enforce laws and regulations, and to inform the public and decision-makers about the state of the environment and the performance of specific facilities. The preparation of the action plan has been notably weakened in key areas because of the lack of appropriate, consistent and reliable data. No adequate baseline exists against which to measure progress towards the action plan's objectives. The current monitoring systems in nearly all the basin countries are not able to support more effective integrated management systems.
- Integrated regional or river basin planning. Central planning and resource allocation were an important feature of water and environmental management in most of the central and eastern European countries in the basin. This has been abandoned for a decentralised approach without the benefit of sufficient time to develop and strengthen institutions at the district and local level so that they are able to carry out such planning. This is not just a problem of creating local capacity, however, because the nature of planning itself must change. The new approach must stress the integration of all sectors and objectives. It must be based on the application of benefit–cost or least-cost analysis, and it must rely on much greater participation on the part of the general public and other concerned groups.

Public and private enterprises both play an important role in some of the water resources management problems. Environmental audits, waste minimisation and demand management can provide a basis for the preparation of step-by-step, low-cost programmes for environmental improvement in industrial facilities. Waste minimisation could be aided by a programme of environmental improvement that includes training, changes in technology to avoid the generation of waste, and "win–win" solutions to reuse wastes. Demand management will decrease considerably water consumption and will provide the basis for increasing water use efficiency.

The general public and NGOs will play an important role in raising public awareness of, and participation in, governmental decisions about resource management and land use. These are vital to sustaining a political commitment to sound water policies. A strong base of support for the action plan will depend on developing mechanisms for the effective participation of the general public and concerned groups in the policies to be developed and the actions to be carried out under the action plan.

IX.7   Conclusions

Four key activities are required to support the proposed action in this large international river basin which covers countries with different economical, political and social conditions. These activities are aimed at water resources management and consist of:

- Enhancing regional and international co-operation.
- Applying an integrated river basin approach.
- Mobilising national financial and human resources.
- Obtaining support from international organisations and financial institutions.

The management process will be complex because it will be necessary to integrate capacity building at the same time as the operations are being improved, operators are being trained, and new equipment is being procured and installed. Furthermore, the action will need to be sustained over at least 20 years.

The cost of the programme is likely to be very large. This fact alone will challenge the political decision-makers in each country who must decide priorities. Much has to be done and the programme has given the impetus required to make a good start. It is hoped that the national action plans will facilitate the commissioning of new wastewater treatment plants and the development of policy, legislature and enforcement mechanisms, as well as encouraging attention to diffuse pollution in order to protect the Black Sea, which is badly needed.

IX.8   References

Task Force 1994 *Strategic Action Plan for the Danube River Basin*, December, 1994. The Task Force for the Programme.

IUCN 1994 *Analysis and Synthesis of National Reviews*. IUCN European Programme. Final report. The World Conservation Union, Gland.

Haskoning 1994 *Danube Integrated Environmental Study*. Final Report.

MOSCOW REGION, RUSSIA

X.1   Introduction

The Russian Federation state report "Drinking Water" issued in 1994, highlighted among other things the ongoing deterioration in water quality and in the reliability of the Moscow region drinking water supply (State Report, 1994). Almost all studies undertaken in the region in recent years, have indicated that the inadequate technical and sanitary condition of the water sources could lead to risks to human health for the population in this vast and important area (Anon, 1992). These potential problems with the Moscow region drinking water supply are of great concern to the Government of the Russian Federation, the authorities of Moscow City, Moscow, Smolensk and Tver Oblasts, to the mass media and to non-governmental organisations (NGOs). However, it was recognised that the policy and strategy for improvement of the region's water supplies should be based on a comprehensive and environmentally-sound approach.

The Ministerial Conference on Drinking Water and Environmental Sanitation "Implementing UNCED Agenda 21" (held in Noordwijk, the Netherlands, March 1993) set forth guiding principles for safe drinking water supply schemes, thereby providing the basis for immediate action by national government and supporting agencies and institutions (UNDP, 1994). In formulating the "Program of Water Quality Improvement in the Sources of Moscow Drinking Water Supply", which was completed in 1994, some efforts were made in the Moscow region to apply those principles in order to attain overall environmental quality and sustainable development objectives.

X.2   Description of the region

The Moscow region (Figure X.1) is located in the western part of the Upper Volga river basin and encompasses the catchment areas of the Volga river with its tributaries (from headwater down to the Ivankovskoye Reservoir

---

*   *This case study was prepared by V.A. Vladimirov*

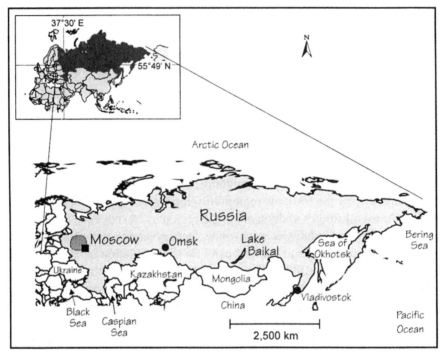

Figure X.1 Location map of the Moscow region

dam) and the Moscow river (from headwater down to Moscow City). Admin-istratively, the area comprises part of the territories of Moscow, Smolensk, Tver Oblasts and Moscow City and covers about 55,000 km². The region has a moderately continental climate with an average annual air temperature of 3–4 °C and an average annual precipitation of 720–800 mm. The evaporation rate in the area is 550–575 mm a^{-1} and is estimated to be 70–80 per cent of total precipitation.

Geologically, the catchment area is composed of mineral coal and Permian and Jurassic bedrocks (limestones, sands and clays). The surface layers consist of glacial and Holocene formations, i.e. boulder loams and sandy loams, as well as lacustrine and fluvioglacial sands and clays. The soils are predominantly of the soaic/podzolic type with medium and light sand loams being prevalent. In depressions in the landscape, marshy and peat bog soils occur. The landscape is mainly low hills of 150–300 m above sea level. The areas of small, flat-top hills (Valdai, Smolensk-Moscow) and ridges (Gzhatsk-Mozhaisk, Klin-Dmitrov) are separated by shallow river valleys and plains. Forest cover in the watersheds varies from 66 per cent of the area at the Volga River headwater to 39 per cent at the site of the Ivankovskoye

Reservoir dam. Native birch and aspen tree forests are dominant but pine tree forests cover the junction area of the Volga, Tma and Tvertsa rivers.

Within the basins of the Vazuzskaya and Moskvoretskaya Water Systems there are 10 administrative districts, nine cities and towns and 19 settlements. The total population is 922,100 people, including 576,000 urban and 346,100 rural residents. In the basin of the Volzhskaya Water System there are 17 administrative districts, 11 cities and towns and 22 settlements with a total population of 1,176,500 people, of which 885,900 are urban and 290,600 are rural residents. The average population density in those basins is 71 persons per km^2 but in the rural area it is only 27 persons per km^2. In total, the water systems of the region provide drinking water to approximately 14 million people, including those of Moscow City.

The economic infrastructure of the region is well developed with numerous types of industry (ferrous and non-ferrous metallurgy, metal works, machinery, electronics, construction, chemical/petrochemical, power generation, mining and textiles), agriculture (crops, livestock, processing of agricultural products), and transport (automotive, railways, inland navigation, airways).

X.3 Water systems

The drinking water supply of Moscow City and the adjacent vicinity is maintained mostly by surface water conveyed from the Vazuzskaya, Moskvoretskaya and Volzhskaya water systems, located within the territories of the Smolensk, Moscow and Tver Oblasts (Figure X.2). The water systems comprise rivers, lakes, reservoirs, canals and hydraulic units. The major water bodies are listed in Table X.1.

The Vazuzskaya water system augments the water supplies and provides water for inter-basin transfer to the Moskvoretskaya water system and to the Volga river. The system incorporates linked canals, pumping and hydropower plants. The reliable water supply (i.e. 95 per cent probability for the year) is 19 m^3 s^{-1} for the Moskvoretskaya water system, 5 m^3 s^{-1} for the Volga river and 1 m^3 s^{-1} for local water use. The Moskvoretskaya water system comprises the Moscow river and its tributaries, with storage reservoirs and diversion dams, and provides a reliable water supply of 51 m^3 s^{-1} (with 95 per cent probability for the year) and 46 m^3 s^{-1} (with 97 per cent probability for the year). The system conveys water for the Rublevskaya and Western water supply plants with a total diversion of 96.7 m^3 s^{-1}.

The Volzhskaya water system includes the Volga river headwater, linked lakes, reservoirs and the Moscow Canal with a series of tandem reservoirs, pumping plants, navigation locks and other structures. The reliable water

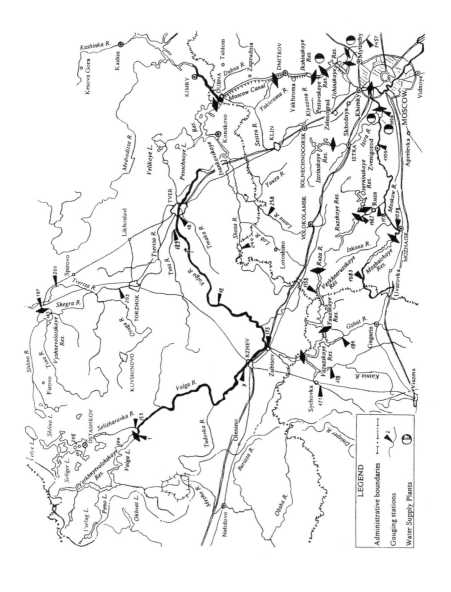

Figure X.2 Map of the basins of the Moscow region that are used for drinking water supplies

Table X.1 The major water bodies of the Moscow region water systems

Vazuzskaya System		Moskvoretskaya System		Volzhskaya System	
Water body	Location (Oblast, City)	Water body	Location (Oblast, City)	Water body	Location (Oblast, City)
Rivers					
Vazuza	Smolensk/Tver Oblasts	Ruza	Moscow Oblast	Volga	Tver Oblast
Osuga	Tver Oblast	Moscow	Smolensk/Moscow Oblasts	Donkhovka	Moscow/Tver Oblasts
Kasnya	Smolensk Oblast	Lusyanka	Moscow Oblast	Doibitsa	Tver Oblast
Gzhat	Smolensk Oblast	Koloch	Moscow Oblast	Shosha	Tver/Moscow Oblasts
Ruza	Smolensk Oblast	Ozerna	Moscow Oblast	Kotlevlya	Tver Oblast
Yauza	Smolensk Oblast	Istra	Moscow Oblast	Lama	Moscow/Tver Oblasts
				Iksha	Moscow Oblast
				Ucha	Moscow Oblast
				Klyazma	Moscow Oblast
				Tvertsa	Tver Oblast
				Shlina	Tver Oblast
Lakes					
				Dolgoye	Tver Oblast
				Vitbino	Tver Oblast
				Seliger	Tver Oblast
				Shlino	Tver Oblast
				Velikoye	Tver Oblast
				Pesochnoye	Tver Oblast
Reservoirs					
Vazuzskoye	Smolensk/Tver Oblasts	Ruzskoye	Moscow Oblast	Verkhnevolzhskoye	Tver Oblast
Yauzskoye	Smolensk Oblast	Mozhaiskoye	Moscow Oblast	Vyshnevolotskoye	Tver Oblast
Verkhneruzskoye	Moscow Oblast	Istrinskoye	Moscow Oblast	Ivankovskoye	Tver Oblast
		Ozerninskoye	Moscow Oblast	Yakhromskoye	Moscow Oblast
				Pestovskoye	Moscow Oblast
				Ikshinskoye	Moscow Oblast
				Pyalovskoye	Moscow Oblast
				Klyazminskoye	Moscow Oblast
				Khimkinskoye	Moscow Oblast/Moscow City
Canals					
Yauza-Ruza Canal	Smolensk/Moscow Oblasts			Moscow Canal	Tver/Moscow Oblasts
Gzhat-Yauza Canal	Smolensk Oblast				

supply from the system is 82 m^3 s^{-1} with 95 per cent probability for the year and 78 m^3 s^{-1} with 97 per cent probability for the year. The total diversion to the northern and eastern water supply plants from the reservoirs of the Moscow Canal is 36.6 m^3 s^{-1}. The major facilities of the water system were completed in the period 1935–67 and they now require technical restoration and/or remodelling.

According to the latest inventory, 92 wastewater treatment plants are located in the area, 17 of which use mechanical, four use physical/chemical and 71 use biological technology. Beginning in 1978, biological wastewater treatment facilities were constructed in almost all cities and towns but their combined total capacity is only about 75 per cent of that required and the treatment efficiency does not comply with the existing standards for water sources of Class II (all drinking water supply sources are sub-divided into three classes depending on their quality). The wastewater of Moscow City is treated at the Kuryanovskaya and Luberetskaya secondary biological treatment stations which discharge treated effluents to the Moscow river downstream of the city.

In the period 1940–84 numerous statements were issued by the Council of Ministers of the Russian Federation devoted to the problems of setting up water protection zones and zones of sanitary protection, of alleviating wastewater pollution, and of improving the technical and sanitary condition of reservoirs and water systems. At present, the legal acts most applicable for enforcing the improving and maintaining of the drinking water supply to the required quality are as follows:

- Existing laws of the Russian Federation, namely "On Protection of the Natural Environment", "On Sanitary and Epidemiological Welfare of the Population", "On Protection of Consumers Rights" and "On Local Administration".
- Draft laws, namely "Water Code of the Russian Federation".
- The Land Code of the Russian Federation.

The operation and maintenance of the Vazuzskaya and Moskvoretskaya systems and water supply plants are carried out by systems operation offices which come under the Moscow municipal enterprise "Mosvodokanal". The Moscow Canal is operated by "Rechflot". The overall organisational management of water users, and of enforcement and compliance related to water use in the area is the responsibility of the Moscow Oka Basin Water Management Office which comes under the Committee on Water Management. The Oblast Committees on Environmental Protection under the Ministry of Environmental Protection and Natural Resources, and the local offices of the State Committee on Sanitary and Epidemiological Survey are responsible for compliance with existing regulations and standards related to wastewater discharges, water quality and other environmental and human health issues. The water quantity and quality monitoring networks are operated by the Federal Survey on Hydrometeorology and Environmental Monitoring and by Mosvodokanal.

X.4   Water resources assessment

The river network of the area is well developed with a density of 0.12–0.35 km km^{-2}. Snowmelt is the principal source of stream flow. The annual average run-off is 6.5–9.0 l s^{-1} km^{-2}, with the spring flood flows accounting for 40–60 per cent and summer flows for 10–20 per cent of the annual streamflow. The existing stationary hydrological network consists of gauging stations maintaining regular hydrometric and hydrochemical observations. Water quality studies are also carried out using surveillance and monitoring data. The assessment of water resources in the basins of the Moscow City drinking water supply is carried out by approved water management subregions using selected control points (Figure X.3).

The total water resources allocated for Moscow City and diverted from surface water sources amounts to 124 m^3 s^{-1} (with 97 per cent probability for the year), including the conveyance of 73 m^3 s^{-1} of water to the Mosvodokanal system, 45 m^3 s^{-1} released to the Moscow, Yauza, Klyazma and Ucha rivers, 3 m^3 s^{-1} released to the Cherkizovskaya system and conveyance losses from the Moscow Canal of 3 m^3 s^{-1}. Of the total amount of water diverted to the Mosvodokanal system, 35.1 m^3 s^{-1} (55 per cent) is used for domestic needs, 22.5 m^3 s^{-1} (17 per cent) is used for industry and 4.5 m^3 s^{-1} is used for miscellaneous needs. In industry, water is used for domestic and drinking purposes (5 m^3 s^{-1}) and for technological needs (9.7 m^3 s^{-1}).

Apart from water diversions by the Water Supply Plants for the Moscow City water supply, some 40–51 per cent of available resources in the basins are released for ecological, power, navigation and other uses, 24–32 per cent for inter-regional water transfers, 2 per cent for consumptive water use and losses, and 1–2.6 per cent for filling reservoirs.

At present, the total water withdrawals from groundwater sources (the Moscow artesian basin) are about 5 m^3 s^{-1}. With the aim of improving the reliability of the Moscow City drinking water supply, the feasibility of using groundwater is being studied, based on the following groundwater withdrawal sites: northern (Klin/Dmitrov/Dubna), western (Ruza/Zvenigorod), southern (Oka) and eastern (Shatura). The total groundwater potential of the

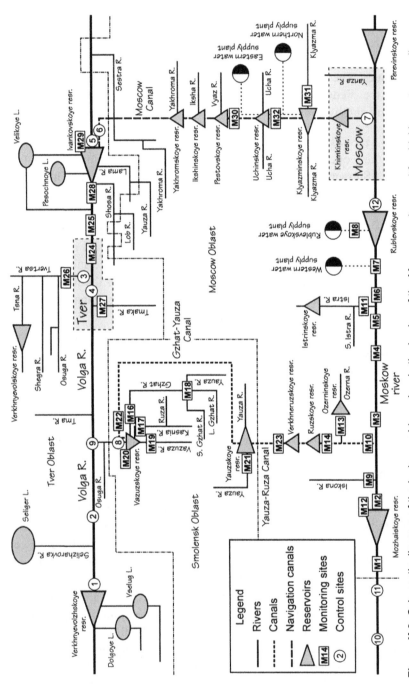

Figure X.3 Schematic diagram of the hydrographic regime in the basins of the Moscow region used for drinking water supplies

sites has been assessed as $41.8 \text{ m}^3 \text{ s}^{-1}$. As a first step, it was recommended to commence use of the southern and northern groundwater sites with prospective resources of 8.5 and $9.3 \text{ m}^3 \text{ s}^{-1}$, respectively.

Water quality is assessed using sampling data for a defined set of physical and biological indicators using the appropriate water quality standards of the Russian Federation. It is, however, noticeable that assessments made by different agencies and institutions sometimes differ as a result of the uncoordinated sampling and methods applied. The conclusions of the sanitary and epidemiology survey was that the Vazuza system water conformed to Class II.

Water quality trend analyses compiled for the Moskvoretskaya system showed that in the 14 years prior to 1992 the average annual concentrations of heavy metals increased 2–5 times and nitrates by 5 times. When compared with maximum allowable concentrations (MAC) the following increases were also observed at the Rudlevskaya Plant site: phenols (8–12 times the MAC), oil products (2–5 times the MAC), and severe microbial pollution (coliform index of 100,000). The water source shows extensive eutrophication, with permanent odour and colour, especially in the spring period, that excludes it even from Class III.

The water of the Volzhskaya system normally does not exceed MACs but elevated concentrations of metals and phosphorus-based organic pesticides have occurred during floods and during the growing season. An integrated assessment puts the Volzhskaya system in Class II. According to the data obtained from the system by Mosvodocanal and the sanitary and epidemiology survey, an integrated toxicity indicator for Class 1 and 2 hazardous substances exceeds the prescribed standards for all water treatment and supply plants.

X.5   Pollution sources

Serious anthropogenic impacts on the water bodies and watersheds of the region imply increasing concentrations of contaminants in the sources of drinking water supplies. Point sources of pollution in the basins of the Moscow City drinking water supplies come mainly from industrial, municipal and agricultural wastewater discharges. According to the state water use accounting data, in 1992 $1{,}917.3 \times 10^6 \text{ m}^3$ of wastewater effluents were discharged into surface water bodies in the area, including $147.5 \times 10^6 \text{ m}^3$ of untreated and inadequately treated wastewater. Diffuse sources of pollution arise mostly from:

- Contaminated precipitation falling within watersheds.
- Soil leaching and erosion.
- Run-off containing fertilisers, pesticides and herbicides.

Table X.2   Total pollution loads for selected variables arising from all sources in the Moscow City drinking water supply systems in 1992

Variable	Total pollution load (t a^{-1})
BOD$_{total}$	1,440,000
Chlorides	105,031
Chromium	90
Grease/oils	1,950
Hydrogen sulphide	87
Iron	900
Potassium	4,850
Nitrogen, ammonium	16,012
Nitrogen, nitrate	1,657
Nitrogen, nitrite	2,501
Oil products	4,452
Phosphorus, total	6,012
Sulphates	90,698
Suspended solids	330,015
Synthetic surface active substances	331
Zink, nickel, cadmium, copper	291

- Run-off, for example, from construction sites, dumps, mining pits, solid waste disposal sites, fertiliser storage, toxic chemical warehouses and leaks from oil and gasoline storage.
- Nutrients in drainage from livestock farms and poultry factories.

Existing pollutant loads from all sources are given for major water quality variables in Table X.2.

X.6   Major problems

The problems affecting the reliability of the quality of the drinking water supply for Moscow City and the surrounding area are as follows:

- Inadequate enforcement and inadequate legal acts and regulations relating to water, including economic instruments.
- Weak institutional and organisational infrastructure for the efficient operation of water systems in relation to environmental and human health issues.
- Inadequate technical and sanitary conditions of the water systems.
- Lack of compliance with the required controls on human activities in water protection zones, riparian belts and sanitary zones.
- Lack of contemporary wastewater treatment facilities for industries, municipal storm sewer systems and other problem areas.
- Improper operation of livestock and poultry farms, and agricultural processing plants that are inappropriate for the watershed environment.

- Current agricultural practices involving the widespread application of mineral fertilisers and toxic chemicals.
- Unsatisfactory condition of the existing water quantity and quality monitoring and assessment network.

X.7   The programme

In October 1993, the Moscow City Government, the Administration of Moscow, Smolensk and Tver Oblasts, the Ministry of Environment Protection and Natural Resources and the Committee on Water Management concluded the Agreement on Joint Water Resources Use and Conservation in the Basins of Moscow City Drinking Water Supply in the Territories of Moscow, Smolensk and Tver Oblasts. Clause 5 of the Agreement stated that *"... a long-term planning document shall be in the form of 'Federal Program of Water Quality Improvement in the Sources of Moscow City Drinking Water Supply', formulated on the basis of regional programmes proposed by Moscow, Smolensk, Tver Oblasts and the Moscow City"*. The Program of Water Quality Improvement in the Sources of Moscow City Drinking Water Supply was also initiated in accordance with the Environmental Action Plan of the Government of the Russian Federation for 1994–95, approved by Government Statement No. 496 of 18 May 1994.

The Program of Water Quality Improvement in the Sources of Moscow City Drinking Water Supply was prepared in 1994 by the Committee on Water Management, the Moscow City Government and the administrations of Moscow, Smolensk and Tver on a collaborative basis as a sub-programme of the Federal programme Water Resources Conservation and Rational Use in Moscow City and Enhancement of its Water Supplies for the Period up to 2010. The Moscow-Oka Basin Water Management Office of the Committee on Water Management of the Russian Federation will be responsible for the general management of the programme. The general manager, jointly with the regional managers of Moscow City, takes responsibility for the implementation and co-ordination of the programme under the supervision of the Expert Council organised in accordance with the Clauses 8 and 10 of the above mentioned agreement.

X.7.1 Programme objectives and scope of activities

The major objectives of the programme comprise the development of efficient measures on:

- Protection of drinking water sources from pollution.

- Restoration and management of the water quality of water supplies, with the aim of reliable delivery of safe drinking water to the populations of Moscow, Smolensk and Tver Oblasts.

The programme activities are grouped into the following categories:

- Measures to protect Moscow City's drinking water sources from pollution, i.e. planning and setting up water protection zones, including the relocation and remodelling of livestock farms and poultry factories, mineral fertiliser and toxic chemical warehouses and other agricultural units, the introduction of new agricultural practices for the rational application of fertilisers and pesticides, and the construction and rehabilitation of wastewater treatment facilities.
    - Water protection measures such as enforcing compliance with regulations by economic enterprises in water protection zones, riparian belts and sanitary zones.
    - Control of wastewater pollution to drinking water sources arising from (a) industrial, agricultural and municipal wastewater, and (b) stormwater from urban and other residential areas.
- A water quality monitoring system: 10 monitoring sites in the Vazuzskaya system, 19 in the Moskvoretskaya system, 11 in the Volzhskaya system, 10 additional hydrometric gauging stations and a water quality centre.
- An automated management system for water conservation: telemetry, computer networks, data banks, simulation modelling and decision support systems.

X.7.2 Implementation and estimated cost and efficiency

The total cost of implementing the programme was estimated as 666.94×10^9 roubles (at 1994 exchange rates) of which 375.25×10^9 roubles would be allocated from the Federal budget and 291.69×10^9 roubles would be allocated from the Oblasts and Moscow City budgets. The remainder would come from enterprise funding and non-budgetary sources. In an evaluation of economic efficiency, the investment return period was estimated at four and a half years.

The implementation of the programme was envisaged for the period 1995–2000. A set of priority measures were included in an immediate action plan comprising reduction of wastewater pollution loads from municipal sewerage works, industrial and agricultural plants and other point sources, and planning of water protection zones. The implementation period for this plan was 1995–97.

In assessing the efficiency of the proposed programme activities two water quality scenarios were used:

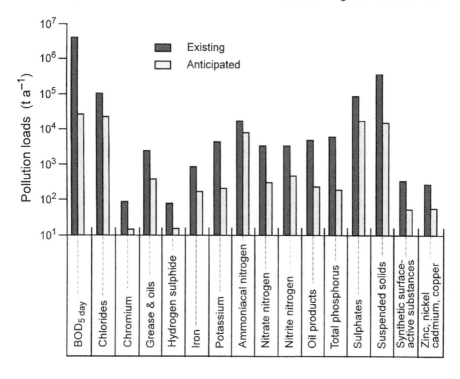

Figure X.4 Existing and anticipated pollution loads following the implementation of the Immediate Action Plan in the basins of the Moscow Region used for drinking water supplies

- Retention of existing water use and conservation trends and practices at the present level.
- Integrated approach to water quality and watershed management.

Water quality forecasts compiled for both alternatives clearly identified that the second scenario could provide a viable basis for attaining the programme objectives in a definite time-frame and for reducing contamination by 40–50 per cent. Existing and anticipated (target) pollution loads from all sources are illustrated in Figure X.4.

X.8   International co-operation

A co-operative programme Improved Drinking Water Protection and Management for the Moscow Region is being implemented as a partnership between Russia and the USA under the auspices of a Joint Commission of V. Chernomyrdin of the Russian Federation and A. Gore of the USA which was established in December, 1993. This programme has two major pilot projects which focus on the protection and enhancement of drinking water supplies in the Moscow region:

- Small River Watershed Management, Moscow River Basin, Istra District.
- Improved Wastewater Compliance and Enforcement, Moscow, Tver and Smolensk Oblasts.

The first project is mostly orientated towards reducing pollution from agricultural and rural land uses which are causing contamination of drinking water sources from the Istra River located in the Istra District of the Moscow Oblast. It will introduce and disseminate low-cost technology and management practices for controlling agricultural and other rural point and diffuse sources of contamination, i.e. large poultry factories and livestock farms, run-off containing sediments from cultivated land, pesticides and fertilisers, and small settlements and recreational facilities constructed without appropriate sewerage and waste treatment capacities. The second project is focused on the control of point-source pollution from certain facilities in Dmitrov, Tver and Gagarin cities.

The projects are funded through an inter-agency agreement between the US Environmental Protection Agency (EPA) and the United States Agency for International Development (USAID) and are implemented from the USA by EPA Regions 5 and 7, the Iowa State University, the US Department of Agriculture, the US Geological Survey and the Minnesota Pollution Control Agency. The Russian counterparts include the Ministry of Environment Protection and Natural Resources, the Committee on Water Management, the Federal Survey for Hydrometeorology and Environmental Monitoring, the State Sanitary and Epidemiology Survey, and the Ministry of Agriculture and Regional Committees on Water Management and Nature Protection.

Major activities under the programme started in 1994 with agreements formulated for a three-year period. In line with the project objectives, an Agreement on Co-operation in the Istra River Basin Small Watershed Management was signed in 1994 between the EPA, USAID and the involved Russian parties. In order to support programme activities, some additional efforts were made by the EPA to provide water quality laboratory assistance through an application to the USAID Commodity Import Program, filed by the Russian Ministry of Environment Protection and Natural Resources in August 1994. Assistance with the microbiological analysis of drinking water is planned by the USEPA.

Further activities on environmental economics and policy are also underway within the Moscow region. The government of the USA is assisting Russian policymakers with environmental policy issues and sustainable development during the country's transition to a market economy. Initial efforts focus on environmental priority-setting based on:

- Economic incentives for private enterprises.
- The use of cost-effectiveness analyses.
- Techniques for identifying the lowest unit-cost options for reducing risks.

Subsequently, policies and programmes will be developed and carried out based on this priority setting approach. In the meantime, efforts should be made to achieve closer co-ordination between the technical assistance from the USA and the programme.

X.9   Conclusion

The Program of Water Quality Improvement in the Sources of Moscow City Drinking Water Supply could be considered as an effort to create, collaboratively, an instrument for integrated environmental and socio-economic management in an important region of Russia. The programme has been reviewed by the Government of the Russian Federation and is in the early stages of implementation.

X.10   References

Anon. 1992 *Ekologicheskie Issledovaniya v Moskve i Moskovskoi Oblasti. Sostoyanie vodoykh sistem.* Otdelenie obshei biologii RAN, Institut vodnykh problem RAN, Tsentr Ekologicheskikh Proektov, MosvodokanalNIIproekt, (*Ecological Studies in Moscow City and Moscow Oblast, Status of Water Systems*, Department of General Biology of RAS, Institute of Water Problems of RAS, Centre of Ecological Projects, MosvodocanalNIIproekt), Moscow.

State Report 1994 *Voda pityevaya*, Ministerstvo okhrany okruzhayushei sredy i prirodnykh resursov Rossiiskoi Federatsii, Gosudarstvenny Komitet Sanitarno-Epidemiologicheskogo Nadzora Rossiiskoi Federatsii, Komitet Rossiiskoi Federatsii po Vodnomu Khozyaistvu, Gosudarstvenny Komitet Rossiiskoi Federatsii po voprosam arkhitektury i stroitelstva, Komitet Rossiiskoi Federatsii po Geologii i ispolzovaniyu nedr, Federalnaya sluzhba Rossii po gidrometeorologii i monitoringu okruzhayushei sredy (*Drinking Water*, Ministry of Environment Protection and Natural Resources of the Russian Federation, State Committee on Sanitary and Epidemiological Survey of the Russian Federation, Russian Federation Committee on Water Management, Russian Federation State Committee on Architecture and Construction, Russian Federation Committee on Geology and Underground Resources Use, Federal Survey of Russia on Hydrometeorology and Environmental Monitoring), Moscow.

UNDP 1994 *Water and Sanitation for All; A World Priority*. United Nations Development Programme, New York.

Case Study XI[*]

CYPRUS

XI.1   Introduction

Cyprus is situated in the north-eastern part of the Mediterranean Sea, 33° East of Greenwich and 35° North of the Equator (Figure XI.1), and is the third largest island in the Mediterranean with an area of 9,251 km^2, of which 1,733 km^2 are forested, 216,000 ha are cultivated and 38,000 ha are irrigated. Irrigated agriculture contributes more than 50 per cent of the value of the total crop production.

The *de jure* population of Cyprus in 1993 was 722,000 with an annual rate of growth of 1.7 per cent. The economically active population is 46 per cent of the total. Employment in agriculture is continuously declining and in 1993 the proportion of the population engaged in agriculture had fallen to 11.9 per cent. Registered unemployment in 1993 was 2.6 per cent. Life expectancy for males is 74.6 years and for females is 79.1 years (Department of Statistics and Research Development, 1995).

The gross national product (GNP) per capita in 1995 was 6,107 Cyprus pounds (US\$ 14,045) with a rate of increase of 5.6 per cent. The contribution of different sectors to total production is given in Figure XI.2 (Department of Statistics and Research Development, 1995).

XI.2   Water resources

XI.2.1 Surface waters

The availability of water in Cyprus is dependant on the annual rainfall, which varies from 340 mm in the coastal plains to 1,100 mm in the Troodos mountains. The average annual rainfall throughout the island is about 500 mm, equivalent to $4,600 \times 10^6$ m^3. About two thirds of the rainfall occurs during the winter months, December to February. It is estimated that about 80 per cent of the rainfall is lost to the atmosphere by direct evaporation and from the remaining 900×10^6 m^3, about 300×10^6 m^3 enrich the aquifer and

*   *This case study was prepared by I. Papadopoulos*

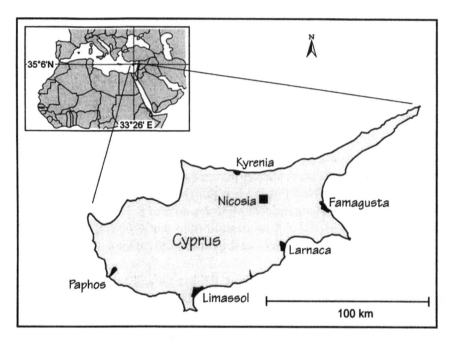

Figure XI.1 Location map of Cyprus

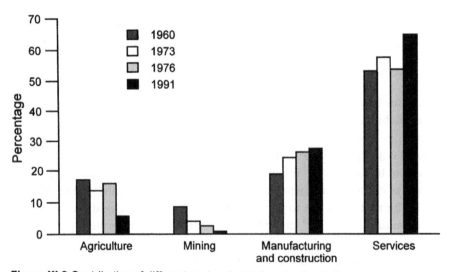

Figure XI.2 Contribution of different sectors to total production in Cyprus

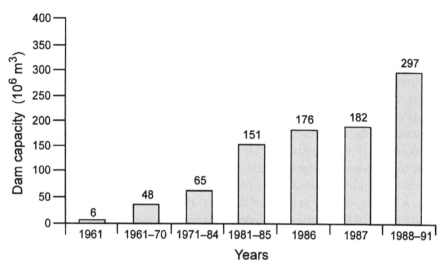

Figure XI.3 Increases in the capacity of dams in Cyprus between 1961 and 1991 (Data supplied by the Water Development Department)

60×10^6 m^3 result in surface run-off. Part of this run-off is used for direct irrigation or is collected in dams and about 260×10^6 m^3 is lost to the sea (Water Development Department, 1989). Projects are underway to divert part of the latter run-off to the dams.

The annual variations in rainfall and snowfall are quite large leading to deficits in water supplies during low rainfall and to floods during high rainfall. When rainfall is only about 360 mm a^{-1} or less, drought conditions occur with negligible run-off and groundwater replenishment. This occurs about once every 16 years. During drought conditions, river flow is drastically reduced thereby affecting available domestic and irrigation water supplies. As a result, Cyprus has embarked on and completed a costly storage dam programme for 297×10^6 m^3 of water (Figure XI.3) which, when considered per unit area of population, is one of the most intensive in the world. Most of the storage dams are integrated into the southern conveyor system which interconnects all important surface water resources from west to east across the island of Cyprus.

XI.2.2 Groundwater

Groundwater is a very important source of water for Cyprus. Water infiltrates directly from rainfall (there is no inflow from outside the island) into confined or unconfined aquifers and can be extracted and used either by pumping or sometimes by gravity feeds in the form of springs. Recently, an aquifer west of the city of Limassol (Akrotiri aquifer) was identified as suitable for partial recharge with treated municipal wastewater produced in the city. Precautionary legal and regulatory actions have been taken to protect the quality of the groundwater, the environment and public health.

Conservation and use of groundwater resources has to be carried out in parallel, and integrated, with surface water resources. Already many aquifers in Cyprus have been seriously over-pumped and their reliable yield has decreased; in many cases the quality of the water has deteriorated and in coastal areas salt water intrusion has occurred.

Due to its extensive storage period, groundwater is ideal as a supplementary water resource in situations of low rainfall and run-off conditions and also as a standby supply in cases of drought. The Government of Cyprus has recently adopted this policy. In order to harmonise the situation in Cyprus with that of the European Union (EU), legal and institutional regulations to control groundwater and its quality are in the final stages of formulation.

The main aquifers of Cyprus are the Western Messaoria, South Eastern Messaoria, Akrotiri and the Kyrenia limestone range. In addition, there are a few minor coastal and some restricted river-valley aquifers. The water from these aquifers is pumped through about 10,000 boreholes and several thousand shallow wells. The total extraction is about 260×10^6 m^3 for irrigation, domestic and industrial purposes. The sustainable yield of the aquifers is about 370×10^6 m^3 and over-pumping occurs in some areas and losses to the sea occur in other areas. The problem of over-pumping has arisen because there are numerous illegal boreholes and uncontrolled withdrawals but in certain areas, such as Famagusta, Morphou, Limassol and Larnaca, salt water intrusion has became a serious problem. In some locations the saline groundwater is not suitable even for the most salt-tolerant crops. The over-pumping has also resulted in most springs drying up.

XI.3   Measures to conserve and replenish groundwater

There are many different groundwater conservation and replenishment measures that can be carried out, many of which have been used in Cyprus, although some of them have not been successful because of inadequate management.

XI.3.1 Conservation and control measures

Aquifers can be declared as conservation areas where laws are enforced for appropriate management action such as:

- Regulation of pumping and the introduction of water meters and efficient conveyance and water application systems.
- Charging of water rates per cubic metre based on the type of crop and quantity extracted.

- Regulation of well drilling, the distance between boreholes and their depth.
- Controlling the water quality of the aquifer.

In order to control water quality, precautionary protection measures and prohibited activities are enforced by law (69/91). The framework of the law defines which substances and/or chemicals are considered toxic or dangerous, as well as pollution by nitrates and other sources (industrial and municipal). Protection measures are proposed for areas recharging groundwater intended for human consumption. These sensitive areas are divided into three zones where different restrictions are imposed. The aim of the law has been to protect groundwater and positive results have been obtained.

XI.3.2 Groundwater recharge

In Cyprus, where favourable conditions occur, groundwater recharge with river flow has been used efficiently for the replenishment of depleted aquifers. As a result, a number of recharge projects have been carried out in Cyprus including:

- Recharge dams, such as at Morphou, Famagusta and Kyrenia.
- Percolation areas downstream of recharge dams such as at Morphou Serrachis valley.
- A recharge canal downstream of the recharge dams and lake at Paralimni.
- An infiltration gallery such as the one traversing the Famagusta aquifer.
- Recharge with treated municipal wastewater.

Recharge of groundwater with reclaimed wastewater not used for direct irrigation is a new concept for Cyprus. Based on the "Cyprus approach" that no water should be allowed to reach the sea, it has been decided that all properly treated wastewater should be used either for irrigation or for groundwater recharge. Moreover, it has been realised that in order to protect the marine environment from eventual pollution and particularly from eutrophication, treated wastewater should not be discharged to the sea. The government of Cyprus has attached particular importance to this because the economy of the country is largely dependent on its tourist industry.

In line with this decision, no effluent from the city of Limassol can be disposed to the sea. Part of the reclaimed wastewater not used for direct irrigation is expected to be used for recharging the Akrotiri aquifer, which is considered to be the third most important in Cyprus. This water could be used subsequently for irrigation in the area. This concept of wastewater use in Limassol has been accepted and commenced in 1996. Under this project, part of the treated effluent (about 20×10^6 m^3 a^{-1} by the completion of the project) will be used for direct irrigation and part will be conveyed to the aquifer for recharge in constructed infiltration basins. Water will be extracted from wells

Table XI.1   Projected water characteristics for reclaimed wastewater from Limassol

Variable	Concentration	Variable	Concentration
BOD_5	2–5 mg l^{-1}	TSS	2–5 mg l^{-1}
NH_3-N	0.5–2 mg l^{-1}	NO_3-N	10–15 mg l^{-1}
Total N	10–17 mg l^{-1}	Phosphorus	5–10 mg l^{-1}
Total coliform[1]	< 2 per 100 ml	Sodium	140–170 mg l^{-1}
Calcium	31–38 mg l^{-1}	Potassium	1–4 mg l^{-1}
Magnesium	34–55 mg l^{-1}	Chloride	34–55 mg l^{-1}
Bicarbonate	239–282 mg l^{-1}	Sulphate	58–64 mg l^{-1}
TDS	300–350 mg l^{-1}		

BOD	Biochemical oxygen demand	[1]	Most probable number
TSS	Total suspended solids		
TDS	Total dissolved solids		Source: CHM HILL, 1992

located downgradient of the basins and will be transferred to the existing irrigation distribution systems. The recharge basins can be used throughout the year for reclaimed water but the recharged water need only be recovered by pumping as required for crop irrigation. Additional water quality variables have been set for the treated effluent used for groundwater recharge with a particular emphasis on nitrogen removal. The treatment plant, therefore, has been designed to provide nitrification and denitrification by which the level of nitrogen in the reclaimed water can be controlled. The projected characteristics of the reclaimed water from the treatment plant are given in Table XI.1. The projected quality of the water should be generally comparable to the quality of the groundwater in the area and is expected to be suitable for the anticipated irrigation and aquifer recharge.

XI.4   Direct use of treated wastewater for irrigation

Water resources in Cyprus are limited and, with the rapid development of urban and rural domestic supplies, conventional water resources have been seriously depleted. As a result, the reclamation and use of wastewater has become a realistic option for providing reliable sources of water to meet shortages and to cover water needs, as well as for meeting wastewater disposal regulations aimed at protecting the environment and public health. However, the use of wastewater could itself be associated with severe environmental and health impacts. Therefore, a multidisciplinary research programme was initiated in 1984 to study the agronomical, environmental and health aspects associated with the use of treated wastewaters for irrigation. Most of the chemical and physico-chemical variables associated with wastewaters have been extensively studied by the Agricultural Research Institute and useful results have been obtained for the rational and environmentally-sound use of these waters (Papadopoulos, 1995). Recently, priority has been given to research on animal feeding and human health aspects. The results indicate that with the treatment level required in Cyprus, with the irrigation technology available and with the code of practice suggested, the health and environmental risks fall within acceptable levels (Jenkins *et al.*, 1994; Papadopoulos *et al.*, 1994).

XI.4.1 Regulatory considerations
In Cyprus, in order to control the treatment and use of wastewater and thereafter to safeguard the environment and public health, very strict guidelines have been formed relating to the quality and the use of treated wastewater (Table XI.2) (Kypris, 1989). In order to allow for the specific situation of Cyprus, these guidelines are more strict than those proposed by the World Health Organization (WHO). In addition, the guidelines are followed by a code of practice intended to ensure protection of public health and the environment even further and they should be considered to be part of the guidelines (Box XI.1).

It is important to stress that when the guidelines were being formulated, the Technical Committee specifically recognised that the conditions affecting the acceptable risk for reuse of reclaimed water may change, that knowledge of real risk may be improved, and that treatment technologies may also be improved in future. Therefore, the Technical Committee considers the guidelines and the code of practice to be open for further modifications based on the latest knowledge and experience gained from ongoing actual use, and from research.

XI.5 Pollution of water resources
Potential pollution of water resources in Cyprus is related to groundwater over-pumping and the intrusion of sea water to aquifers as already discussed above, to wastewater use and to intensive agriculture.

Industrial activities are rather limited in Cyprus and therefore the main sources of pollutants are, and will increasingly be, urban sewage plants and the use of the effluents. For this reason guidelines and a code of practice have been formulated and legally enforced to protect human health and the environment (Table XI.2 and Box XI.1).

Recently, there has been a considerable increase in the application of fertilisers and pesticides in order to enhance agricultural production. In Cyprus, because of the limited amount of agricultural land and the high cost

Table XI.2 Guidelines for the quality of wastewater used for irrigation in Cyprus

Irrigation area	BOD (mg l⁻¹) 80% limit[1]	BOD Max. allowed	SS (mg l⁻¹) 80% limit[1]	SS Max. allowed	Faecal coliforms (No. per 100 ml) 80% limit[1]	Faecal coliforms Max. allowed	Intestinal worm (No. per litre)	Treatment required
Amenity areas of unlimited access	10	15	10	15	50	100	Nil	Secondary, tertiary and disinfection
Crops for human consumption; amenity areas of limited access	20	30	30	45	200	1,000	Nil	Secondary, storage > 1 week and disinfection, or tertiary and disinfection
	na	na	na	na	200	1,000	Nil	Stabilisation using maturation ponds with a total retention time > 30 days or secondary and storage > 30 days
Fodder crops	20	30	30	45	1,000	5,000	Nil	Secondary and storage > 1 week or tertiary and disinfection
	na	na	na	na	1,000		Nil	Stabilisation using maturation ponds with total retention time > 30 days or secondary and storage > 30 days
Industrial crops	50	70	na	na	3,000	10,000	na	Secondary and disinfection
	na	na	na	na	3,000	10,000	na	Stabilisation using maturation ponds with a total retention time > 30 days or second-ary and storage > 30 days

BOD   Biochemical oxygen demand
SS   Suspended solids
na   Not applicable

[1] These values must not be exceeded in 80% of samples per month
Irrigation of vegetables is not allowed. Irrigation of ornamental plants for trade purposes is not allowed.

No substances accumulating in the edible parts of crops and proved to be toxic to humans or animals are allowed in the wastewater effluents.

Box XI.1 Code of practice for treated domestic sewage effluent used for irrigation in Cyprus

1. The sewage treatment and disinfection plant must be kept and maintained continuously in satisfactory and effective operation for as long as treated sewage effluent is intended for irrigation.
2. Skilled operators should be employed to attend the treatment and disinfection plant, following formal approval by the appropriate authority that the persons are competent to perform the required duties, necessary to ensure that the conditions of clause 1 are satisfied.
3. The treatment and disinfection plant must be attended every day and records must be kept of all operations performed.
4. All outlets, taps and valves in the irrigation system must be secured to prevent their use by unauthorised persons. All such outlets must be coloured and clearly labelled to warn the public that the water is unsafe for drinking.
5. No cross-connections with any pipeline or works conveying potable water are allowed. All pipelines conveying sewage effluent must be satisfactorily marked with red tape to distinguish them from domestic water supply. In unavoidable cases where sewage effluent and domestic water supply pipelines must be laid close to each other, the sewage or effluent pipes should be buried at least 0.5 m below the domestic water pipes.
6. The irrigation methods allowed and the conditions of application differ between different plantations as follows:
   a. Park lawns and ornamental gardens in amenity areas of unlimited access:
   - Subsurface irrigation methods.
   - Drip irrigation.
   - Pop-up, low angle, low pressure and high precipitation rate sprinklers. Sprinkling should preferably be practised at night and when people are not around the amenity areas.
   b. Park lawns and ornamental gardens in amenity areas of limited access, industrial and fodder crops:
   - Sub-surface irrigation methods.
   - Drip irrigation.
   - Surface irrigation methods.
   - Spray or sprinkler irrigation is allowed with a buffer zone of about 300 m.
   For fodder crops, it is recommended to stop irrigation at least one week before harvesting. No animals supplying milk should be allowed to graze on pastures irrigated with sewage.
   c. Vines:
   - Drip irrigation.
   - Minisprinklers and sprinklers (irrigation should stop two weeks before harvesting).

Continued

Box XI.1   Continued

d. Trees with fruits eaten raw without peeling:
- Drip irrigation.
- Hose basin irrigation.
- Bubbler irrigation.

No fruits to be collected from the ground.

e. Trees with fruits eaten after peeling, nuts and similar fruits:
- Drip irrigation.
- Minisprinklers (stop irrigation one week before harvesting).

No fruits to be collected from the ground except nuts. Other irrigation methods could also be considered.

7. In order to meet the required standards, tertiary treatment is essential when treated effluent is intended for irrigation of fruit trees and amenity areas of unlimited access. The following tertiary treatment methods are acceptable:
- Sedimentation by storage for not less than 30 days in open basins without agitation.
- Coagulation plus flocculation followed by rapid sand filtration.
- Any other method which may secure the total removal of helminth eggs and reduce faecal coliforms to acceptable levels.

8. Appropriate disinfection methods should be applied when sewage effluents are to be used for irrigation. In the case of chlorination the residual free or total level of chlorine in the effluent should be equal to or more than 0.5 and 2 mg l^{-1} respectively at the point of use.

9. Suitable facilities for monitoring the essential water quality variables, should be available at the treatment site.

of labour and water, increases in production through the application of fertilisers and pesticides have become very important, but have also resulted in some groundwater pollution. Pollution of groundwater by nitrates is becoming a serious problem in areas of intensive agriculture. Measures are taken by the Ministry of Agriculture to minimise the application of fertilisers and pesticides and a code of practice concerning fertilisers and other chemicals, similar to the code of practice for wastewater use, has been formulated. Farmers are advised to apply suitable fertilisers and other chemicals using an appropriate method at the best time of year and to keep their fertiliser and pesticide activities away from rivers and open wells.

XI.6 Conclusions and recommendations

Based on experimental data and on the application of wastewater for irrigation in Cyprus, the following conclusions and recommendations can be put forward:

- Wastewater, when properly treated, used and managed, could be considered as an additional, innovative and reliable water resource with particular application in agriculture.
- With the use of all treated wastewater for direct irrigation or irrigation following groundwater recharge, the irrigated land in Cyprus will be expanded by 6 per cent. Similarly, the equivalent amount of freshwater could be saved for other purposes.
- The treatment and use of wastewater for irrigation, within the acceptable level of risk for the environment and for public health, is considered, under the conditions found in Cyprus, to be the best option for long-term sustainable agriculture with a sound environmental basis.
- Wastewater reclamation and use may contribute to the protection of the environment but inappropriate treatment and use may also adversely affect the environment and human health. Therefore, the formulation of guidelines and a code of practice concerning treatment and use of wastewater are essential.
- In order to be effective, the guidelines and code of practice should be followed by legal enforcement.

XI.7 References

Department of Statistics and Research Development 1995 *Statistical Abstract 1993*. Ministry of Finance, Nicosia.

CHM HILL 1992 *Limassol Sewage Effluent and Sludge Reuse Study*. Final Report. CHM HILL, Nicosia.

Jenkins, C.R., Papadopoulos, I. and Stylianou, Y. 1994 Pathogens and wastewater use for irrigation in Cyprus. In: *Land and Water Resources Management in Mediterranean Region*, Volume IV. Proceedings of a conference held in Bari, Italy, 4–8 September 1994. CIHEAM, 979–989.

Kypris, D. 1989 Considerations for the quality standards for the reuse of treated effluent. In: *Wastewater Reclamation and Reuse*. Proceedings of a conference held in Cairo, Egypt, 11–16 December 1988, United Nations Food and Agriculture Organization, Rome.

Papadopoulos, I. 1995 Non conventional water resources: Present situation and perspective use for irrigation. In: *International Seminar on Economic*

Aspects of Water Management in the Mediterranean Area. Proceedings of a seminar held in Marrakech, Morocco, 17–19 May, 1995.CIHEAM, 54–76.

Papadopoulos, I., Economides, S., Stylianou, Y., Georgiades, E. and Koumas, A. 1994 Use of treated wastewater for irrigation of sudax for animal feeding. In: *Land and Water Resources Management in Mediterranean Region, Volume IV*. Proceedings of a conference held in Bari, Italy, 4–8 September 1994. CIHEAM, 991–8.

Water Development Department 1989 *Fifty Years of Water Development, 1939–1989, in Cyprus*. Water Development Department, Nicosia.

KINGDOM OF JORDAN

XII.1 Introduction

This case study focuses on the management and control of wastewaters and water pollution sources in the Hashemite Kingdom of Jordan in order to increase the available supply of waters of suitable quality on a sustainable basis. Although applicable to the whole of Jordan, special emphasis is placed on the Amman-Zarqa region because of its high level of population and economic activity. Due to low rainfall and increasing water supply demands, Jordan has to consider all possible methods of water conservation and reuse.

This case study presents an analysis of the water pollution problems in Jordan and identifies some solutions. The basic information and data presented here were gathered by the author, with the assistance of others, during a consulting assignment in Jordan under a contract through the United States Agency for International Development (USAID) in 1992.

XII.2 General information on Jordan and Greater Amman

Jordan is typical of countries in the Middle-East (Figure XII.1) facing population and development growth, while still limited by their water resources. Figure XII.2 shows rainfall distribution in Jordan as isohyets for a normal year. The more acute water problems occur in the more highly popu-lated areas of Jordan, including Amman and Zarqa (see Figures XII.1 and XII.2). Figure XII.3 shows monthly and mean annual rainfall data for an Amman rainfall station. Typically, rainfall occurs from October to April, with over 75 per cent occurring during the four months of December to March.

According to a Greater Amman Planning Report published in 1990, the total population of Jordan was estimated at 3,112,000, of which 2,177,000 (70 per cent) were urban (10 per cent in refugee camps and informal areas) and 935,000 (30 per cent) were rural. Assuming an increasing urban

---

*   This case study was prepared by Herbert C. Preul

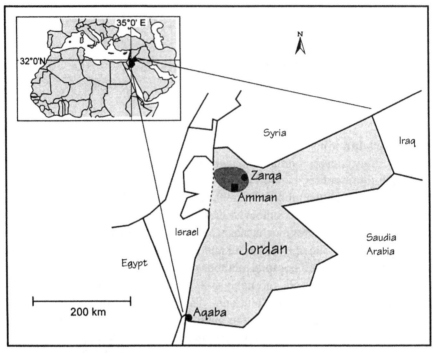

Figure XII.1 Location map of Jordan, indicating Amman and Zarqa where some of the more acute water shortages occur

population, the total population in the year 2005 is projected to be 4,139,000, of which 3,158,000 would be urban and 981,000 would be rural.

In 1985, the population of Greater Amman was 900,990. There were 144,708 households of which 141,000 occupied dwellings and 16,000 buildings were vacant. Low-rise apartment buildings accounted for 60 per cent, one or two story villas and houses accounted for 30 per cent and single story buildings dwellings accounted for 10 per cent of dwellings. The projected population for the year 2005 is 2,000,000.

Figure XII.4 shows the projections for water supply and demand in Jordan between 1990 and 2015 as determined in a water management study for USAID (USAID/Jordan, 1992). The projected shortage represents a formidable deficit. The study concluded that no single supply management method could solve this shortage, but that a combination of management alternatives would probably prove to be the best solution. Some of the wastewater control alternatives considered are discussed in this case study.

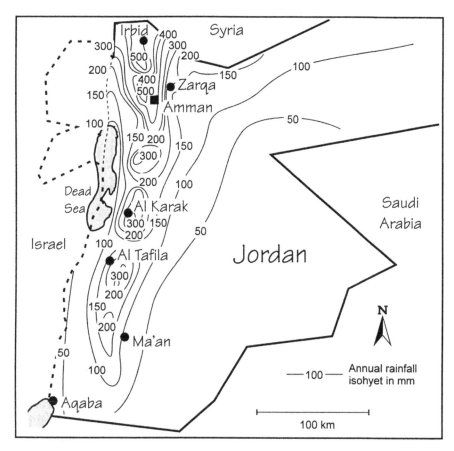

Figure XII.2 Rainfall distribution as isohyets for a normal year (long-term average) for Jordan

XII.3 Wastewaters and water pollution control

The major discharges of wastewaters are from municipal treatment plants and industrial and commercial operations. The largest contributors are concentrated in the Zarqa River Basin, including the Amman-Zarqa region. There are 14 major wastewater treatment plants (WWTPs) operating in Jordan. The As-Samra plant, serving Amman and Zarqa, has the greatest capacity with a current flow of about 100,000 m^3 d^{-1}. Other existing and proposed plants include a range of treatment processes, but waste stabilisation ponds are the most common method used.

There are more than 100 major wet-type industrial operations (i.e. those industries which use water in some form of processing and which produce

Rainfall station: AL0023   Amman/J.Amman (3rd Circ.)

Annual mean	488.8 mm							
Absolute max.	39.3	157.6	243.0	230.4	293.2	236.6	256.2	10.0
Mean monthly	10.3	41.0	80.7	120.4	97.5	104.8	32.7	1.4
Absolute min.	0.0	4.3	1.6	16.9	23.7	23.2	0.0	0.0

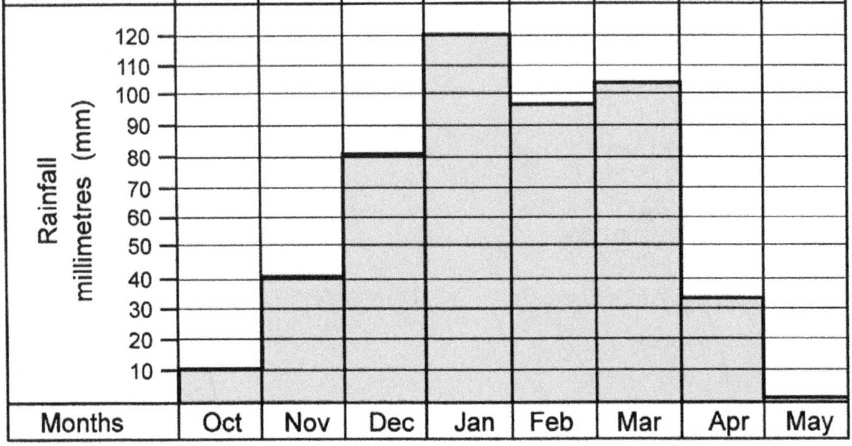

Months	Oct	Nov	Dec	Jan	Feb	Mar	Apr	May

Figure XII.3   Monthly and mean rainfall data for the Amman rainfall station during the period 1965/66–1984/85 (Data from *Amman-Zarqa Basin Water Resources Study Report*, November 1989, North Jordan Water Resources Investigation Project)

wastewater, such as the chemical industry, pulp and paper mills and food and drink processing) in the Amman-Zarqa region, as well as hundreds of additional smaller industrial operations and commercial shops which discharge small amounts of wastewaters. Of a total of 108 major wet-type industrial operations, 55 are connected to the Amman-Zarqa sanitary sewers and 53 discharge to surface water bodies (mostly wadis).

XII.4   Existing major wastewater management problems and needs
Most of the major problems with wastewater management are concentrated in the Amman-Zarqa region. The major problem in this basin concerns the wastewater handling and treatment facilities, known as the Ain-Ghazal/As-Samra system. Although currently being upgraded, these facilities have been grossly overloaded with an effluent exceeding prescribed limits. When completed in 1985, the As-Samra pond system was adequate for the intended treatment. Since that time, however, both the hydraulic and organic loads discharging to this system have dramatically increased, due to:
- Large increases in population.

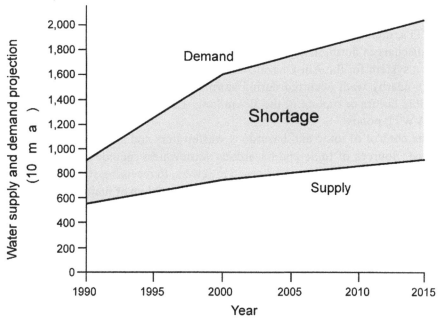

Figure XII.4 Water supply and demand for Jordan projected from 1990 to 2015 (After USAID/Jordan, 1992)

- The Ain Ghazal WWTP being taken out of service and its load being transferred to As-Samra.
- Increased wastewater loads and diversions to As-Samra.
- Increased septic tank dumpage (sewage pumped from septic tanks and dumped into the pond influent for further treatment).

The original wastewater treatment ponds were designed to handle an average of 68,000 m^3 d^{-1} but current flows are about 100,000 m^3 d^{-1} or greater. In 1991 the average annual flow to the As-Samra ponds was 97,471 m^3 d^{-1}. The chemical oxygen demand (COD) of the influent was 1,574 mg l^{-1} and the biochemical oxygen demand (BOD) was 703 mg l^{-1}. The effluent had 180 mg l^{-1} of suspended solids and a BOD of 104 mg l^{-1} effluent (equivalent to an 85 per cent removal). The effluent is usually high in nutrients (as ammonia nitrogen and phosphorus) and high in coliform bacteria (total and faecal). Consequently, downstream water quality, in the Wadi Zarqa, River Zarqa and King Talal Reservoir, has been deteriorating continuously. Studies by Engineering-Science, Inc. (1992) have shown that nutrients in the Wadi Zarqa averaged 4 mg l^{-1} N and 0.3 mg l^{-1} P during the one year period, 1989–90.

Emergency standby handling and containment facilities are needed at all WWTPs, including municipal and industrial plants, in order deal with spills and discharges during equipment failures. There is also an urgent need for such a system for the Ain-Ghazal/As-Samra siphon-pump where overflows into a nearby wadi occurred during storms in 1992. A further threat is the possible failure or rupture of the 39 km long, 1,200 mm diameter, siphon to the WWTP ponds.

The control of toxic and hazardous wastewaters and sludges is urgently needed. Sources of toxic and hazardous wastewaters include WWTPs and industrial wastes which are discharged to sewers, to receiving streams and to stormwater run-off as a result of spills. This is a problem of major concern in the Zarqa river basin where contamination in the food chain exceeds acceptable health limits. Studies on hydrochemical pollution of the Amman-Zarqa basin by Hanaineh-Abdeinour *et al.* (1985) during 1979–81 showed an *"obvious increase in trace elements"*. The study classified the Amman-Zarqa waters at that time as *"weakly to heavily polluted"*. Heavy pollution was mainly caused by: Cd, NO_3, SO_4, Cl, K and Na. Several trace elements were also observed to be increasing, including Fe, Pb, Mn, Zn, Cu and Cr.

Significant increases in elements normally associated with industrial discharges were also identified as follows: Cl showed a 6.5 fold increase, NO_3 showed a 2.2-fold increase, SO_4 showed a 5.0-fold increase and TDS showed a 2.2-fold increase. Although these results do not present a complete inventory of elements in all the possible toxic and hazardous industrial wastes being discharged, they do show an emerging pattern of concern. It is expected that these concentrations will have continued to increase since the study was carried out. A central toxic and hazardous waste treatment facility is needed for the handling and disposal of these wastes.

Inadequate on-site, pre-treatment of industrial wastewaters is a prevalent problem. Although many industries have on-site treatment facilities, they are generally inadequate as indicated by the discharges being directed to the As-Samra WWTP. Data show that the COD and the total suspended solids (TSS) concentrations in the influents are extremely high at all of the 14 major plants in Jordan, largely due to the discharge of industrial wastes. Ordinary domestic sewage in Jordan typically has BOD values in the range of 600–700 mg l^{-1}, but industrial discharges may drastically increase these values, such as at the Irbid plant where the influent has a BOD of around 1,140 mg l^{-1}. Available data show that all of the 14 major WWTPs are receiving industrial discharges, and for nine out of the 14 plants the treatment efficiencies are reasonable, giving 90 per cent BOD removal or more. Nevertheless, the discharges are still exceeding desired limits. Effluents should have less than 30 mg l^{-1} BOD, 30 mg l^{-1} TSS and 60–100 mg l^{-1} COD. Several of the plants are achieving these results but most are not, particularly the As-Samra waste stabilisation ponds at their current load.

Government instructions for discharging industrial and commercial wastewater into public sewers, as published in the official newspaper of the HKJ on 17 September 1988, Edition No. 3573, prescribe the following limits: 800 mg l^{-1} BOD, 1,100 mg l^{-1} TSS, 2,100 mg l^{-1} COD, 50 mg l^{-1} P and 50 mg l^{-1} fat, oil and grease (FOG). Although these are relatively lenient limits and regulations, a survey of municipal WWTP concentrations indicated that a large number of industries were not complying with them. In order to bring WWTP effluents into a desired range of compliance, there is a need for much higher level of on-site, pre-treatment by all industries, together with consistent monitoring.

Waste minimisation measures are needed. Although certain private organisations, such as the Chamber of Industries, are available to promote the activities of industries and commercial operations, there is a lack of effort to minimise waste discharge in an organised way.

A more direct and effective method of technical assistance to industries in relation to WWTP requirements is needed. In most cases, managers and WWTP operators are willing to provide proper treatment facilities, but are uncertain about the actual treatment facilities required. Industries in the same proximity should also be encouraged to combine their needs into a mutual WWTP for greater efficiency.

A more effective and responsive approach is needed for monitoring and compliance. At present, industries may be informed of non-compliance by the discharges from their WWTP effluents, but they need further information on the proper technical approach for rectifying the problem. There is a need for a more responsive "link" between monitoring and compliance.

Basin-wide comprehensive water quality management programmes and an environmental protection agency are needed in order to cross environmental boundaries and to follow the effects of a range of environmental emissions, not only in water but also in other media such as air, solid waste, soil and sediments. Table XII.1 gives, as an example, the trends between 1987 and 1989 in average values for selected toxic elements in the reservoir sediments of the King Talal Reservoir. The results were reported by Gideon (1991) from data compiled from reservoir suspended sediment annual reports.

In the same study, selected boreholes (water wells) in the Amman-Zarqa catchment area in 1990 showed heavy contamination with TDS, Na, Cl and

Table XII.1   Average concentrations of toxic elements in sediments of the King
Talal Reservoir, 1987–89

Variable	1987	1988	1989
Iron (mg kg^{-1})	17,392	19,094	25,110
Aluminium (mg kg^{-1})	12,275	17,869	22,077
Arsenic (mg kg^{-1})	2.80	1.53	4.36
Cadmium (mg kg^{-1})	11.80	6.66	8.78
Chromium (mg kg^{-1})	36.0	36.0	42.3
Lead (mg kg^{-1})	35.0	41.0	44.0
Manganese (mg kg^{-1})	362	413	442
Zinc (mg kg^{-1})	90	97	108

Source: Gideon, 1991

NO_3. Although polluted water discharges are largely responsible for this gross contamination of resources, there are associated emissions in other media (e.g. air) which should be investigated in a co-ordinated way.

Training programmes in basic water pollution control awareness and WWTP operation and maintenance are needed immediately. Although many water pollution control professionals within the government and involved in WWTP operation and maintenance have impressive educational backgrounds, there is a need to focus more closely on practical problems in the field. For example, although university educated engineers are expected to be capable in the basic aspects of wastewater management, they very often lack practical experience, especially where financial resources are extremely limited. Seminars and symposiums are excellent for drawing attention to problems. In addition, continuous workshop-type training is needed for all operational personnel in both government and industry.

XII.5   Management solution alternatives

In this section management alternatives for solutions to the problems discussed above and their associated needs are considered in the same order as above. Water conservation and sustainable quality effects are also noted.

Expansion and improvements in the Ain-Ghazal/As-Samra wastewater treatment system are believed to be in progress in order to alleviate the major problems in this area. This expansion should meet all current and future effluent requirements through to the year 2015. Assuming that the existing As-Samra waste stabilisation pond system will be expanded and improved, there will be some increase in evaporation losses from the ponds. These losses could be partially off-set by covering the anaerobic ponds with floating

Styrofoam sheets or other floating material. These ponds do not need to be open to the atmosphere. Based on an area of 18 ha of anaerobic ponds with an evaporation rate of approximately 2.0 m a^{-1}, covering the ponds would save approximately 360,000 m^3 a^{-1}. Covering the other ponds, i.e. aerated, facultative and maturation ponds, is not recommended because it would interfere with the treatment processes and because the costs of such untried methods would be uncertain. The bottoms of the ponds can be sealed thereby eliminating seepage losses equivalent to about 5 per cent of the pond inflow. Seepage losses for a flow of 100,000 m^3 d^{-1} a^{-1} at 5 per cent loss would be 1.8×10^6 m^3 a^{-1}. Such a water loss is worth recovering using a low cost method such as bottom sealing.

An alternative also worth investigating is the possible development of a small hydro-power station using the flow and head of the pond effluent. A suitable site could be downstream on the Zarqa river where heads in the range of 50–100 m may be available. Based on a flow of 100,000 m^3 d^{-1}, the following power generation could be possible:

- For a head of 50 m: approximately 600 horsepower or 400–500 kW.
- For a head of 100 m: approximately 1,200 horsepower or 800–1,000 kW.

Although the power that could be generated is not great, there would also be some water quality benefits downstream. In fact, the most important effect of the As-Samra treatment system improvements will be realised in downstream water quality improvements in a range of water resources.

Emergency standby handling and containment facilities for all WWTPs and industrial plants are needed to contain spills and accidental discharges. The Ain-Ghazal siphon-pump system is currently causing the most concern. The benefits of installing such facilities include the prevention of water quality degradation in rivers and streams. These benefits could be quantified using risk analysis techniques.

Control of disposal of WWTP sludges and industrial toxic and hazardous materials is required. Municipal WWTP sludges are normally not considered to be hazardous and therefore may be used as a soil conditioner in certain restricted areas. Although they have some fertiliser value, it is generally not worth further processing to market as a cost-recovery product. Waste stabilisation pond systems produce very little sludge, which is one of their major advantages. The existing As-Samra anaerobic ponds require de-sludging only after intervals of several years of operation. In addition, the sludge quantities produced are relatively small. The other ponds, employing facultative and maturation processes, never need to be de-sludged if properly operated.

The disposal of industrial sludges, including toxic and hazardous materials, is a much more difficult problem requiring special handling and disposal methods. A hazardous waste treatment facility for the Amman-Zarqa industrial complex is currently in the planning stage through the World Bank Industrial Waste Unit. This will allow industries to use a central service and should prevent indiscriminate disposal and miscellaneous discharges into the sewers and streams. Similar facilities in other governorates may be needed as industrial development increases.

As far as possible, all industries should be required to connect to the sewer system and to provide on-site, pre-treatment which will control effluents according to standards. As an economy measure certain industries in close proximity could combine their discharges for treatment in a common facility. An industrial waste discharge fee system, based on quantity and quality, would also encourage on-site pre-treatment and compliance because of the costs incurred for violations. However, this approach must be combined with an efficient monitoring and enforcement mechanism.

By instituting a fee system, based on quantity and quality, it is expected that industries will be much more responsive to reductions in water use and waste disposal, mainly because of the possible cost associated with non-compliance. Coupling this system with an industrial waste minimisation programme is expected to reduce industrial water demand by 50 per cent within an 8-year period. Vast improvements in water quality control could also be expected. Further, the collection of fees would help to fund better monitoring and enforcement.

Industrial waste minimisation is the application of low-cost, low-risk alternatives for reducing and reusing waste materials. A broad range of cost savings is possible for conservation of water as well as for conservation of other valuable materials. A typical industrial waste minimisation programme should include the following management initiatives: waste audits, improved housekeeping, substitute materials, and recycling and re-using wastes.

In wet-type industries, water savings can be dramatic in well-managed programmes, with savings in water consumption up to 70 per cent or more in certain industries over an 8-year period (Center for Hazardous Materials Research, 1991). Although difficult to quantify, improvements in the water quality of industrial effluents can be expected to be even more dramatic than those achieved in water conservation, especially for toxic discharges. Many of the industrial chemicals in waste streams can be recovered and reused, e.g. chrome in tannery wastes, with considerable cost recovery benefits to the industry. Benefits may also occur in reduced wastewater effluent charges under the industrial waste discharge fee system.

Industrial managers have expressed the need to be more closely advised on their WWTP requirements so as to be more responsive to the discharge regulations. An alternative approach to this problem would be to arrange for direct technical assistance through existing private industrial support agencies in close co-ordination with the governmental ministries in charge of monitoring and compliance. This technical assistance should be closely coupled with monitoring results obtained by the appropriate Ministry. Although not possible to quantify, long-term improved technical assistance should accrue significant benefits.

Consistent and effective monitoring is fundamental to the enforcement of compliance with effluent standards. Currently, the system only identifies non-complying WWTPs and industries sporadically and often problems are not corrected. Therefore, in order to be more effective in correcting problems, it has been suggested that non-compliance notifications should be coupled with immediate technical guidance either from the appropriate ministry or from a private industrial support agency, together with a deferred time period in which to make corrections and to achieve compliance. Although such measures can be expected to enhance water quality, the benefits cannot be measured directly.

Comprehensive water quality management programmes are required through river basin authorities. A wide range of environmental emissions occur, particularly in industrial areas such as the Zarqa river basin, and therefore it has been suggested that water quality management and monitoring should be co-ordinated to trace contamination in the full range of water resources and environmental media. This would include flowing surface waters, impoundments, water supplies, drinking waters, irrigation waters, groundwaters, wells, soil contamination, irrigation use, pesticide applications, pollution from urban run-off, non-point pollution sources, air pollution and solid waste disposal. Such a basin-wide programme is best accomplished through river basin authorities or through an environmental protection agency which would cross ministerial boundaries but could still integrate the efforts of various ministries. Through this approach, problems can be traced and corrected more responsively. These new authorities or the environmental protection agency should have certain enforcement powers.

River basin authorities have been highly successful for water pollution control in various developed countries; examples include Ruhr Verbands in Germany and River Commissions in the USA. The expected benefits include enhancement of water quality and enforcement efforts that will be more responsive and better co-ordinated.

Certain training programmes have been recommended as being required immediately and could be the key to most of the problems discussed above. The most immediate need is for the training of appropriate government engineers and scientists, WWTP managers and operators of municipal and industrial plants. Beyond this initial need, a broader training programme should include other government water resource control management personnel, private sector industrialists, selected consultants and industrial service company principals. The subjects that could be included in the training programme, depending on the personnel to be trained and their needs, are as follows:

- Basic water pollution control.
- Point-source pollution.
- Non-point source pollution.
- Pollution prevention and waste minimisation.
- Pollution measurement and monitoring.
- Industrial water conservation.
- Pollution control audits and feasibility studies.
- WWTP design and equipment requirements.
- WWTP operation and maintenance.
- Equipment requirements, costs and project financing.

Along with the proposed training programmes, two demonstration facilities should be set up for use in connection with the training programme. These would be a typical industrial plant with a WWTP and a typical municipal WWTP.

The overall objective of the broader training concept programme is to produce an environmental awareness which will form the basis for establishing higher priorities in water conservation and quality control throughout the country. Although the benefits of these training programmes are not directly measurable, they will be immediate and far reaching.

XII.6 Recommendations and possible results

The major discharges of wastewaters in Jordan are from municipal and industrial WWTPs, with the largest plants located in the Amman-Zarqa region. The effluents from the As-Samra waste stabilisation pond system and from over 100 wet-type industries in this region constitute by far the largest portion of the total available wastewater flows that require water conservation and quality management. The most immediate priority recommendations for achieving benefits in water conservation and water quality are:

- An improved Ain-Ghazal/As-Samra treatment system.
- Implementation of an industrial waste discharge fee system.

- Implementation of an industrial waste minimisation programme.
- Training programmes in water pollution control and WWTP operation and maintenance.
- Investigation into a small power station using the As-Samra effluent.

Longer-term water conservation and water quality effects will result from the following actions:

- Basin-wide water quantity and quality management through river basin authorities or an environmental protection agency.
- Effective water quality monitoring and compliance.
- Technical assistance to industrial waste dischargers.
- A central toxic and hazardous waste handling and treatment facility.
- Emergency handling and containment facilities for all WWTPs and industrial waste dischargers.

The above recommendations will result in significant water conservation savings, but the greatest effects are expected to be achieved in water quality enhancement. Although the benefits of water quality improvements are difficult to quantify, the effects of the improvements become quantifiable in terms of water available for reuse for a variety of purposes. Thus water quality improvements will have far reaching benefits for overall water use throughout Jordan.

XII.7 References

Center for Hazardous Materials Research 1991 *Industrial Waste Minimization Manual for Small Quantity Generators*. University of Pittsburgh Applied Research Center, Pittsburgh.

Engineering-Science Inc. 1992 *Effects of Nutrient Removal at the As-Samra Waste Stabilization Ponds on the Quality of King Talal Reservoir*. Engineering-Science Inc., Pasadena, California.

Gideon, Raja 1991 *The Potential Impact of Industrial Wastes on Water Resources in Amman-Zarqa Basin*. Proceedings of the Second Environmental Pollution Symposium, 1990, Friederich Ebert Stiftung Goethe-Institut, Amman Water Research and Study Center. University of Jordan, Amman.

Hanaineh-Abdeinour, L., Fayyad, M., and Tutingi, M. 1985 *Hydrochemical Pollution of the Amman-Zarqa Basin*. Dirasat Vol. XII No. 7. University of Jordan, Amman.

USAID/Jordan 1992 *A Water Management Study for Jordan*. Project in Development and the Environment, Technical Report No. 4, USAID/Jordan Project No. 398-0365. Chemonics International Consulting Division, Inc., Washington, D.C.

Case Study XIII[*]

SANA'A, YEMEN

XIII.1   Introduction

The Republic of Yemen (Arabia felix) is located in the south and south-eastern part of the Arabian Peninsula and covers an area of 555,000 km^2 (Figure XIII.1). The country is surrounded from the west and south by the Red and the Arabian Seas. To the east and north it is bordered by the Sultanate of Oman and the Kingdom of Saudi Arabia respectively. In addition to Sana'a city, which is the capital, the country consists of 17 governorates of which 11 are located in the north (prior to 1990 known as North Yemen) and six in the south (prior to 1990 known as South Yemen). According to the High Water Council (HWC, 1992a) the total population was estimated to be 12.4 million in 1990 and 14 million in 1992. Eighty per cent are thought to live in the central and southern highlands which receives most of the erratic, limited rainfall. It is projected that the country's population will reach 23.4 million by the year 2010. Increasing water demand in recent years and the limited availability of surface water resources have increased the pressure on the available, mostly non-renewable, groundwater resources.

According to the *World Development Report* (World Bank, 1993), the per capita gross national product (GNP) of Yemen in 1991 was US$ 520. The major sectors that play important roles in the country's economy are agriculture, industry, services and mining. HWC (1992b) summarised the share of those sectors in the Gross Domestic Product (GDP) in 1990 as 20.6, 12.9, 58.1 and 8.4 per cent respectively. Although agriculture is not the largest contributor to the national economy, it employs around 60 per cent of the active labour force. In 1990, the total cultivated agricultural land was estimated to be 1.12×10^6 ha of which 61 per cent was rain-fed, 28 per cent was irrigated with groundwater, 2 per cent was irrigated with permanent springs and the remaining 9 per cent was cultivated by spate irrigation. In 1992, irrigated agriculture consumed about 90 per cent of the total water

[*]   *This case study was prepared by Mohamed Al-Hamdi*

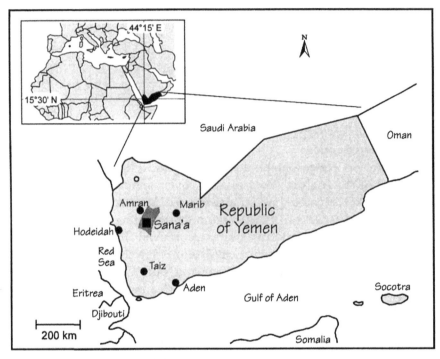

Figure XIII.1 Location map of Yemen indicating the Sana'a basin

demand and accounted for about 50 per cent of the value of agricultural production. While total exports in 1990 amounted to YR 8.3×10^9 (the 1995 official exchange rate was US\$ $1 =$ YR 12 and the parallel market rate for January 1995 was US\$ $1 \approx$ YR 100), of which crude oil and agricultural products had the largest shares (87 and 10 per cent respectively), agricultural trade registered a deficit of 88 per cent. Inflation in 1988 was around 16 per cent, but as a result of the Gulf crisis and the return of more than a million labourers from the Gulf states, who previously provided hard currency, inflation increased to 50 per cent between 1990 and 1991.

Yemen depends mainly on external borrowing to implement its development programmes. As of 1990 the total debt stood at US\$ 7.1×10^9, which was about 85 per cent of the GDP; 12 per cent of the debt comes from short-term commercial sources, 16 per cent from long-term multilateral sources, and the remaining 72 per cent from bilateral sources.

XIII.1.1   Structure of the water sector
The two main institutions responsible for water in Yemen are the Ministry of Electricity and Water (MEW) and the Ministry of Agriculture and Water

Resources (MAWR). The MEW is in charge of water supply and wastewater collection and treatment in urban centres, in addition to water supply in rural areas. Three organisations are directly attached to the MEW: the National Water and Sewerage Authority (NWSA), the General Directorate of Rural Water Supply (RWSD) and the High Water Council (HWC). The NWSA is a financially autonomous authority in charge of water supply and wastewater collection and treatment for the urban areas. Since the establishment of the authority in 1973, its jurisdiction has expanded to cover 12 cities in addition to Sana'a. The minister of MEW chairs the board of directors that runs the authority. The RWSD is mainly in charge of the rural water supply. The main role of this directorate has been the construction of small-scale water supply projects (mostly funded by external donors), which are usually handed to local councils for operation and maintenance. So far, rural sanitation has not received much attention, and on-site disposal facilities are the most common approach in the rural communities. The HWC was established under the same legislation that established the MEW in 1981, and its role is to co-ordinate the activities of all agencies in the water sector. The main task of the Council was to formulate national water plans and strategies and to prepare national water legislation. The Council consisted of deputy ministers of concerned ministries and was chaired by the Minister of Electricity and Water. As a result of under-staffing, the council was reformulated in 1986 to consist of concerned ministers and chaired by the Prime Minister. The Technical Secretariat of the HWC was also established in 1986 to assist the Council in the performance of its duties. Currently, no law had been passed to support the formulation of the Council as an independent agency and, therefore, it had been facing difficulties in meeting its obligations and duties.

After reunification of North and South Yemen in May 1990, the MAWR was formed from the previous Ministry of Agriculture and Fisheries in the north and the Ministry of Agriculture and Agrarian Reform in the south. These ministries had been in charge of development of water resources for agricultural purposes. However, since May 1990 the MAWR has been given the responsibility of managing national water resources, i.e. it has become a water manager and a major water user at the same time.

XIII.1.2 Legislative framework

At present, there exists no national water legislation. Prior to May 1990, the HWC had prepared draft national water legislation and, because of the seriousness of groundwater depletion, the HWC also drafted a by-law on regulating groundwater extraction and a law to establish a National Water Authority. In the drafted law, the proposed National Water Authority was given the responsibility of allocating available water resources, specifying water use priorities and controlling annual consumption in order to ensure the sustainability of economic and social development. Due to the altered responsibilities for water resources management that occurred after May 1990, the MAWR drafted, independently, a second national water legislation in 1992 with a law to establish a National Water and Irrigation Authority.

However, neither of these laws were passed and the lack of water legislation has subsequently created an atmosphere of uncoordinated water use which is evident from the continuous decline of groundwater levels nation-wide. In short, the seriousness of the present water situation highlights the immediate need for water legislation and the establishment of a national agency to manage the scarce water resources in Yemen.

XIII.2   Water issues

The Sana'a basin is located in the central highlands (Figure XIII.2) and covers approximately 3,200 km^2, ranging from less than 2,000 m to more than 3,200 m above sea level. The climate of the basin area is characterised by a low and erratic rainfall pattern with an average of 250 mm a^{-1}. Sana'a, the capital of Yemen, is located in the Sana'a plain (Figure XIII.2) at an elevation of about 2,200 m above sea level. According to the first national census in 1975, the population of the city was 134,588 inhabitants and it had increased more than three-fold to 424,450 by 1986. Although the national population growth rate was around 3 per cent, the population of the city grew at an annual rate of 11 per cent and was then projected to continue at a similar rate. This rapid growth is mainly attributed to improved economic conditions which stimulated internal migration from the rural areas. At present, the population of the city is estimated to be over 1 million and is projected to increase to over 3.4 million by the year 2010.

XIII.2.1   Water resources

The principal source of water in the region is groundwater from three aquifer layers, namely alluvial deposits, volcanic units and the Tawilah sandstone. Of the three aquifers, the Tawilah is considered to be the most productive and has the best water quality. The capacity of the Tawilah is estimated at 2,230 × 10^6 m^3 (total storage) of which only 50 per cent is considered withdrawable. In addition to low recharge as a result of low rainfall in the recent past, increased extraction (mainly for agriculture) has resulted in a substantial drop in groundwater levels (3–4 m a^{-1}). It is important to realise that while the total water demand in the Sana'a basin area was estimated to be 220 × 10^6 m^3 a^{-1} in 1995, recharge estimates for the Tawilah aquifer vary

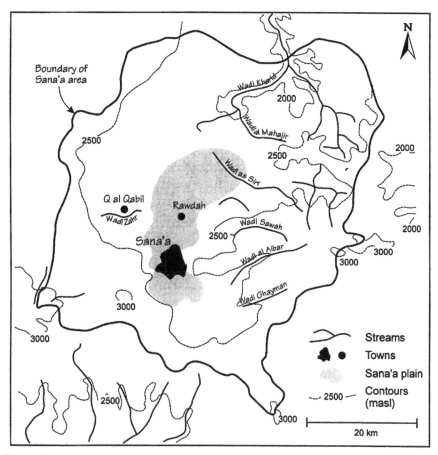

Figure XIII.2 Map showing the major features of the Sana'a basin

between only 27×10^6 and 63×10^6 m³ a⁻¹. The large difference between consumption and recharge is being filled with water from long-term storage, referred to as groundwater mining. The present pattern of water use in Sana'a is clearly unsustainable and, if allowed to continue, depletion of this valuable and scarce resource is inevitable.

XIII.2.2 Water use

Groundwater in the region is used exclusively to satisfy the water needs of the different water-using sectors, namely irrigated agriculture, municipal use and industrial use.

Prior to the Yemeni revolution in 1962, agriculture in the Sana'a basin area depended on dry farming practices and spate irrigation. The introduction of drilled boreholes in the 1970s, and the identification of the Tawilah as a highly productive aquifer, encouraged farmers to use groundwater for irrigation. Having realised the importance of the Tawilah, the government tried to regulate agricultural water use in the area by passing a law in 1973 which identified a local protection zone around the NWSA wellfields and prohibited further drilling of new wells or cesspits unless permitted. At present, agriculture in the basin area consumes about 175×10^6 m^3 a^{-1}, which accounts for 80 per cent of the total water demand in the basin area. Moreover, qat (a tree from which the leaves are chewed as a stimulant in Yemen) and grapes (a cash crop) are estimated to consume around 40 and 25 per cent respectively of the agricultural water demand in the region. The main reasons behind the over-use of groundwater for irrigation can be summarised as:

- Unclear water rights and thus unregulated extraction.
- Fuel subsidies and low import duties on agricultural equipment.
- High returns on cash crops.
- Inefficient irrigation practices.

Within the Sana'a basin, it is estimated that the present population is about 2.34 million, of which 1.4 million live in urban areas. Although the per capita consumption rate varies, it is estimated that the total municipal water demand in 1995 was 36.9×10^6 m^3 a^{-1}, of which about 29×10^6 m^3 a^{-1} was consumed in the urban areas. It was also projected that the total yearly municipal water demand would increase to 138×10^6 m^3 a^{-1} by the year 2010 (HWC, 1992c). The industrial water demand was estimated at 4.7×10^6 m^3 a^{-1} in 1990 and was projected to increase to 6.2×10^6 m^3 a^{-1} in 1995. Van der Gun et al. (1987) reported that the government of the Yemen Arab Republic (North Yemen prior to reunification in 1990) took measures to prevent the further establishment of major water-consuming industries in the Sana'a area and this could explain the low rate of increase in water use compared with the other sectors.

XIII.2.3 Sources of groundwater pollution

In the Sana'a basin area, unregulated direct disposal underground of municipal and industrial wastewater by means of on-site disposal facilities (cesspits) presents a potential threat of groundwater contamination. The thick, unsaturated zone, resulting from deep groundwater levels (100–170 m below ground level) suggests that groundwater pollution is unlikely. However, the complex geological structure and the presence of rock fractures could reduce the travel time of pollutants through this layer. The use of pesticides and chemical fertilisers in agriculture in Yemen is, however, still at a relatively low level and therefore groundwater contamination from this source is not of major concern at present.

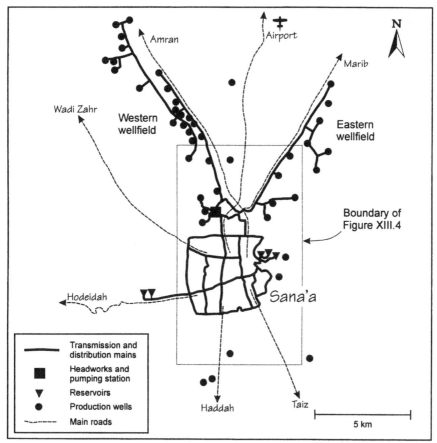

Figure XIII.3   Map of the Sana'a area showing the location of pumping stations, reservoirs and the NWSA wells (After Al-Hamdi, 1994)

XIII.2.4   Water and wastewater in Sana'a city

In the city of Sana'a, the municipal water supply consists of both public and private water supplies. In 1993, the public water supply produced around 17.8×10^6 m^3 providing 43 per cent of the city's population with a per capita consumption of about 120 l d^{-1}, including 35 per cent that was not accounted for. Groundwater from the NWSA wellfields (Figure XIII.3) is of good quality and meets the World Health Organization (WHO) drinking water guidelines. Nevertheless, chlorination is usually applied as a safety measure in the distribution network. Private water supplies, which depend on unmonitored private boreholes in the city, some of which also draw from the Tawilah, were estimated to have produced 6.7×10^6 m^3 in 1993. Although the private water supply is supposed to cover 57 per cent of the city's population, the high price of the water is suspected to reduce the per capita consumption to about 35 l d^{-1}.

As of 1993, only 12 per cent (10,000–12,000 m^3 d^{-1}) of the city was connected to the sewerage system which conveys wastewater to stabilisation ponds in Rowdda, north of Sana'a, for treatment (see Figure XIII.4). The rest of the city (35,000 m^3 d^{-1}) depended on cesspits with infiltration as the main mechanism of wastewater disposal. Al-Eryani *et al.* (1991) concluded that domestic wastewater had produced some changes in the quality of ground-water under the heavily populated area of the city and around the stabilisation ponds at Rowdda. Al-Shaik (1993) summarised an investigation of the water quality of some wells along the path of the effluent from the stabilisation ponds north of Rowdda. The study identified a contaminated area along the effluent channel and recommended continuous monitoring of the investi-gated area, as well as the NWSA wellfields. Al-Hamdi (1994) investigated the quality of groundwater in the city of Sana'a and classified the city into three quality zones: north, middle and south (Figure XIII.4). Groundwater in the middle zone contained more nitrate and chloride than the other zones, suggesting that wastewater disposal in this zone has had a negative effect on the quality of the groundwater. Furthermore, a polluted sub-area (sub-middle) was identified within the middle zone, which was characterised by NO_3^- concentrations within the range 100–160 mg l^{-1}, Cl$^-$ concentrations within the range 220–400 mg l^{-1} and electrical conductivity within the range 975–2,045 mS cm^{-1}. It was argued that the present pollution could be attrib-uted to wastewater disposal and that the polluted zone would expand towards the north, because the general direction of groundwater flow in the area is from south to north. No immediate risk was thought to exist for the NWSA wellfields but more than 50 per cent of the city's population depend on unmonitored private wells scattered within the city's perimeter.

The use of cesspits in the eastern and western parts of the city (Nokom and Allakama) has resulted in an overflow of wastewater to the ground surface because the local geology infiltration rates are very low. In addition to the poten-tial health hazards resulting from direct human exposure, Al-Hamdi (1994) has suggested that intermittent de-pressurisation of the drinking water distribution network could induce some suction of wastewater into the network.

Based on groundwater samples taken near industrial activities, mainly large factories located outside the city, Al-Eryani *et al.* (1991) concluded that industrial wastewater in the Sana'a area was not presenting an immediate threat to the quality of the groundwater; however, no detailed information about the waste disposal methods and the characteristics of the industrial

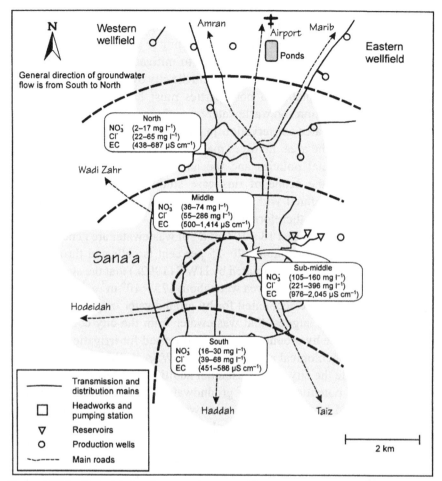

Figure XIII.4 Map showing the groundwater quality variation in the city of Sana'a. The general direction of groundwater flow is from South to North (After Al-Hamdi, 1994)

wastewaters was given. In addition to large factories, which are mostly located outside the city, many small workshops, oil-changing garages and car washes are located within the city. The results presented by Al-Hamdi (1994) suggest that direct disposal of wastewater from these activities could lead to serious groundwater contamination.

From the above discussion, it is evident that groundwater depletion is currently taking place, while at the same time the quality of groundwater under the city is threatened by extensive wastewater disposal. Water rights have not been settled with farmers and, therefore, they consider groundwater to be communal property whereby they have the right to fulfil their domestic and agricultural water needs. Competition for groundwater extraction could increase the rate of depletion of the aquifer leading to a subsequent decrease in irrigated agriculture in the area. In order to mitigate the possible future conflicts that could arise between farmers and the city over water resources, a management plan acceptable to both parties must be concluded. In this context water conservation and wastewater reuse for irrigation could prove to be two key issues. Water conservation in irrigated agriculture, the largest groundwater user in the area, involves many aspects, including agricultural economy, governmental policies and the national legal conditions. Such aspects are beyond the scope of this case study. Wastewater reuse is, however, closely integrated with groundwater management and pollution control and this aspect is therefore discussed below.

Current estimates show that 18×10^6 m^3 a^{-1} of wastewater are generated by the city of Sana'a, of which about 20–25 per cent is collected through the sewerage system. It has been estimated by HWC (1992c) that the agricultural water requirements in the basin area were about 175×10^6 m^3 a^{-1} in 1995, of which 160×10^6 m^3 were accounted for by groundwater irrigation for cash crops. These estimates suggest that wastewater from the city could reduce agricultural water use by around 12 per cent if reused for irrigation at properly selected hydrogeological areas, i.e. at the NWSA wellfield region. This reuse could provide the city with substantial additional water supplies while also reducing the potential threat of groundwater contamination under the city. Farmers could be convinced to reuse wastewater because it would be cheaper than groundwater (collection and treatment would be paid for by the Government and by consumers) and more reliable (especially with the continuous decline in groundwater levels and the threat of complete exhaustion of the aquifer). Such reuse should be constrained by legal agreements where treated wastewater (the property of the city) is traded for undefined groundwater rights. Thus farmers involved in these agreements would receive treated wastewater, in addition to possible privileges, such as extra attention from relevant governmental agencies, awareness programmes for wastewater irrigation and certain financial incentives (i.e. loans and subsidies), in return for discontinuing groundwater irrigation. The increasing scarcity of groundwater in the area could make such agreements attractive to farmers especially when long-term (sustainable) agriculture in the area is most likely to be wastewater-irrigated.

With respect to pollution control, wastewater reuse could serve three objectives simultaneously:

- It would eliminate all adverse health effects that could result from drinking contaminated groundwater, from direct exposure to overflowing wastewater, and from direct contamination of the drinking water distribution network.

- The private sector could continue to provide part of the population with safe drinking water.
- The increased groundwater supplies, as a result of less groundwater irrigation, would allow the NWSA to increase the coverage of the regulated and monitored public water supply.

However, the absence of a co-ordinating agency and the present divided responsibilities for water resources are major constraints to the implementation of such management options.

XIII.2.5   Critical water issues

As indicated above Yemen in general, and Sana'a in particular, are facing a critical water shortage due to unregulated and uncoordinated water use. Moreover, there is a potential risk of groundwater contamination as a result of unregulated wastewater disposal. The risk of groundwater pollution could incur serious health problems because more than 50 per cent of the city's population rely on private wells for their water needs. In addition to adverse health effects, polluted groundwater becomes very costly to treat.

XIII.3   Planned interventions

The government of Yemen realised that there was a critical water shortage in Sana'a and initiated, with the assistance of the Dutch government, a project in the late 1980s to look for alternative water sources for the city, i.e. a supply orientated approach. The government also realised the need for water legislation and for a national agency to manage, regulate and co-ordinate the use of water resources in a manner that will ensure sustainable development.

With regard to the risk of groundwater contamination in the Sana'a area, the NWSA has appreciated that direct wastewater disposal and the overloaded stabilisation ponds are the main contributors to changing groundwater quality in certain areas of the city. Thus collection and proper treatment of wastewater is viewed as the key to protect the Tawilah aquifer from further quality degradation. If the sewerage system is expanded to cover the entire city and if wastewater is adequately treated so that it can be re-used in agriculture, the quality of the groundwater will be protected and some of the agricultural water demand should be reduced. Recently, land has been acquired for a new activated sludge treatment plant, but funds still need to be allocated for its construction. In response to continuous public complaints, the NWSA intends, in an emergency programme, to connect the eastern and western parts of the city (Nokom and Allakama) to the sewerage system in order to eliminate the overflow of wastewater and to reduce the threat of drinking water contamination in the distribution network.

XIII.4   Lessons learned and conclusions

In an effort to manage the current unsustainable use of the groundwater resources in the Sana'a area, the Government has focused on a supply orientated approach with a project to evaluate different water sources. At the same time, the Government has failed to address demand management measures as a viable option in water resources management. Importing water from other regions to Sana'a, given the scarcity of water nationwide, would be very costly and could face strong local resistance in the supplying regions. Implementation of demand management in Yemen requires an in-depth understanding of water rights. Settlement of those rights would become essential if the Government wished to set water-use priorities and to control the (re)allocation of water resources.

The 1973 law to protect the NWSA wellfields from depletion and from deterioration in water quality can be considered ineffective for the following reasons:

- Small ratio of protection zone to total basin area.
- Such regulations are difficult to monitor and to enforce.
- There was no other alternative for wastewater disposal and therefore permits for cesspits were always granted.

The quality of groundwater under the central part of the city of Sana'a and around the stabilisation ponds has deteriorated as a result of unregulated direct disposal of wastewater. Although immediate action is required, the availability of financial resources to expand the sewerage system and to construct proper treatment facilities seems to be the major constraint. To date, economic and financial incentives have been neglected in water management and pollution control in Yemen.

Five main points have been highlighted by this case study:

- Unregulated disposal of municipal and industrial wastewater could cause serious changes in the quality of groundwater and therefore could have the potential to result in adverse health effects and high treatment costs. Reuse of wastewater in a water-scarce regions like Sana'a can be considered as an attractive and effective opportunity because it reduces the threat of groundwater contamination while also providing a water source with a high nutrient content for irrigation. However, the success of a wastewater reuse programme depends on several conditions:
  - The sewerage system should expand to cover the entire city (very costly).

- Reclaimed wastewater for irrigation should be free of toxic substances that may arise from industrial discharges, and the hygienic and agronomic quality of the water should be suitable for irrigation.
- Farmers should be amenable to the use of reclaimed wastewater for irrigation (wastewater irrigation of cash crops could reduce the market price of those crops).
- The present institutional arrangement of the water sector in Yemen, where there is no proper co-ordination in the use of scarce water resources or effective management of pollution control, can be viewed as a prime factor leading to the unsustainability of those water resources.
- A demand-orientated approach should be considered as an important element in water resource management. This is particularly important in arid and semi-arid areas where water resources are limited although demand, due to increased populations needing water and food, is always increasing.
- Economic and financial incentives should be considered seriously in water management and pollution control. Pricing could play an important role in demand reduction and pollution prevention.
- Sustainable use of scarce water resources should be included in the regional and national economic and social development plans and strategies.

XIII.5   References

Al-Eryani, M., Ba-issa, A. and Al-Shuibi, Y. 1991 *Groundwater Pollution in the Sana'a Basin: a Preliminary Appraisal*. Environmental Protection Council, Sana'a, Republic of Yemen.

Al-Hamdi, M. 1994 Groundwater Pollution due to Municipal Wastewater Disposal. M.Sc. thesis, IHE, Delft, The Netherlands.

Al-Shaik, H. 1993 *Report on the Extent of Groundwater Pollution due to the Effluent of the Sana'a Stabilisation Ponds at Rowdda*. National Water and Sewerage Authority, Sana'a, The Republic of Yemen.

Van der Gun, G., Trietsch, R. And Uneken, H. 1987 Sources for Sana'a Water Supply. Unpublished mission report, Sana'a, The Republic of Yemen.

HWC 1992a *National Water Legislation and Institution Issues*. Final report, Volume II, UNDP/DESD project YEM/88/001. The Technical Secretariat of the High Water Council, Sana'a, Republic of Yemen.

HWC 1992b *Water Resources Management and Economic Development*. Final report, Volume I, UNDP/DESD project YEM/88/001. The Technical Secretariat of the High Water Council, Sana'a, Republic of Yemen.

HWC 1992c *Water Resources Management Options in Sana'a Basin*. Final report, Volume IX, UNDP/DESD project YEM/88/001. The Technical Secretariat of the High Water Council, Sana'a, Republic of Yemen.

World Bank 1993 *World Development Report 1993, Investing in Health*. Oxford University Press, New York.

Appendix

PARTICIPANTS IN THE WORKING GROUP

M. Adriaanse, Information and Developments, Institute for Inland Water Management and Waste Water Treatment (RIZA), Ministry of Transport, Public Works and Water Management, The Netherlands

G. Alabaster, United Nations Centre for Human Settlements (UNCHS/Habitat), Nairobi, Kenya

M. Al-Hamdi, Sana'a University, Faculty of Engineering, Sana'a, Yemen

W. Ankersmit, Technical Advice Department, Ministry of Foreign Affairs/DGIS, The Hague, The Netherlands

G.J.F.R. Alaerts, Professor of Sanitary Engineering and Vice Rector, International Institute for Infrastructural, Hydraulic and Environmental Engineering (IHE), Delft, The Netherlands

L. Anukam, Federal Environmental Protection Agency (FEPA), Abuja, Nigeria

C. Bartone, Senior Environmental Specialist, Urban Development Division, The World Bank, Washington, D.C., USA

J. Bartram, Manager, Water and Wastes, European Centre for Environment and Health, World Health Organization, Rome, Italy

J. Bernstein, Environment Specialist, The World Bank, Washington, D.C., USA

S.A.P. Brown, Wates, Meiring & Barnard, South Africa

P. Chave, Head of Pollution Control, National Rivers Authority, Bristol, UK

R. T. Cruz, Assistant Project Director, River Rehabilitation Secretariat, Pasig River Rehabilitation Program, Department of Environment and Natural Resources, Carl Bro International a/s, Metro Manila, Philippines

R. Enderlein, Environment & Human Settlement Division, United Nations Economic Commission for Europe, Geneva, Switzerland

U. Enderlein, Urban Environmental Health, Division of Operational Support in Environmental Health, World Health Organization, Geneva, Switzerland

R. Helmer, Chief, Urban Environmental Health, Division of Operational Support in Environmental Health, World Health Organization, Geneva, Switzerland

R.M. Hermann, Head, Hydraulic and Sanitary Engineering Department, Escola Politecnica da Universidade de Sao Paulo, Sao Paulo, Brazil

I. Hespanhol, Urban Environmental Health, Division of Operational Support in Environmental Health, World Health Organization, Geneva, Switzerland

A. Kandiah, Senior Officer, Water Resources Development and Management Services, Food and Agriculture Organization of the United Nations, Rome, Italy

H. Larsen, Water Quality Institute (VKI), Danish Academy of Technical Sciences, Horsholm, Denmark

B. Locke, Deputy-Executive Secretary, Water Supply & Sanitation Collaborative Council, World Health Organization, Geneva, Switzerland

P. Marchandise, World Health Organization/Nancy Project Office, European Centre for Environment & Health, Vandoeuvre-les-Nancy, France

A. Milburn, Executive Director and Managing Editor, International Association on Water Quality (IAWQ), London, UK

I. Natchkov, Ministry of Environment, Sofia, Bulgaria

P. J. Newman, WRc plc., Medmenham, UK

I. Papadopoulos, Agricultural Research Institute, Ministry of Agriculture, Natural Resources and Environment, Nicosia, Cyprus

R. Pors, DGIS/Environment Programme, The Hague, The Netherlands

H.C. Preul, International Water Resources Association (IWRA), University of Cincinnati, Department of Civil and Environmental Engineering, Cincinnati, USA

J. Rasmussen, Deputy Director, Water Quality Institute, Danish Academy of Technical Sciences, Horsholm, Denmark

V.A. Rezepov, Deputy Director, Centre for International Projects (CIP), Moscow, Russian Federation

D.W. Rodda, Team Leader, Danube Programme Coordination Unit, Vienna International Centre, Vienna, Austria

H. Romero-Alvarez, Instituto Mexicano de Tecnologia de Agua, Oficinas Consejo Nacional de Agua, Mexico City, Mexico

J. Schwartz, Chief Environmental Adviser, River Rehabilitation Secretariat, Pasig River Rehabilitation Programme, Carl Bro International a/s, Metro Manila, Philippines

Y. Sharma, Additional Director, National River Conservation Directorate, Ministry of Environment and Forests, New Delhi, India

E. Skarbovic, Water and Lithosphere Unit, United Nations Environment Programme (UNEP), Nairobi, Kenya

J. Smet, International Reference Center, The Hague, The Netherlands

G.E. Stout, Executive Director, International Water Resources Association, IWRA, University of Illinois, Urbana, USA

L. Ulmgren, Director, International Department, Stockholm, Sweden

V. Vladimirov, c/o Center for International Projects (CIP), Moscow, Russia

R. Wirasinha, Executive Secretary, Water Supply and Sanitation Collaborative Council, World Health Organization, Geneva, Switzerland

T.F. Zabel, Senior Consultant, Environment, WRc plc., Medmenham, UK

C. Zhang, The World Bank, Washington, DC, USA

Index

Printed and bound by CPI Group (UK) Ltd, Croydon, CR0 4YY

01/11/2024

01782626-0019